AAPG FOREIGN REPRINT SERIES NO. 2

Geology and Productivity:

Arabian Gulf

Selected papers reprinted from
AAPG Bulletins and Special Publications

compiled by
Anthony E. L. Morris

Published November 1978 by
The American Association of Petroleum Geologists
Tulsa, Oklahoma, U.S.A.

Published November 1978

Library of Congress Catalog Card No. 78-65773
ISBN: 0-89181-801-4

Printed by
Edwards Brothers, Inc.
Ann Arbor, Michigan

Contents:

Preface and Bibliography:

Geology and Productivity of Arabian Gulf Geosyncline: Preface

The Arabian Gulf geosyncline, as used here, is that portion of the Tethys sedimentary belt between southeast Turkey and Oman (see index map). Its boundaries are geologic: the Arabian shield and the Dead Sea fault on the west, the southeastern overthrust belt of Turkey on the north, the Zagros thrust (suture) and the Oman line on the east, and the geanticline which forms the hills of the southern Arabian Peninsula on the south. The region has an area of 450,000 sq mi (1,165,000 sq km). A milestone was reached on January 1, 1978, when cumulative production from the region exceeded 100 billion bbl. Stated reserves are on the order of 360 billion bbl. Current production is 21 million bbl per day.

Standard Oil of California through a Canadian subsidiary, Bahrain Petroleum Co., was the first United States corporation to establish production in the region. Concession options covering the sheikdom of Bahrain Islands were acquired in 1928 and, after years of international political wrangling, the first well was completed as a discovery on June 1, 1932.

Prior to that, United States oil interests had been in the Middle East since 1908 when Rear Admiral Colby M. Chester negotiated with the Turkish government for mineral and oil rights in Mesopotamia. World War I prevented the formal ratification of the agreements but provided the basis for a claim after the Armistice when the victors were "dividing the spoils." So it came about that, on July 31, 1928, the Near East Development Corp., consisting of Standard of Jersey, Socony Oil Co., Atlantic Refining Co., and Mexican Petroleum Co. (Doheny), became a 24 ¾%-interest holder in the Iraq Petroleum Co., the successor to the Turkish Petroleum Co. This contract later became known as the Red Line Agreement because the parties to it drew a red line around a mutual area of interest (the entire Arabian Peninsula) in which all agreed that none would take a petroleum concession without offering the others a pro rata share and, if any of the parties did not approve, the offering party could not take the concession on its own. It was thus that Gulf Oil, who had acquired the Bahrain concession option from a New Zealander, Major Frank Holmes, was obliged to divest itself of it and offered it to Standard of California at cost ($50,000).

The first AAPG presence was a brief review of Persian (as then known) oil fields in 1922 (AAPG Bull., v. 6, p. 383). The Association has since published over 1,000 pages of studies on this prolific region.

Selection of articles from this array of fine material required a Draconian approach which finally yielded 3 overview articles and 6 regional papers covering structure, stratigraphy, and productivity in Iraq (1), Iran (2), Saudi Arabia (2), and the southern Arabian Gulf (1).

Discussions of such fascinating problems and conditions as the tectonic history of the Oman Mountains, the diverse nature of fracture porosity in carbonate reservoirs, and the olistostromes or nappes of northern Iraq and southeast Turkey unfortunately could not be included but can be easily located in the bibliography of AAPG papers covering the Arabian Gulf geosyncline and nearby Middle East areas included at the end of the volume.

Anthony E. L. Morris

Petroleum Consultant
Los Angeles, California

July 1, 1978

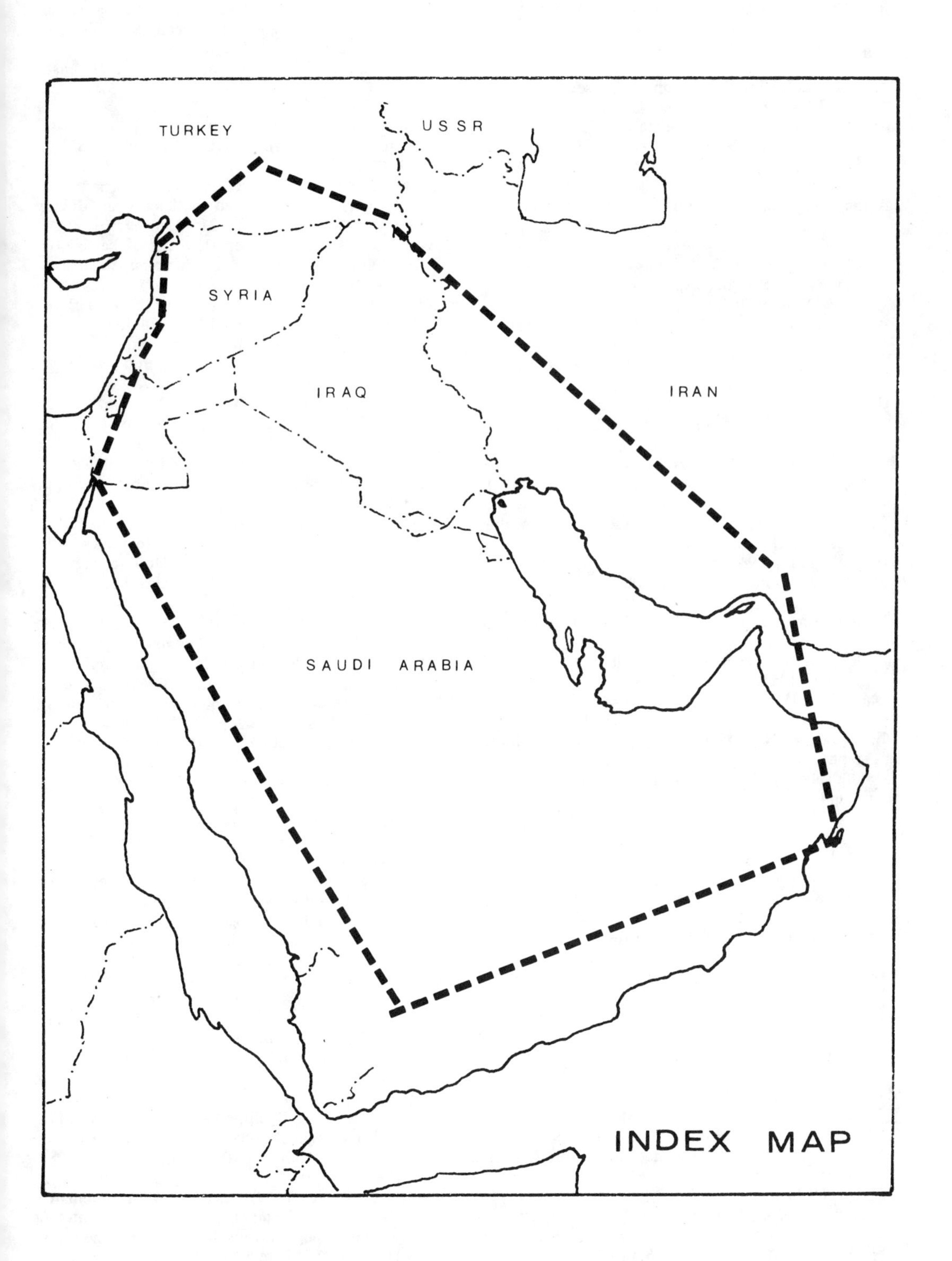
TURKEY
USSR
SYRIA
IRAQ
IRAN
SAUDI ARABIA
INDEX MAP

BIBLIOGRAPHY OF AAPG PUBLICATIONS[1,2]
(Arabian Gulf Region)

Ala, M. A., 1974, Salt diapirism in southern Iran: AAPG Bull., v. 58, p. 1758-1770.

Amiel, A. J., and G. M. Friedman, 1971, Continental sabkha in Arava Valley between Dead Sea and Red Sea: significance for origin of evaporites: AAPG Bull., v. 55, p. 581-592.

Arabian American Oil Company Staff, 1959, Ghawar oil field, Saudi Arabia: AAPG Bull., v. 43, p. 434-454.

Baker, N. E., and F. R. S. Henson, 1952, Geological conditions of oil occurrence in Middle East fields: AAPG Bull., v. 36, p. 1885-1901.

Banner, F. T., and G. V. Wood, 1964, Lower Cretaceous-Upper Jurassic stratigraphy of Umm Shaif field, Abu Dhabi Marine Areas, Trucial Coast, Arabia: AAPG Bull., v. 48, p. 191-206.

Clapp, F. G., 1936, Geology and bitumens of the Dead Sea area, Palestine and Transjordan: AAPG Bull., v. 20, p. 881-909.

Colman-Sadd, S. P., 1978, Fold development in Zagros simply folded belt, southwest Iran: AAPG Bull., v. 62, p. 984-1003.

Daniel, E. J., 1954, Fractured reservoirs of Middle East: AAPG Bull., v. 38, p. 774-815.

De Golyer, E., 1944, Preliminary report of the technical oil mission to the Middle East: AAPG Bull., v. 28, p. 919-923.

Dunnington, H. V., 1958, Generation, migration, accumulation, and dissipation of oil in northern Iraq, *in* L. G. Weeks, ed., Habitat of oil: AAPG, p. 1194-1251.

Elder, S., 1957, Sediments and water of Persian Gulf: discussion: AAPG Bull., v. 41, p. 332-333.

El-Naggar, Z. R., and I. A. Al-Rifaiy, 1972, Stratigraphy and microfacies of type Magwa Formation of Kuwait, Arabia; pt. 1: Rumaila Limestone Member: AAPG Bull., v. 56, p. 1464-1493.

—— —— 1973, Stratigraphy and microfacies of type Magwa Formation of Kuwait, Arabia; pt. 2: Mishrif Limestone Member: AAPG Bull., v. 57, p. 2263-2279.

Emery, K. O., 1956, Sediments and water of Persian Gulf: AAPG Bull., v. 40, p. 2354-2383.

Eyoub, D., 1930, Prospective oil territory in Turkey: note: AAPG Bull., v. 14, p. 1088-1089.

—— 1931, Petroleum possibilities of Turkey: AAPG Bull., v. 15, p. 629-669.

Falcon, N. L., 1958, Position of oil fields of southwest Iran with respect to relevant sedimentary basins, *in* L. G. Weeks, ed., Habitat of oil: AAPG, p. 1279-1293.

Fohs, F. J., 1927, Geology and the petroleum and natural gas possibilities of Palestine and Sinaitic Peninsula: AAPG Bull., v. 11, p. 135-149.

—— 1947, Oil-reserve provinces of Middle East and southern Soviet Russia: AAPG Bull., v. 31, p. 1372-1383.

Gill, W. D., and M. A. Ala, 1972, Sedimentology of Gachsaran Formation (Lower Fars Series), southwest Iran: AAPG Bull., v. 56, p. 1965-1974.

Glennie, K. W., et al, 1973, Late Cretaceous nappes in Oman Mountains and their geologic evolution: AAPG Bull., v. 57, p. 5-27.

Greig, D. A., 1958, Oil horizons in the Middle East, *in* L. G. Weeks, ed., Habitat of oil: AAPG, p. 1182-1193.

Hart, E., and J. T. C. Hay, 1974, Structure of Ain Zalah field, northern Iraq: AAPG Bull., v. 58, p. 973-981.

Henson, F. R. S., 1950, Cretaceous and Tertiary reef formations and associated sediments in Middle East: AAPG Bull., v. 34, p. 215-238.

Howard, W. V., 1933, Reservoir rocks of Persian oil fields and other limestone reservoirs: discussion: AAPG Bull., v. 17, p. 563-565.

Hull, C. E., and H. R. Warman, 1970, Asmari oil fields of Iran: AAPG Mem. 14, p. 428-437.

Ilhan, E., 1967, Toros-Zagros folding and its relation to Middle East oil fields: AAPG Bull., v. 51, p. 651-667.

James, G. A., and J. G. Wynd, 1965, Stratigraphic nomenclature of Iranian oil consortium agreement area: AAPG Bull., v. 49, p. 2182-2245.

Kamen-Kaye, M., 1970, Geology and productivity of Persian Gulf synclinorium: AAPG Bull., v. 54, p. 2371-2394.

—— and H. H. Wilson, 1970, Late Cretaceous eugeosynclinal sedimentation, gravity tectonics, and ophiolite emplacement in Oman Mountains, south Arabia: discussion and reply: AAPG Bull., v. 54, p. 536-538.

Kendall, C. G. St. C., and P. A. d'E. Skipworth, 1969, Holocene shallow-water carbonate and evaporite sediments of Khor al Bazam, Abu Dhabi, southwest Persian Gulf: AAPG Bull., v. 53, p. 841-869.

Kent, P. E., 1958, Recent studies of south Persian salt plugs: AAPG Bull., v. 42, p. 2951-2972.

—— and H. D. Hedberg, 1976, Salt diapirism in southern Iran: AAPG Bull., v. 60, p. 458.

Law, J., 1957, Reasons for Persian Gulf oil abundance: AAPG Bull., v. 41, p. 51-69.

Lebküchner, R. F., F. Orhun, and M. Wolf, 1972, Asphaltic substances in southeastern Turkey: AAPG Bull., v. 56, p. 1939-1964.

Lees, G. M., 1933, Reservoir rocks of Persian oil fields: AAPG Bull., v. 17, p. 229-240.

Mason, S. L., 1930, Geology of prospective oil territory in Republic of Turkey: AAPG Bull., v. 14, p. 687-704.

McQuillan, H., 1973, Small-scale fracture density in Asmari Formation of southwest Iran and its relation to bed thickness and structural setting: AAPG Bull., v. 57, p. 2367-2385.

—— 1974, Fracture patterns on Kuh-e Asmari Anticline, southwest Iran: AAPG Bull., v. 58, p. 236-246.

Metwalli, M. H., G. Philip, and M. M. Moussly, 1974, Petroleum-bearing formations in northeastern Syria and northern Iraq: AAPG Bull., v. 58, p. 1781-1796.

Moody, J. D., K. W. Glennie, and B. M. Reinhardt, 1974, Late Cretaceous nappes in Oman Mountains and their geologic evolution: discussion and reply: AAPG Bull., v. 58, p. 889-898.

Omara, S., 1964, Diapiric structures in Egypt and Syria: AAPG Bull., v. 48, p. 1116-1125.
Owen, R. M. S., and S. N. Nasr, 1958, Stratigraphy of the Kuwait-Basra area, *in* L. G. Weeks, ed., Habitat of oil: AAPG, p. 1252-1278.
Powers, R. W., 1962, Arabian Upper Jurassic carbonate reservoir rocks: AAPG Mem. 1, p. 122-192.
Powers, S., 1926, Reflected buried hills in oil fields of Persia, Egypt, and Mexico: AAPG Bull., v. 10, p. 422-442.
Renouard, G., 1955, Oil prospects of Lebanon: AAPG Bull., v. 39, p. 2125-2169.
Rigo de Righi, M., and A. Cortesini, 1964, Gravity tectonics in foothills structure belt of southeast Turkey: AAPG Bull., v. 48, p. 1911-1937.
Seyed-Emami, K., 1975, Jurassic-Cretaceous boundary in Iran: AAPG Bull., v. 59, p. 231-238.
Steineke, M., and R. A. Bramkamp, 1952, Mesozoic rocks of eastern Saudi Arabia (abs.): AAPG Bull., v. 36, p. 909.
—— —— and N. J. Sander, 1958, Stratigraphic relations of Arabian Jurassic oil, *in* L. G. Weeks, ed., Habitat of oil: AAPG, p. 1294-1329.
Stöcklin, J., 1968, Structural history and tectonics of Iran: a review: AAPG Bull., v. 52, p. 1229-1258.
Sugden, W., 1962, Structural analysis, and geometrical prediction for change of form with depth, of some Arabian plains-type folds: AAPG Bull., v. 46, p. 2213-2228.
Tasman, C. E., 1939, Oil possibilities in southern Turkey: note: AAPG Bull., v. 23, p. 690-691.
—— 1949, Stratigraphy of southeastern Turkey: AAPG Bull., v. 33, p. 22-31.
—— 1950, Stratigraphic distribution of evidences of bituminous substances in Turkey: AAPG Bull., v. 34, p. 1718-1728.
Temple, P. G., and L. J. Perry, 1962, Geology and oil occurrence, southeast Turkey: AAPG Bull., v. 46, p. 1596-1612.
Thode, H. G., and J. Monster, 1970, Sulfur isotope abundances and genetic relations of oil accumulations in Middle East basin: AAPG Bull., v. 54, p. 627-637.
Tleel, J. W., 1973, Surface geology of Dammam dome, Eastern Province, Saudi Arabia: AAPG Bull., v. 57, p. 558-576.
Welland, M., and K. W. Glennie, 1975, Late Cretaceous nappes in Oman Mountains and their geologic evolution: discussion and reply: AAPG Bull., v. 59, p. 1686-1688.
Wilson, H. H., 1969, Late Cretaceous and eugeosynclinal sedimentation, gravity tectonics, and ophiolite emplacement in Oman Mountains, southeast Arabia: AAPG Bull., v. 53, p. 626-671.
—— K. W. Glennie, and M. W. Hughes Clarke, 1973, Late Cretaceous nappes in Oman Mountains and their geologic evolution: discussion and reply: AAPG Bull., v. 57, p. 2282-2290.

[1] Does not include annual "Foreign Developments" papers, which (since 1970) include several excellent bibliographies.
[2] Through June 1978.

SELECTED BIBLIOGRAPHY FROM NON-AAPG SOURCES

Adasani, M., 1965, The Greater Burgan field: 5th Arab Petroleum Cong. (Cairo), Paper 35 (B-3), 27 p.
Akkus, M. F., 1971, Geologic and stratigraphic investigation of the Darende-Balaban basin (Malatya, ESE Turkey): Turkey Mineral Research and Explor. Inst. Bull. 76, p. 1-54.
Al-Ani, A. A., 1975, The geology and structure of Buzurgan field—southeast Iraq: 9th Arab Petroleum Cong. (Dubai), Paper 122 (B-3), p. 1-7.
Allemann, F., and T. Peters, 1972, The ophiolite-radiolarite belt of the North-Oman Mountains: Eclogae Geol. Helvetiae, v. 65, p. 657-697.
Al-Malek, A. K., 1975, Evolution sedimentologique et paleogeographique du nord ouest Syrien (Kurd-Dagh) pendant le Cretace: 9th Int. Sediment. Cong. (Nice), Theme 5, v. 2, p. 243-250.
Al Naqib, K. M., 1959, Geology of the southern area of Kirkuk Liwa, Iraq: 1st Arab Petroleum Cong. (Cairo), Proc., p. 1-50.
—— 1967, Geology of the Arabian Peninsula, southwestern Iraq: U.S. Geol. Survey Prof. Paper 560-G, 54 p.
Al-Omari, F. S., 1972, Bibliography on the geology of Iraq (pt. I): Iraq Geol. Soc. Jour., v. 5, p. 173-178.
—— and A. Sadek, 1974, New contributions to the upper-Senonian stratigraphy in northern Iraq: Geol. Rundschau, v. 63, p. 1217-1231.
Altinli, I. E., 1966, Geology of eastern and southeastern Anatolia: Turkey Mineral Research and Explor. Inst. Bull 66, foreign ed., p. 35-76.
Andersen, K. H., R. I. Baker, and J. Raoofi, 1964, Development of methods for analysis of Iranian Asmari reservoirs: Iran Petroleum Inst. Bull. 16, 17, December.
Arkell, W. J., R. A. Bramkamp, and M. Steineke, 1952, Jurassic ammonites from Jebel Tawaiq: Royal Soc. London Philos. Trans., ser. B., v. 236, p. 241-313.
Baker, B. H., 1970, The structural patterns of the Afro-Arabian rift system in relation to plate tectonics: Royal Soc. London Philos. Trans., ser. A, v. 267, p. 383-391.
Bender, F., Geology of Jordan: Berlin, Gebruder Borntraeger, 207 p.
—— et al, 1969, Beitrage zur geologie Jordaniens: Geol. Jahrb. Beihefte, heft 81, 247 p.
Beydoun, Z. R., 1966, Geology of the Arabian Peninsula, eastern Aden Protectorate and part of Dhufar: U.S. Geol. Survey Prof. Paper 560-H, 49 p.
—— and H. V. Dunnington, 1975, The petroleum geology and resources of the Middle East: Beaconsfield, Bucks., U.K., Scientific Press Ltd., 99 p.
Bizon, G., et al, 1974, New stratigraphic data on the Tertiary basins of southern Turkey: Inst. Francais Petrole Rev., v. 29, p. 305-326 (in French).
Bramkamp, R. A., and R. A. Powers, 1958, Classifications of Arabian carbonate rocks: Geol. Soc. America Bull., v. 69, p. 1305-1317.

Braud, J., and L. E. Ricou, 1971, The Zagros fault or main thrust, an overthrust and a strike-slip fault (in French): Acad. Sci. Comptes Rendus, ser. D., v. 272, p. 203-206.

British Petroleum Company, 1964, Geologic maps, columns, and sections of the High Zagros of southwest Iran: London, British Petroleum Co., Ltd.

Brown, G. F., 1972, Tectonic map of the Arabian Peninsula: (Arab) Ministry Petroleum and Mineral Res. (Riyadh), Map A-2, scale 1:4,000,000.

Cordey, W. G., 1971, Stratigraphy and sedimentation of the Cretaceous Mardin Formation in southeastern Turkey, *in* A. S. Campbell, ed., Geology and history of Turkey: Petroleum Explor. Soc. Libya, 13th Ann. Field Conf.

Cornelius, P. F. S., et al, 1973, The Musandam expedition 1971-72 scientific results: pt. 1: Geog. Jour., v. 139, pt. 3, p. 400-445.

Danninger, H. P., 1974, The salt domes of Hormuz: Kosmos, v. 70, p. 388-392 (in German).

Dubertret, L., E. J. Daniel, and F. Bender, 1963, Liban, Syrie, Jordanie: Paris, Lexique Strat. Internat., v. III, Asie, fac. 10c, 437 p.

Dunnington, H. V., 1967, Stratigraphical distribution of oilfields in Iraq-Iran-Arabia: Inst. Petroleum Jour., v. 53, no. 520, p. 129-161.

—— et al, 1959, Iraq: Paris, Lexique Strat. Internat., v. III, Asie, fac. 10a, 333 p.

El Aouar, M. A., 1960, Reservoir performance under gas and water injection, Abqaiq field: 2nd Arab Petroleum Cong. (Beirut), v. 2, p. 290-307.

Elder, S., 1963, Umm Shaif oil field: history of exploration and development: Inst. Petroleum Jour., v. 49, no. 478, p. 308-315.

—— and K. F. C. Grieves, 1965, Abu Dhabi marine areas geology: 1st Internat. Cong. "Le Petrole et la Mer," (Monaco), sec. 1, no. 127, 8 p.

Evans, G., et al, 1969, Stratigraphy and geologic history of the sabkha, Abu Dhabi, Persian Gulf: Sedimentology, v. 12, p. 145-159.

Falcon, N. L., 1961, Major earth-flexuring in the Zagros Mountains of southwest Iran: Geol. Soc. London Quart. Jour., v. 117, p. 367-376.

—— 1967, The geology of the north-east margin of the Arabian basement shield: Adv. Sci. (London), v. 24, no. 119, p. 31-42.

—— 1969, Problems of the relationship between surface structure and deep displacements illustrated by the Zagros Range (with discussion), *in* Time and place in orogeny: Geol. Soc. London Spec. Pub. 3, p. 9-22, 292-293.

—— 1973, The Musandam (northern Oman) expedition, 1971-72: Geog. Jour., v. 139, p. 1-19.

—— 1974, Southern Iran: Zagros Mountain, in Mesozoic-Cenozoic orogenic belts: data for orogenic studies: Alpine-Himalayan orogens: Geol. Soc. London Spec. Pub. 4, p. 199-211.

Fleck, R. J., et al, 1976, Geochronology of the Arabian shield, western Saudi Arabia: K-Ar results: Geol. Soc. America Bull., v. 87, p. 9-21.

Fox, A. F., 1959, Some problems of petroleum geology in Kuwait: Inst. Petroleum Jour., v. 45, no. 424, p. 95-110.

—— 1961, The development of the south-east Kuwait oilfields: Inst. Petroleum Review, December, p. 373-379.

—— and R. C. C. Brown, 1968, The geology and reservoir characteristics of the Zakum field, Abu Dhabi: AIME-SPE, Saudi Arabia Sec., 2d Regional Tech. Symposium (Dhahran), Proc., p. 39-66.

Gidon, M., et al, 1974a, Synsedimentary Tertiary sliding and movements in Burujird region, Zagros, Iran: Acad. Sci. Comptes Rendus, ser. D, v. 278, p. 421-424 (in French).

—— et al, 1974b, Late Cretaceous tectonics in the Burujird region, eastern Zagros, Iran: Acad. Sci. Comptes Rendus, ser. D, v. 278, p. 577-580 (in French).

—— et al, 1974c, The character and amplitude of displacements of the Main fault in the Burujird-Durud region, eastern Zagros, Iran: Acad. Sci. Comptes Rendus, ser. D, v. 278, p. 701-704 (in French).

Glennie, K. W., et al, 1974, The geology of the Oman Mountains: Royal Geol. and Mining Soc. (Netherlands) Trans., v. 31, 423 p.

Hajash, G. M., 1967, The Abu Dhabi sheikhdom—the onshore oilfields history of exploration and development: 7th World Petroleum Cong. (Mexico), Proc. P.D. 2A, no. 5, p. 165-188.

Harris, J. T., J. T. C. Hay, and B. N. Twombley, 1968, Contrasting limestone reservoirs in the Murban field, Abu Dhabi: AIME-SPE, Saudi Arabia Sec., 2d Regional Tech. Symposium (Dhahran), Proc., p. 149-187.

Harrison, J. V., 1930, The geology of some salt plugs in Laristan (southern Persia): Geol. Soc. London Quart. Jour., v. 86, p. 463-522.

—— and N. L. Falcon, 1969, Is the Zagros fault line of Iran a wrench fault?: Geol. Mag., v. 106, p. 608.

Haven, H., A. Azar-pey, and E. R. Sayah, 1971, Iran Pan-American geological and geophysical methods in the Persian Gulf area: 8th World Petroleum Cong. (Moscow), Proc. v. 2, p. 131-145.

Haynes, S. J., and H. McQuillan, 1974, Evolution of the Zagros suture zone, southern Iran: Geol. Soc. America Bull., v. 85, p. 739-744.

Helal, A. H., 1968, Stratigraphy of outcropping Paleozoic rocks around the northern edge of the Arabian shield (within Saudi Arabia): Deutsch. Geol. Gesell. Zeitschr., v. 117 (1965), p. 506-543.

Hemer, D. O., 1968, Diagnostic palynological fossils from Arabian formations: AIME-SPE, Saudi Arabia Sec., 2d Regional Tech. Symposium (Dhahran), Proc., p. 311-325.

Hill, A. C. C., and E. P. Zomola, 1960, Associated gas in Saudi Arabia: 2d Arab Petrol. Cong. (Beirut), v. II, p. 267-284.

Hosoi, H., and R. Murakami, 1971, Radiated rupture-faults on the Arabian type oil field (in Japanese): Japanese Assoc. Petroleum Technologists Jour., v. 36, no. 3, p. 10-17.

Ilhan, Emil, 1965, Alpine orogeny and the formation of hydrocarbons in southeast Anatolia:

Turkish Petroleum Adm. Pubs. Bull. 10, p. 31-44.
Kassler, P., 1973, The structural and geomorphic evolution of the Persian Gulf, *in* The Persian Gulf: New York, Springer-Verlag, p. 11-32.
Kats, Ya. G., and T. P. Onufriyuk, 1974, Use of cosmic photographs for verifying geological structure of Zagros and for modifying small-scale geological maps of the region: Vyssh. Ucheb. Zavedeny Izv., Geologiya Razved., no. 12, p. 34-43 (in Russian).
Kent, P. E., 1970, The salt plugs of the Persian Gulf region: Leicester Literary and Philos. Soc. Trans., v. 64, p. 56-88.
—— and H. R. Warman, 1972, An environmental review of the world's richest oil-bearing region—the Middle East: 24th Inter. Geol. Cong. (Canada), Proc. sec. 5, p. 142-152.
Khain, V. Ye., Ya. G. Kats, and A. G. Selitskiy, 1973, Tectonic regionalization and main features of modern structures of Alpine belt in the Near and Middle East; pt. 1, Western segment: Internat. Geol. Rev., v. 15, p. 1117-1127.
—— et al, 1973, Tectonic regionalization and main features of modern structures of Alpine belt in the Near and Middle East; pt. 2, Eastern segment: Internat. Geol. Rev., v. 15, p. 1128-1133.
Kiersznowski, S. F., 1968, The performance of the Khursaniyah Arab reservoir system: AIME-SPE, Saudi Arabia Sect., 2d Regional Tech. Symposium (Dhahran), Proc. 557-571.
Kukal, A., and A. Saadallah, 1970, Paleocurrents in Mesopotamian geosyncline: Geol. Rundschau, v. 59, p. 666-686.
Lebkicher, R., et al, 1960, Aramco handbook: New York, Arabian American Oil Co., 343 p.
Loutfi, G., and A. S. Jaber, 1970, Geology of the upper Albian-Campanian succession in the Kuwait-Saudi Arabia, Neutral Zone, offshore area: 7th Arab Petroleum Cong., Kuwait, Preprint 62 (B-3), 22 p.
Makhus, M., 1974, Regularities of structure and location of native sulfur deposits in the Mesopotamian basin: Akad. Nauk SSSR Doklady, v. 218, p. 171-174 (in Russian).
Malhas, A. R., and A. B. Tal, 1974, The tectonic and stratigraphic framework of Jordan and its implications regarding potential hydrocarbon accumulations: Amman, Jordan, Natural Resources Authority.
Metwalli, M. H., G. Philip, and M. M. Moussly, 1972, Oil geology, geochemical characteristics, and the problem of the source reservoir relations of the Jebissa crude oils (Syrian Arab Republic): 8th Arab Petroleum Cong. (Algiers), Paper 66 (B-3), 20 p.
Millon, R., 1970, Geologic structures revealed by interpretation of the aeromagnetic survey of the Arabian shield: (France) Bur. Recherches Geol. et Minieres Bull., ser. 2, sec. 4, p. 15-27 (in French).
Milton, D. I., 1967, Geology of the Arabian Peninsula, Kuwait: U.S. Geol. Survey Prof. Paper 560-F, 7 p.
Mina, P., M. T. Razaghnia, and Y. Paran, 1967, Geological and geophysical studies and exploratory drilling of the Iranian continental shelf-Persian Gulf: 7th World Petroleum Cong. (Mexico), Proc. PD 9, no. 10, p. 179-222.
Mitchell, R. C., 1956, Aspects geologiques du desert occidental de l'Irak: Soc. Geol. France Bull., ser. 6, tome 6, fac. 4-5, p. 391-406.
—— 1957, Physiographic regions of Iraq: Soc. Geog. Egypte Bull., tome 30, p. 75-96.
—— 1959, Stratigraphic and lithologic reconnaissance studies in northern Iraq: Soc. Geog. Egypte Bull., tome 32, p. 291-299.
Morton, D. M., 1959, The geology of Oman: 5th World Petroleum Cong. (New York), Sec. 1, paper 14, 14 p.
National Iranian Oil Company, 1959, Geological map of Iran: Tehran, Iran, National Iranian Oil Co., scale 1:2,500,000.
Nelson, P. H., 1968, Wafra field, Kuwait-Saudi Arabia Neutral Zone: AIME-SPE, Saudi Arabia Sect., 2d Regional Tech. Symposium (Dhahran), Proc., p.101-120.
Nikolaevskii, A. S., 1972, Upper Cretaceous oil-bearing reef complex in northeastern Syria: Geol. Nefti i Gaza, no. 9, p. 71-76 (in Russian).
Orhun, F., 1969, Characteristic properties of the asphaltic substances in southeastern Turkey, their degrees of metamorphosis and their classification problems: Turkey Mineral Research and Explor. Inst. Bull. 72, foreign ed., p. 97-109.
Picard, L., 1939, On the structure of the Arabian Peninsula: 17th Internat. Geol. Cong. (Moscow), v. 2, p. 433-442.
Plumhoff, F., and H. Schumann, 1966, Biostratigraphy of the Upper Cretaceous in Djebel Abd-el-Aziz, northeastern Syria: Neues Jahrb. Geologie u. Paleontologie Abh., v. 125, p. 345-362.
Ponikarov, V. P., 1966, The geological map of Syria: Damascus, Syrian Arab Republic, Ministry of Industry, 111 p. explanatory notes, scale 1:1,000,000.
—— 1967, The geology of Syria, pt. I: Stratigraphy, igneous rocks, and tectonics: Damascus, Syrian Arab Republic, Ministry of Industry, 229 p.
—— E. D. Sulidi-Kondrat'ev, and E. A. Dolginov, 1966, Geologic history of development and tectonics of Arabian-Nubian shield: Sovetskaya Geologiya, no. 8, p. 67-75 (in Russian).
Powers, R. W., 1968, Arabie Saoudite: Paris, Lexique Strat. Internat., v. III, Asie, fac. 10b.
—— et al, 1966, Sedimentary geology of Saudi Arabia: U.S. Geol. Survey Prof. Paper 560-D, 177 p.
Prosdocimo, L., and M. Aftabrushad, 1968, SIRIP exploration in offshore area of Persian Gulf Bahrgan Sar and Nowruz oil fields discoveries: Iran Petroleum Inst. Bull. 32, p. 43-55.

Qatar Petroleum Company, 1966, A review of the geological occurence of oil and gas in Qatar: 2d Arab Petroleum Cong. (Beirut), v. II, Paper 27.

Quennell, A. M., 1951, Geology and mineral resources of (former) Transjordan, *in* Colonial geological and mineral resources: London, Govt. Printing Office, v. 2, p. 85-115.

—— 1958, The structure and geomorphic evolution of the Dead Sea rift: Geol. Soc. London Quart. Jour., v. 114, pt. 1, p. 1-24.

Reinhardt, B. M., 1969, On the genesis and emplacement of ophiolites in the Oman Mountains geosyncline: Schweizer Mineralog. u. Petrog. Mitt., v. 49, no. 1, p. 1-30. (See also Geologie en Mijnbouw, v. 49, p. 161-163.)

Reyre, D., 1971, Ten years petroleum exploration in Middle East (in French): Rev. Assoc. Tech. Pet., no. 209, p. 31-46.

—— and S. Mohafez, 1970, Preliminary contribution by NIOC-ERAP to the geology of Iran, pt. 1 (in French): Inst. Francais Petrole Ref., v. 25, p. 687-713.

—— —— 1970, Preliminary contribution by NIOC-ERAP to the geology of Iran, pt. 2 (in French): Inst. Francais Petrole Rev., v. 25, p. 979-1014.

Richter, M., 1966, The northern end of the African graben system: Neues Jahrb. Geologie u. Palaontologie Monatsh., no. 1, p. 14-21 (in German).

Ricou, L. E., 1971, The ophiolitic peri-Arab crescent, a belt of nappes formed during the Upper Cretaceous: Rev. Geographie Phys. et Geologie Dynamique, v. 13, p. 327-349 (in French).

Rigassi, D., 1971, Petroleum geology of Turkey, *in* A. S. Campbell, ed., Geology and history of Turkey: Petroleum Explor. Soc. Libya, 13th Ann. Field Conf., p. 453-482.

Roberts, D. G., 1969, Structural evolution of the rift zones in the Middle East: Nature, v. 223, p. 55-57.

Rosen, N. C., 1969, Bibliography of geology of Iran: Iran Geol. Survey Spec. Pub. 2, 77 p.

Saidi, A. M., and R. E. Martin, 1966, Application of reservoir engineering in the development of Iranian reservoirs: Iran Inst. Petroleum Bull. 24.

Sandring, A. J., and W. Sugden, 1975, Qatar: Paris, Lexique Strat. Internat., v. III, Asie, fac. 10b.

Sanlav, F., M. Tolgay, and M. Genca, 1963, Geology, geophysics, and production history of the Garzan-Germik field, Turkey: 6th World Petroleum Cong. (Frankfurt/Main), sec. 1, P.D. 3, Paper 35.

Saudi Arabia, Directorate General of Mineral Resources, 1972, Topographic map of the Arabian Peninsula (scale 1:4,000,000), compiled by U.S. Geol. Survey: Directorate General Mineral Resources, Map AP-1.

Sayyab, A., 1973, Stratigraphy of the Cretaceous-Jurassic contacts at Iraq and neighboring area, *in* Colloque du Jurassique a Luxembourg 1967: Stratigraphie general du Jurassique hors de l'Europe: France, Bur. Recherches Geol. Minieres Mem. 75 (1971), p. 695-700.

—— and R. Valek, 1968, Patterns and general properties of the gravity field of Iraq: 23d Internat. Geol. Cong. (Prague) Proc., sec. 5, p. 129-142.

Schmidt, G. C., 1964, A review of Permian and Mesozoic formations exposed at the Turkey-Iraq border near Harbol: Turkey, Mineral Res. and Explor. Inst. (Maden Tetkik ve Arama Enstitusu Yayinlarindan), Bull. 62.

Seibold, E., and K. Vollbrecht, 1969, The bottom relief of the Persian Gulf (in German): Meteor Forschung, (C), no. 2, p. 29-56.

—— and J. Ulrich, 1970, The bottom relief of northwestern Gulf of Oman (in German): Meteor Forschung. (C), no. 3, p. 1-14.

Shirazi, M., and M. P. Ramazanpour, 1973, Natural gas in Iran and its part in the energy pattern, *in* 4th Symposium Development Petroleum Resources of Asia and the Far East, Proc., v. 3, pt. 2, Documentation: U.N. ECAFE, Mineral Res. Development Ser. 41, p. 139-146.

Shurrab, A. H., 1960, Geology of eastern Nejd: 2d Arab Petroleum Cong. (Beirut), v. II, Paper 26.

Slinger, F. C. P., and J. G. Crichton, 1959, The geology and development of the Gachsaran field, southwest Iran: 5th World Petroleum Cong. (New York), sec. 1, Paper 18, 22 p.

Steineke, M., and M. P. Yackel, 1960, Saudi Arabia and Bahrain: Am. Geog. Soc. Spec. Pub. 31, p. 203-229.

Stepanov, D. L., F. Golshani, and J. Stocklin, 1969, Upper Permian and Permian-Triassic boundary in north Iran: Iran Geol. Survey Rept. 12, 72 p.

Stocklin, J., 1961, Lagunare Formationen und Salzdome in Ostiran: Eclogae Geol. Helvetiae, v. 54, nr. 1.

—— 1974, Possible ancient margins in Iran, *in* C. A. Burke and C. L. Drake, eds., The geology of continental margins: New York, Springer Verlag.

—— and A. O. Setudehnia, 1972, Iran: Paris, Lexique Stratigraphique Internat., v. III, fac. 9b, 376 p.

Tayim, H. A., 1960, Utilization of natural gas in Saudi Arabia: 2d Arab Petroleum Cong., v. II, p. 308-321.

Tchalenko, J. S., 1973, Discussion on the paper, by A. A. Nowroozi, "Focal mechanism of earthquakes in Persia, Turkey, West Pakistan, and Afghanistan and plate tectonics of the Middle East": Seismol. Soc. America Bull., v. 63, p. 731-732.

—— 1975, Strain and deformation rates at the Arabia/Iran plate boundary: Geol. Soc. London Quart. Jour., v. 131, pt. 6, p. 585-586.

—— and J. Braud, 1974, Seismicity of the Zagros (Iran): the main recent fault between 33° and 35°N: Royal Soc. London Philos. Trans., ser. A, v. 277, p. 1-25.

Thralls, W. H., and R. C. Hasson, 1956, Geology and oil resources of eastern Saudi Arabia: 20th Int. Geol. Cong. (Mexico), Symposium on oil and gas deposits, tomo II, p. 9-32.

Tikrity, S. S., and A. A. Al-Ani, 1972, Pre-Cambrian basement of Iraq: 8th Arab Petroleum Cong. Proc., sec. Ƅ-3, no. 96, p. 1-6.

Trusheim, F., 1974, The tectogenesis of the Zagros: Deutsch. Geol. Gesell. Zeitschr., v. 125, p. 119-149 (in German).

Tschopp, R. H., 1967a, The general geology of Oman: 7th World Petroleum Cong. (Mexico), Proc. P. D. 2, Paper 15, p. 189-210.

—— 1967b, Exploration history of the Fahud oilfield: 7th World Petroleum Cong. (Mexico), Proc. P.D. 2, Paper 16, p. 211-223.

Turkey, Institute of Mineral Research and Exploration (Maden Tetkik ve Arama Enstitusu Yayinlarindan), 1963, Geological map of Turkey: Ankara, 10 sheets, scale 1:500,000, explanatory notes in English.

Uflyand, A. K., 1965, Tektonika i istoriya razvitiya severo-vostochnogo okonchaniya Pal'mirid (Siriya): Geotektonika, no. 3, p. 20-36.

U.S. Geological Survey, 1963, Geologic map of the Arabian Peninsula: U.S. Geol. Survey Misc. Geol. Inv. Map, I-270A, scale 1:2,000,000.

Van Bellen, R. C., 1956, The stratigraphy of the "Main Limestone" of the Kirkuk, Bai Hassan, and Qarah Chauq Bagh structures in north Iraq: Inst. Petroleum Jour., v. 42, no. 393, p. 233-263.

Von Wissmann, H., C. Rathjens, and F. Kossmat, 1943, Beitrage zur tektonik Arabiens: Geol. Rundschau, Band 33, Heft, 4-6, p. 221-253.

Walther, H. W., 1972, Salt diapirs in southeastern Iran: Geol. Jahrb., v. 90, p. 359-387 (German).

Wells, A. J., 1969, The crush zone of the Iranian Zagros Mountains, and its implications: Geol. Mag., v. 106, p. 385-394.

Wetzel, R., 1974, Etapes de la prospection petroliere en Syrie et au Liban: Paris, Compagnie Francaise des Petroles, Notes and Mem. 11, p. 40-70.

—— and D. M. Morton, 1959, Contribution a la geologie de la Transjordanie: Notes et Mem. Moyen-Orient, t. 7, p. 95-191.

Williams, G. A. D., 1960, Lateral velocity variations across the Qatar Peninsula and their effect on the interpretation of seismic results: 2d Arab Petroleum Cong. (Beirut), v. II, p. 158-163.

Willis, R. P., 1967, Geology of the Arabian Peninsula: Bahrain: U.S. Geol. Survey Prof. Paper 560-E, 4 p.

Wolfart, R., 1967, Geologie von Syrien und dem Lebanon, *in* Beitrage zur regionalem Geologie der Erde, v. 6: Berlin, Gebrüder Borntraeger, 328 p.

—— 1968, Tectonics and paleogeographic development of the mobile shelf in the area of Syria and Lebanon (in German): Deutsch. Geol. Gesell. Zeitschr., v. 117 (1965), p. 544-589.

Yasamanov, N. A., 1970, Paleogeography of the Near and Middle East in the Early Cretaceous Period (in Russian): Vyssh. Ucheb. Zavedeniy Izv., Geologiya Razved., no. 10, p. 70-77.

Zvereva, O. V., and A. G. Selitsky, 1970, Spatial distribution of oil and gas in the hydrocarbon-bearing basin of the Persian Gulf (in Russian): Vyssh. Ucheb. Zavedeniy Izv., Geologiya Razved., no. 10, p. 106-112.

Reprinted for private circulation from
THE BULLETIN OF THE AMERICAN ASSOCIATION OF PETROLEUM GEOLOGISTS
Vol. 36, No. 10, October, 1952

GEOLOGICAL CONDITIONS OF OIL OCCURRENCE IN MIDDLE EAST FIELDS[1]

N. E. BAKER[2] AND F. R. S. HENSON[3]
New York and London

ABSTRACT

The Middle East geological provinces are: (1) the Arabo-Nubian and Arabo-Somali massifs of pre-Cambrian igneous and metamorphic rocks in western and southern Arabia; (2) the foreland shelf, north and east of the massifs; and (3) the orogenic Taurus-Zagros-Oman mountain belt, peripheral to the foreland.

Interpretations of the tectonics are controversial, some authors emphasizing the possible role of taphrogenesis and vertical movements in the basement, while others postulate a long history of intermittent compression (Alpine orogenies) from the north and northeast.

All commercial oil so far discovered occurs in anticlinal traps in the foreland shelf and the orogenic mountain belt; though the possible presence of other types of accumulation is not excluded.

The sedimentational history of the region was conditioned by slow progressive subsidence of the Arabian foreland, interrupted intermittently by epeirogenic uplift movements which are reflected in thickness and facies variations of the sediments.

Deposits of successive marine cycles tongue into a flange of continental sands surrounding the Arabo-Nubian and Arabo-Somali massifs.

Exclusive of non-productive pre-Mesozoic and post-Miocene rocks, the stratigraphy of the oil-field belt consists of: (1) Triassic to Lower Cretaceous—mainly chemical limestone-dolomite deposition; prolific oil producer in eastern Arabia; (2) Middle Cretaceous to Oligocene—(a) globigerinal limestone, chalk, and marl, with oil in associated reef complexes and in fractured limestone in northern Iraq, southwestern Iran, and southeastern Turkey; (b) sands and shales from western shorelines; prolific production in Bahrein, Kuwait, and Basrah; and (3) Miocene—rapidly alternating evaporites, limestones, and clastics; major oil pools in southwestern Iran and Iraq.

Fracturing and availability of plastic or non-fractured cover are important factors in the migration and accumulation of Middle East oil.

INTRODUCTION

The purpose of this paper is to bring together knowledge of known geologic conditions in the Middle East and to show the effect of these conditions on the

[1] Read before the American Association for the Advancement of Science (Section E), and the Geological Society of America, at Philadelphia, December 28, 1951. Manuscript received, January 12, 1952. Published by permission of the Iraq Petroleum Company, Ltd., and associated companies.

[2] Suite 4916, 500 Fifth Avenue, New York City.

[3] Iraq Petroleum Company, Ltd., London, England.

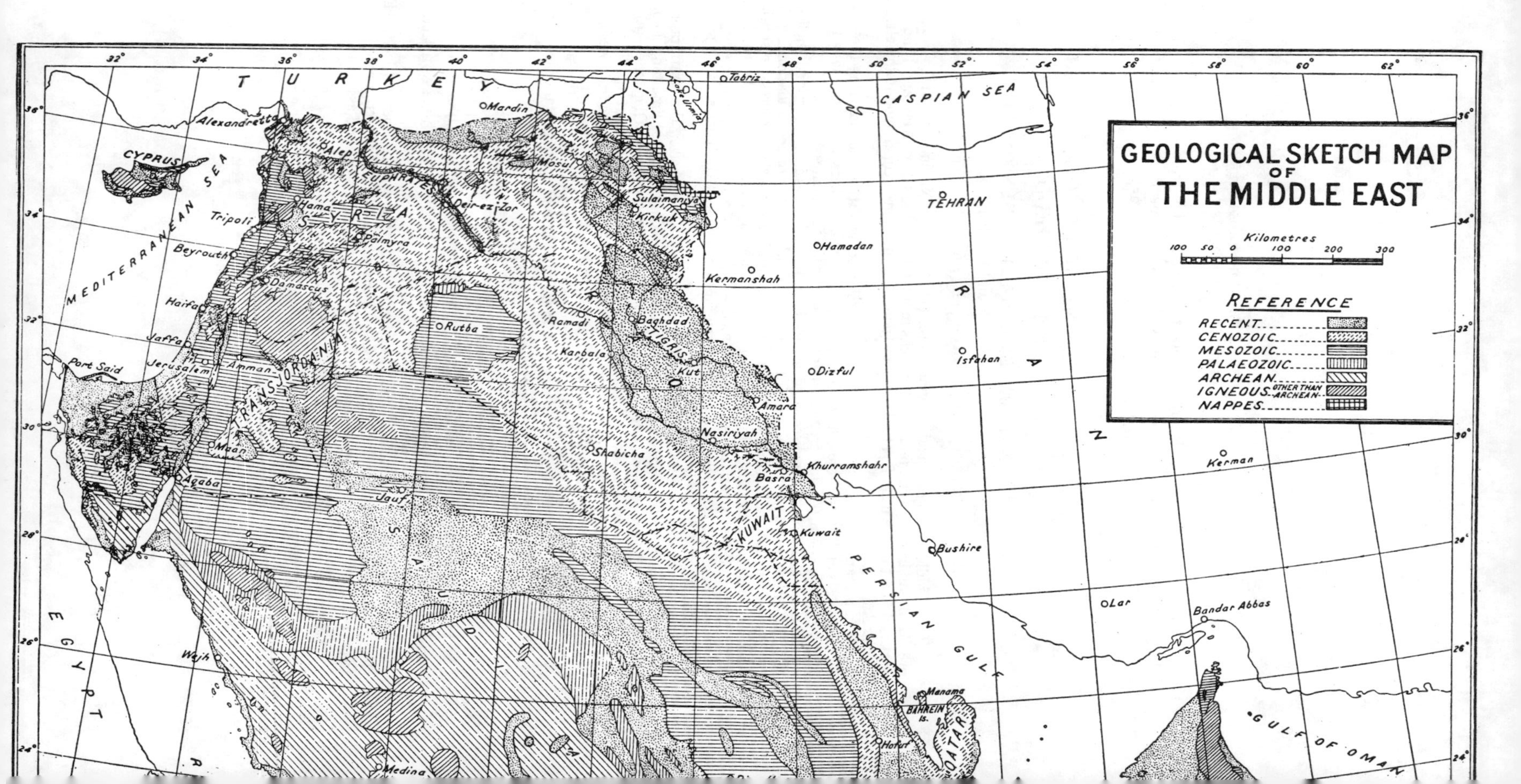
GEOLOGICAL SKETCH MAP
OF
THE MIDDLE EAST
Kilometres
100 50 0 100 200 300
REFERENCE
RECENT
CENOZOIC
MESOZOIC
PALAEOZOIC
ARCHEAN
IGNEOUS OTHER THAN ARCHEAN
NAPPES
TURKEY
IRAN
SYRIA
TRANSJORDANIA
IRAQ
SAUDI ARABIA
KUWAIT
EGYPT
CYPRUS
MEDITERRANEAN SEA
CASPIAN SEA
PERSIAN GULF
GULF OF OMAN
EUPHRATES
TIGRIS
QATAR
BAHREIN IS.
Alexandretta
Alep
Hama
Tripoli
Beyrouth
Palmyra
Damascus
Haifa
Jaffa
Jerusalem
Amman
Port Said
Maan
Aqaba
Wejh
Medina
Jauf
Rutba
Mardin
Mosul
Deir-ez-Zor
Sulaimaniya
Kirkuk
Baghdad
Ramadi
Karbala
Kut
Amara
Nasiriyah
Shabicha
Basra
Khurramshahr
Kuwait
Hofuf
Manama
Tabriz
Lake Urmia
Kermanshah
Hamadan
Dizful
TEHRAN
Isfahan
Bushire
Lar
Kerman
Bandar Abbas

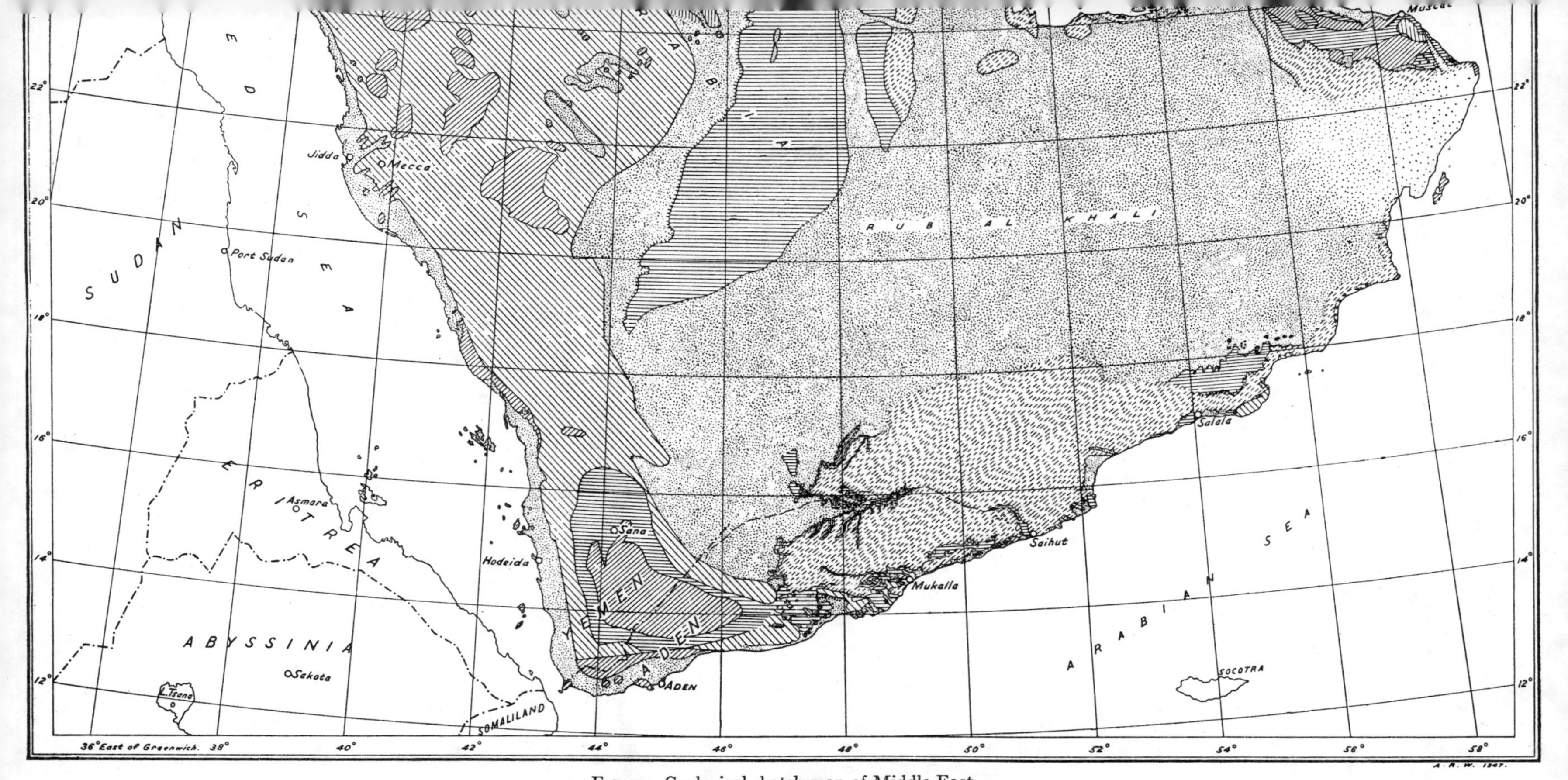

FIG. 1.—Geological sketch map of Middle East.

generation and accumulation of oil. The substance of the paper has been drawn from many sources. However, acknowledgments and bibliographical references are kept to the minimum, since adequate leads to many other sources of information on the Middle East are contained in the works cited in the text and in Figure 2.

The various possibilities of the geologic methods of oil generation and accumulation are suggested; but the main purpose of the paper is to present a concise conception of the fundamental geologic factors which have produced this oil region.

GENERAL GEOLOGY AND TECTONICS (FIG. 1[4] AND FIG. 2)

The foreland of the Middle East region is composed of three geologically distinct provinces.

1. *The Massif zone*, consisting mainly of intensely deformed igneous and metamorphic rocks of pre-Cambrian age. Within this zone, the Arabo-Nubian massif crops out over much of western and central Arabia, and the Arabo-Somali massif, exposed in Somaliland, is believed to extend northeastward off the south coast of Arabia, where peneplaned basement rocks appear locally under sedimentary cover.

Between the Arabo-Nubian and Arabo-Somali massifs, is the Sabea Block of Yemen, a mass of late (?Tertiary-Quaternary) eruptive igneous rocks.

The massif zone determines the western and southern limits of the great sedimentary basins, in which lie the Middle East oil fields.

2. The *Shelf* of Arabia and adjacent lands, extending over a large, now mainly desert, area north and east of the Massif zone. This foreland shelf subsided slowly from Cambrian time onward, beneath shallow epi-continental seas.

Subsidence was interrupted intermittently by slight epeirogenic uplift movements so that the deposits of successive marine cycles tongue westward and southward toward the massifs, overlapping them, or interdigitating with blankets of continental sand which surrounded the massifs from time to time.

As Picard has pointed out,[5] such marine encroachments may have been more extensive than the present evidence suggests, since their deposits have been removed in part during periods of regression.

With this reservation, and with some exceptions, it may be said that successive marine transgressions extended farther west and south until the late Eocene, when the reverse process began with emphasis on regression.

Exclusive of Paleozoic history, which is relatively little known, subsidence and sedimentation in the epi-continental seas of Arabia and its borderlands were remarkably uniform throughout the region considered until the beginning of Cretaceous time, when the Orogenic-Geosynclinal zone began to develop and to define the northern and eastern limits of the foreland proper.

[4] N. E. Baker, *Science of Petroleum*, 2d ed. Published by permission of Oxford University Press.

[5] L. Picard, "Structure and Evolution of Palestine," *Bull. Geol. Dept. Hebrew Univ.*, Jerusalem, Vol. 4 (1943), Nos. 2–4.

3. The *Orogenic-Geosynclinal zone,* caused by tangential compression (Alpine orogenies) from the north, northeast, and east, and now marked respectively by the Taurus, Zagros, and Oman mountain ranges. This orogenic zone is further subdivisible into a *belt of autochthonous folds*, bordering the foreland, and a peripheral *zone of thrusting.*

Known oil fields are confined to the autochthone, and to adjacent parts of the foreland.

The main diastrophisms which have been recognized in the region are indicated by regressions and unconformities (Fig. 3).

In the oil-field areas, most of the recorded movements were gentle, rarely producing any visible angular discordances, until late Tertiary thrusting and folding occurred on a spectacular scale in the Taurus-Zagros-Oman orogenic belt, impressing on the region its present geomorphology.

Before this, broad warpings of the crust occurred, and these produced a changing pattern of gentle basins and swells which may be interpreted either as geanticlines and geosynclines due to incipient "Alpine" compression, or to vertical movements of the basement. Such major features are inferred from thickness and facies variations of the sediments.

In addition, many smaller anticlinal structures rose intermittently in the foreland, some of them at least from the Cretaceous onward; their origin has been ascribed to uplift of basement blocks[6] and/or to the local upthrust of Cambrian salt (Persian Gulf area), but normal folding must also be considered as a possibility.

Deep-seated faulting, blanketed by later sediments, is believed from geophysical evidence to be associated with foreland tectonics, but visible large-scale faulting is rare and of late date.

Figure 2 shows the varied strike of foreland structures, the dominant directions in different areas being north, east, northeast, and northwest, though many intermediate trends occur. The pattern is thought to be more consistent with a history of taphrogenesis than with one of regional compression on a broad front, though opinions differ.

Compression is, however, obvious in the giant late Tertiary anticlines and thrusts of the orogenic zone, with fold-arcs varying in strike, no doubt according to the direction of resistance offered by the jagged edge of the residual Arabian foreland.

Controversy exists about the tectonic history of the Taurus-Zagros-Oman belt before this culminating orogeny.

It is possible that foreland-type tectonics and sedimentation previously predominated in the whole region; indeed, with some exceptions, the orogenic belt is differentiated from the foreland more by its late Tertiary tectonics than by its sediments, and there is evidence, both in its detailed stratigraphy and in

[6] S. W. Tromp, "Block-Folding Phenomena in the Middle East," *Geologie en Mijnbouw,* Vol. 11 (1949), No. 9.

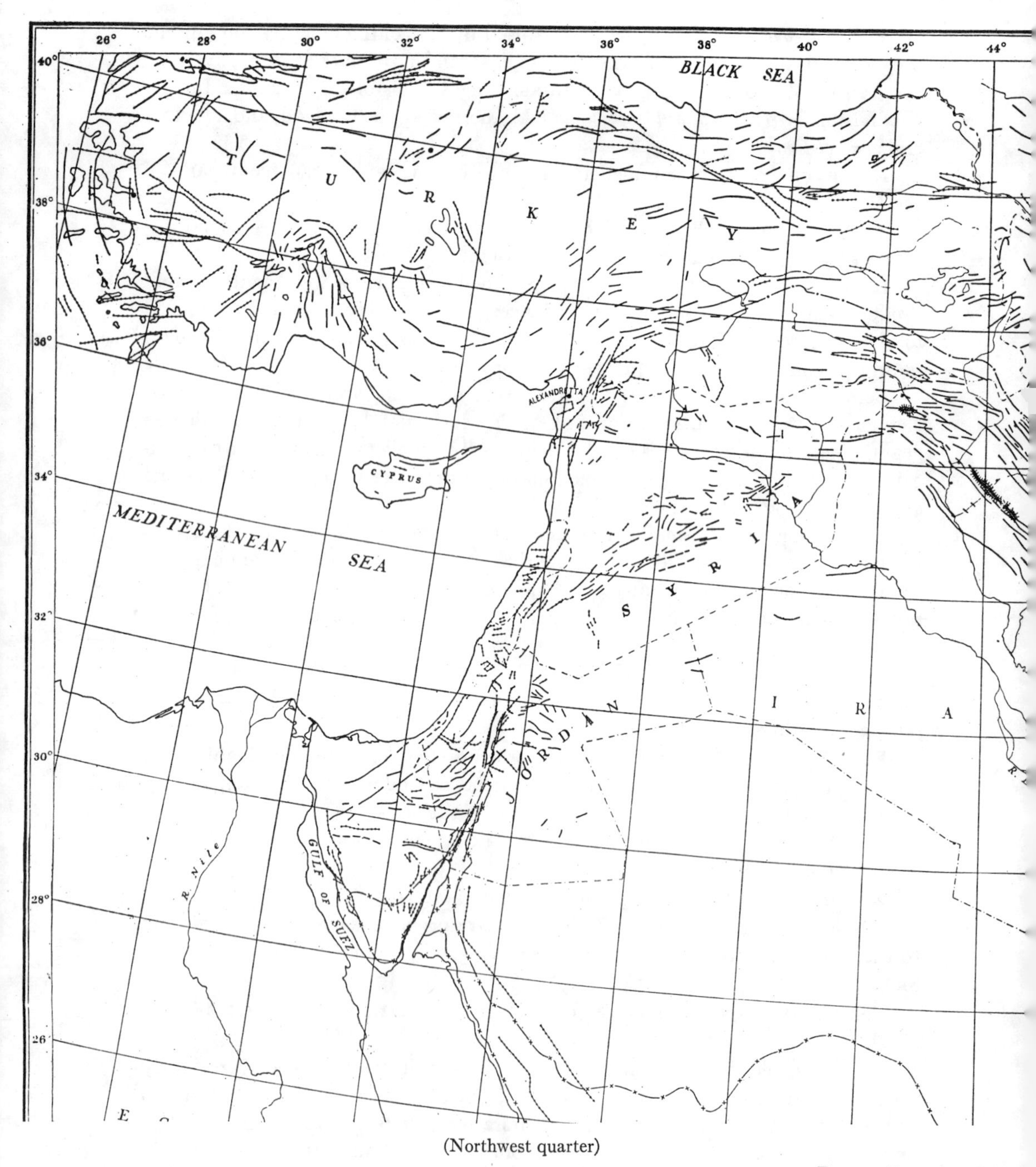

(Northwest quarter)

FIG. 2.—Tectonic map

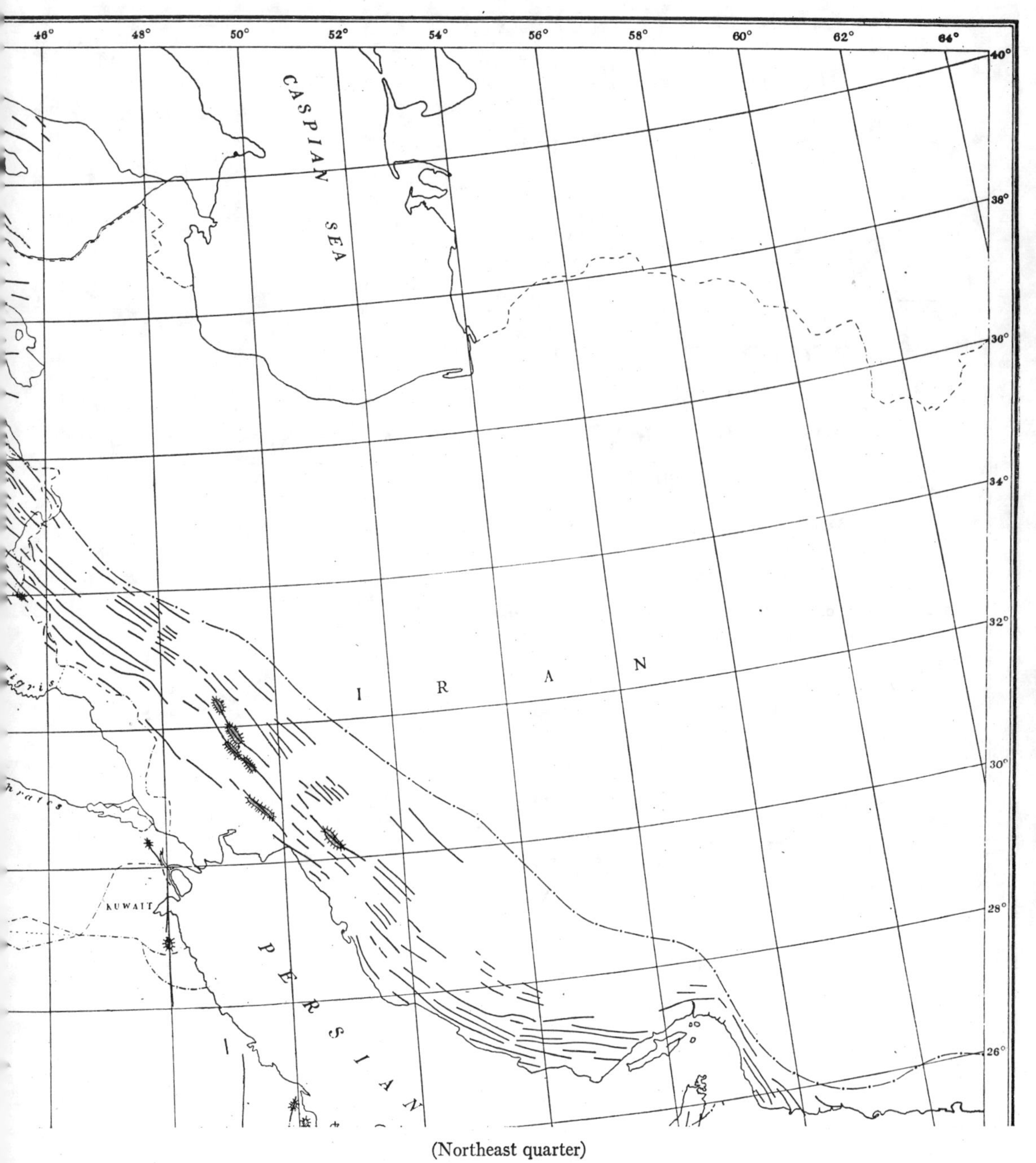

(Northeast quarter)

of Middle East.

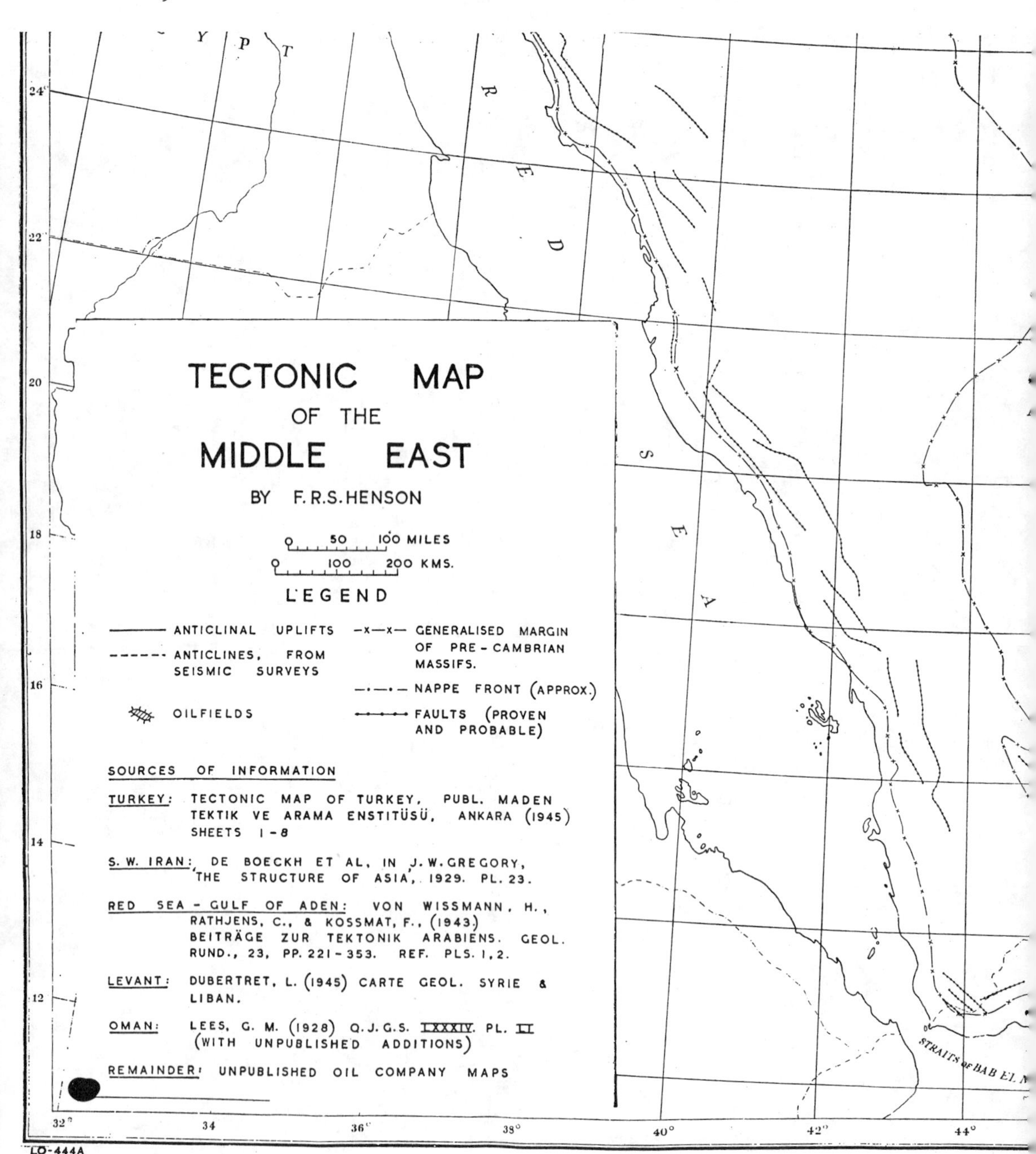

(Southwest quarter)

FIG. 2.—Tectonic map

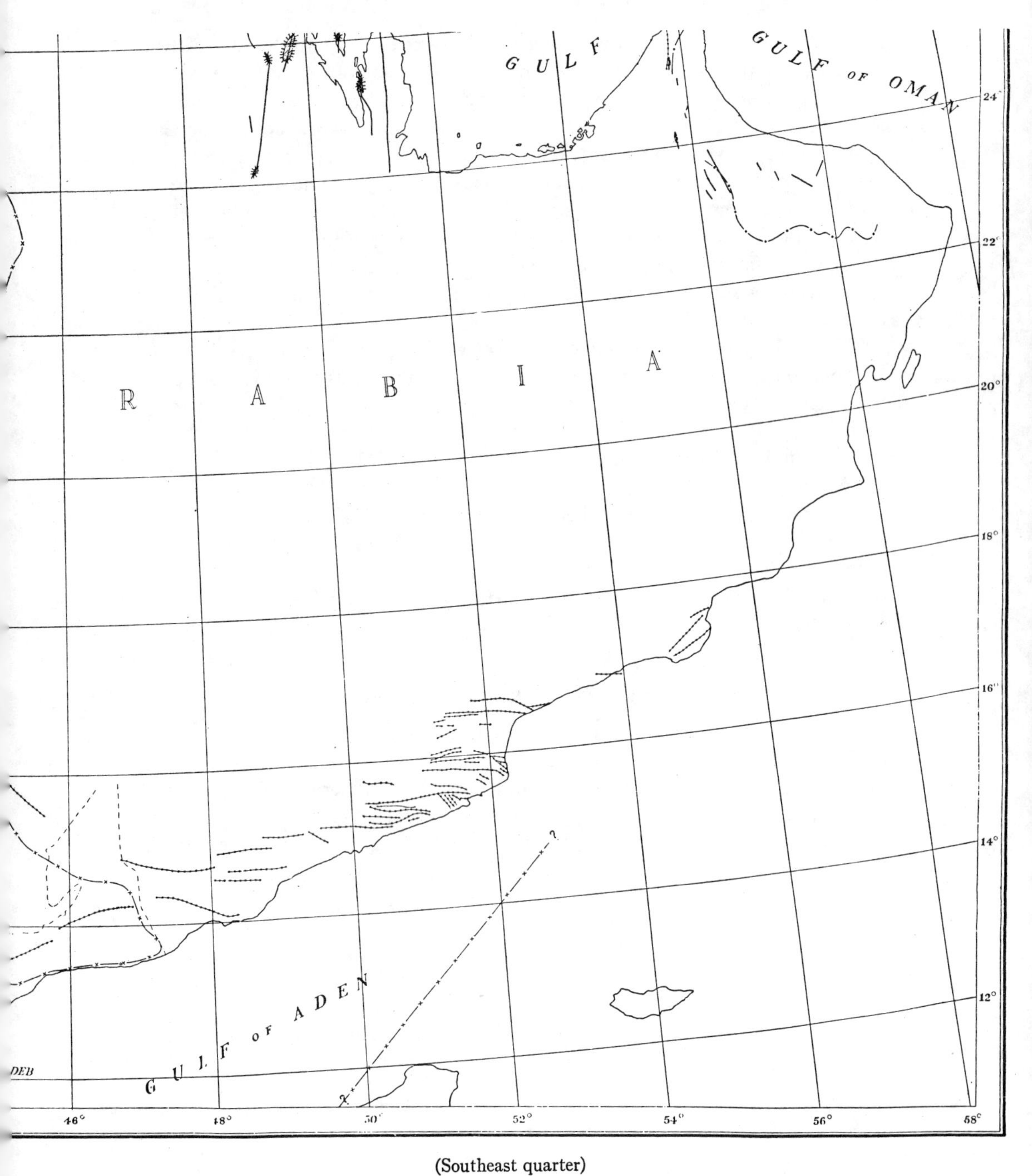

(Southeast quarter)

of Middle East.

the pattern of its folding, that older features may have interfered locally with normal orogenic-geosynclinal development, which was of the complex type.[7]

These questions have been discussed by Henson,[8] who assessed at maximum the possible role of taphrogenesis in the geological history of the Middle East.

Other authors,[9] on the contrary, consider that a dominating part has been played by successive waves of compression, caused by Alpine orogenies from the north (Taurus), northeast (Zagros), and east (Oman). In particular, a Cretaceous phase of powerful thrusting and folding is postulated in the orogenic belt.

Geanticlinal warpings and progressive rise of anticlines alike are attributed to this type of movement, and taphrogenesis is minimized because of the rarity of visible faulting.

No doubt the truth lies somewhere between the two extremes.

Solution of these problems is of obvious interest to the oil prospector, since the evolution of structural patterns and sedimentary environments, hence the origin, migration, and accumulation of oil, are all affected.

No definite answer can be given yet, but some of the considerations involved are presented.

Meanwhile it may be remarked that all commercial oil pools so far discovered in the Middle East occur in anticlinal traps—probable Germano-type or salt-dome structures in the foreland, and normal anticlines in the orogenic belt—but that the possible presence of other (for example, stratigraphic) types of accumulation is not excluded, because no systematic search for them has been made.

STRATIGRAPHY AND OIL OCCURRENCES

A generalized diagram illustrating the nature and variations of the sediments is given in Figure 3. Further discussion is confined to Triassic-Miocene stratigraphy, since all known commercial, and possibly commercial, oil pools are contained in rocks deposited during that time-interval.

The Triassic-Miocene stratigraphy reveals in succession three regionally dominant regimes of sedimentation, with some exceptions. Transition from one regime to the next is not everywhere sharp or exactly synchronous.

TRIASSIC-LOWER CRETACEOUS

Prevalent "chemical" deposits consisting of (a) dense, generally azoic, microcrystalline limestones; (b) oölitic, pseudo-oölitic, and detrital limestones, locally rich in remains of calcareous algae, small mollusks, echinoids, and foraminifera,

[7] F. J. Pettijohn, *Sedimentary Rocks*, p. 447. Harper and Brothers, New York (1949).

[8] F. R. S. Henson, "Observations on the Geology and Petroleum Occurrences of the Middle East," *Proc. Third World Petroleum Congress*, Sect. I, 1951; in press.

[9] G. M. Lees, and F. D. S. Richardson, "The Geology of the Oil-Field Belt of S.W. Iran and Iraq," *Geol. Mag.*, Vol. 77 (1940), No. 3.

P. E. Kent, F. C. Slinger, and A. N. Thomas, "Stratigraphical Exploration Surveys in South-West Persia," *Proc. Third World Petroleum Congress*, Sect. I, 1951; in press.

with sporadic coral and stromatoporoid reef lentils; (c) saccharoidal dolomites, formed mainly by diagenesis of oölitic and saccharoidal types, probably during para-induration; and (d) evaporites, occurring at intervals, with some of considerable lateral extent probably reflecting phases of marine regression.

Subordinate shales occur in some areas.

The oölites and associated detrital limestones vary in proportion to other types in different sectors, and undoubtedly indicate shoaling conditions in shallow water. It is surprising therefore to find that some units of this nature are remarkably widespread and very constant in thickness.

However, it must be remembered that nearly all the evidence comes from the crestal areas of drilled or exposed anticlines; and without information about the nature of the equivalent deposits in synclines, it is impossible to determine whether shoaling was confined to rising anticlinal highs, or whether widespread shallowing alternated with subsidence through intermittent epeirogenic uplifts and/or through basin-filling.

In the outcrops of Nejd and Kurdistan, certain breaks in sedimentation are found within the Triassic-Lower Cretaceous interval, but fossil control is inadequate as yet to correlate these events accurately with facies changes throughout the basins.

Oil occurs prolifically in the porous shoal limestones and dolomites under anhydrite or dense "chemical" limestone cover.

The "Arab zone" (Upper Jurassic) of the Persian Gulf area produces oil from such rocks between persistent anhydrite bands in most, but not all, drilled structures.

Other potentially commercial oil occurrences in limestones of the Triassic (northern Iraq) and Middle Jurassic (eastern Saudi-Arabia; northern Iraq) are more sporadic below limestone cover, possibly because limestone is more liable than anhydrite to fracturing without re-sealing through diagenesis or plasticity under load.

It is believed that all these oil accumulations are indigenous to the formations in which they are found.

The generation of oil may be visualized under stagnant, organically rich bottom conditions, in depressions flanking the shoal areas over which reservoir rocks were formed; or the oil may have formed only during phases of subsidence.

In either occurrence, the alternation of subsidence and shoaling without appreciable current deformation must have provided ideal conditions for retention of the oil until rising anticlines stimulated and directed migration; but the latter may have occurred in several stages, since some of the reservoirs contain bituminous residues as well as live oil.

Notable exceptions to this regime are found where clastics (sands and/or shales) tongue down from littoral zones into the dominantly calcareous deposits of the basins.

Sedimentation of this nature becomes of dominating importance in the Lower

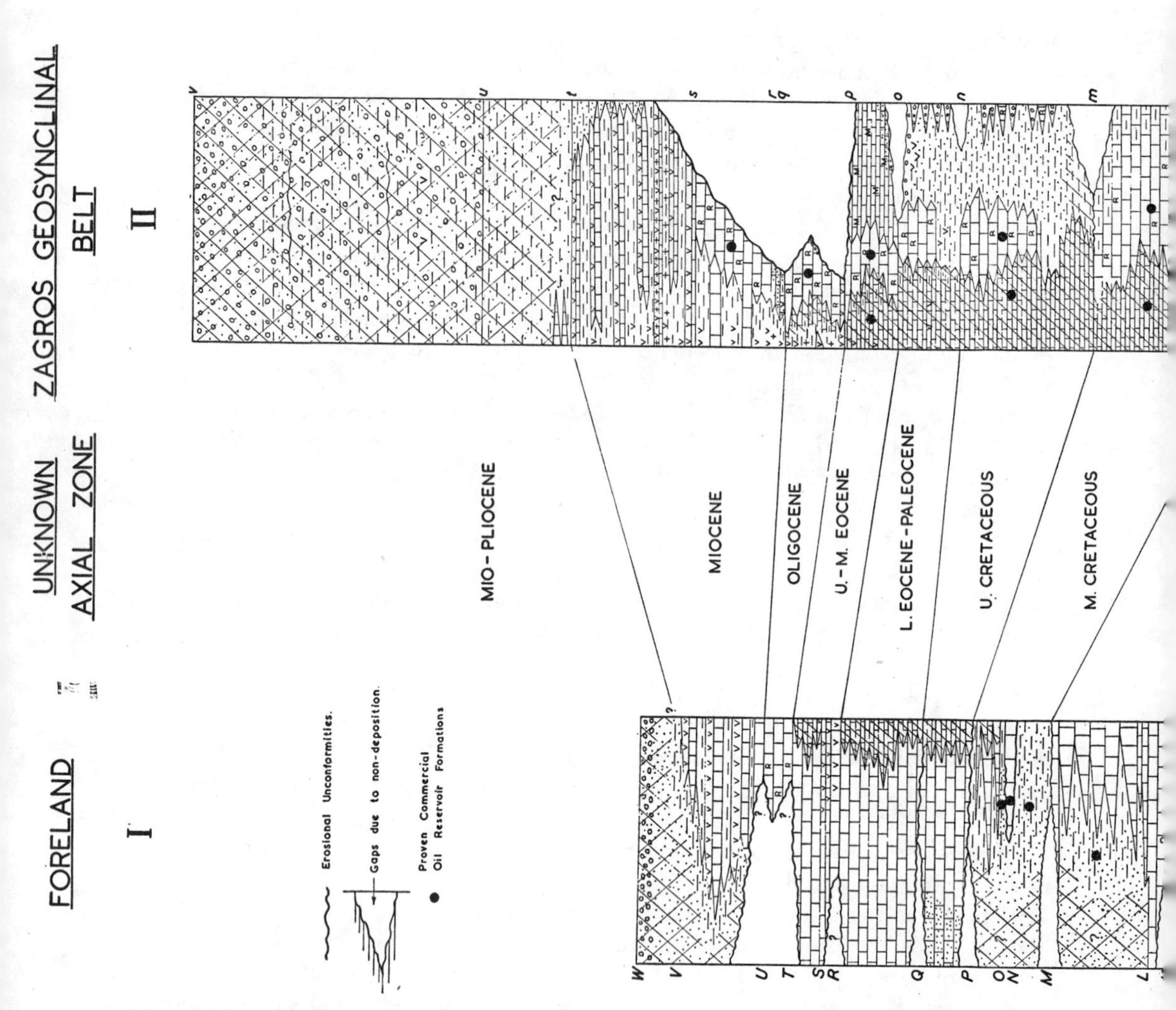

FIG. 3.—Diagram illustrating stratigraphic variations in Zagros-Persian Gulf oil-field belt, by F. R. S. Henson. Not drawn

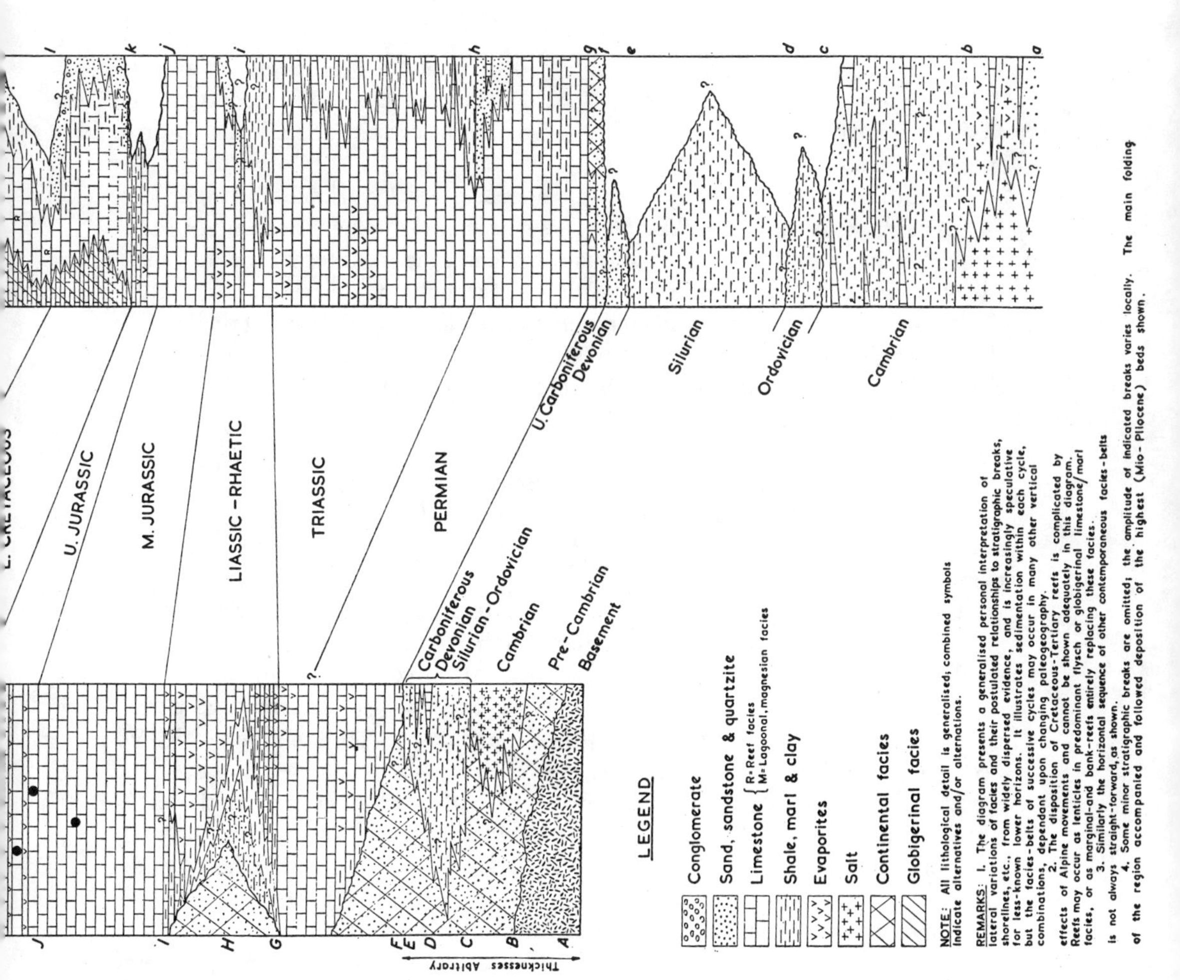

to scale. Thicknesses of intervals are broadly comparative only, representing averages from 35 scattered sections.

Cretaceous of southwestern Iraq, where a sand and shale section largely replaces Neocomian-Aptian limestones and shales of other areas, and has rich oil pools.

Alternations of sand and plastic shale provide ideal reservoir and cover conditions for oil which in part is almost certainly indigenous.

MIDDLE CRETACEOUS-OLIGOCENE

Although the Lower Cretaceous is included in the first regime of sedimentation because of its prevalent facies, it should be remarked that the second regime was introduced exceptionally at this earlier date in certain areas of the Zagros orogenic zone.

Conditions were changing with considerable local variability during the Lower Cretaceous, because of incipient (?synorogenic) Alpine movements which, through succeeding Middle Cretaceous-Oligocene time, resulted in the emphasis and fluctuation of basin and swell structure throughout the region.

Under these controls, characterizing the second regime, globigerinal limestones, chalks, marls, and shales predominated in basinal environments, and passed toward shorelines, and over swells, into reef and shoal limestone complexes.

Further details of this sedimentation and of its possible connections with oil formation and accumulation have been discussed by Henson.[10]

Oil occurs prolifically in an Eocene-Oligocene reef-complex at Kirkuk (Iraq), and relict reef pools, respectively of Middle and Upper Cretaceous age, are exposed nearby.

Upper Cretaceous fore-reef limestones contain potentially commercial oil at Jawan (northern Iraq), and the Raman Dagh (southeastern Turkey) oil field produces from a Middle Cretaceous reef.

Opinion is divided whether these pools are genetically connected with reef conditions or whether the reef limestones serve merely as reservoirs for anticlinal accumulations of oil from other sources.

The latter view receives some support from the fact that commercial oil is obtained also from fractures in Upper Cretaceous limestones of the basinal, globigerinal facies at Ain Zalah (northern Iraq). It is possible that both conditions occur together or separately.

Important exceptions to this regime are the following.

1. The deposition of a widespread sand, shale, and limestone sequence of Middle Cretaceous age in the Persian Gulf region.

Sands are confined to the lower part (Albian-Cenomanian) of the section, tonguing down into the basin from continental sand areas on the west (? and south); they yield the rich oil production of Bahrein, Kuwait, and Basrah (part).

In the latter area, the productive measures repeat (after a limestone intercalation) the conditions of the Lower Cretaceous clastics.

[10] F. R. S. Henson, "Cretaceous and Tertiary Reef Formations and Associated Sediments in the Middle East," *Bull. Amer. Assoc. Petrol. Geol.*, Vol. 34 (1950), No. 2.

There is clear evidence that some of the productive anticlines of the Persian Gulf region were rising at least as early as the Cretaceous, a circumstance very favorable to penecontemporaneous oil accumulation in these sands.

However, only a minority of the Persian Gulf anticlines produce from the Cretaceous sands, and some variations in basinal position and tectonic history may account for this fact through their effects upon oil migration.

2. In certain belts of the Taurus-Zagros orogenic zone, major flysch-filled troughs were developed during Upper Cretaceous and Eocene time, their sediments passing laterally into the reef or globigerinal facies characteristic of the second regime.

Some lenticular sands and conglomerates in the flysch are bitumen-impregnated, but an early test failed to yield production, possibly because the sands contain too much interstitial argillaceous matter.

3. Development in certain areas of basinal evaporites, some of them reflecting marginal uplift movements.

MIOCENE

Just as there is some temporary overlap of the first and second regimes of sedimentation, so are the conditions of the Miocene anticipated in the Oligocene of certain areas, and even in the Eocene of the Persian Gulf region, though with considerable modifications of detail in each.[11]

It has been mentioned that, in the ebb and flow of marine cycles of sedimentation, the emphasis after the Eocene was on regression rather than transgression.

In consequence of the resulting progressive restriction of basins, evaporite deposition became increasingly prevalent in basinal environments.

Thus, in the third regime, a widespread Miocene (lower Fars) formation is found, consisting of more or less cyclical alternations of anhydrite, limestones (some organic), and clastics, with important salt members in some areas.

At the base of the lower Fars, however, is a thick, transgressive limestone (upper Asmari) of neritic facies, which passes basinally into evaporites and shales, at least in the lower Persian Gulf region.

The rich oil fields of southwest Iran produce in part from the upper Asmari limestone, but this Miocene reservoir is in many places superimposed upon Oligocene reef limestones (lower and middle Asmari) which contribute much of the oil.

Both the Miocene and the Oligocene components of the Asmari limestone complex are richly organic and it is possible that some or all of their oil may be indigenous; each is productive in certain areas where the other is absent or barren.

Oligocene reefs are oil-bearing at Kirkuk (Iraq), where the upper Asmari is absent, and bituminous Oligocene shales occur in adjacent Syrian territory.

The upper Asmari is commercially productive at Naft Khaneh (eastern Iraq),

[11] In parts of Syria the conditions of the second regime continue into the Miocene.

where the Oligocene is mainly in evaporite facies; and it is oil-bearing also at Qaiyarah (northern Iraq) where the Oligocene is absent or insignificant.

The lower Fars itself is potentially productive from limestones and basal sands in the Basrah area of Iraq, and contains thin bituminous shales, particularly in the southeastern Persian Gulf region; its oil may be wholly or partly indigenous.

The evidence for indigenous Miocene oil is by no means negligible; but there are definite indications in some areas of the vertical migration of oil from lower sources, through fractures.

Thus a mixed, or allothigenous, origin must also be considered.

This question is referred to again under Structural Conditions of Oil Accumulation.

Deposits following the Miocene lower Fars formation are not of interest here since they are mainly in non-petroliferous, continental facies, expressing further retreat of the sea toward the southeast.

STRUCTURAL CONDITIONS OF OIL ACCUMULATION

Most of the anticlines producing oil in the foreland are considered to be of block-fold[12] and/or salt-plug (Persian Gulf) type. They are relatively gentle, and observed faulting (not everywhere present) is crestal, on a small scale, but increasing in throw at depth.

There is stratigraphic evidence that some, if not all, of these structures rose intermittently during Cretaceous and Eocene time, with possibilities of earlier and later movement.

These conditions were most favorable for normal anticlinal accumulation of oil within migration range of individual structures; and even hard-rock cover has proved effective in most places.

However, the main oil strata are not everywhere productive, a fact which may be explained locally by lack of cover, or regionally, perhaps, by stratigraphic and tectonic factors affecting the primary migration of oil.

In the orogenic zone, conditions are very different, and some views on the southwestern Iranian sector have been published recently by Kent, Slinger, and Thomas.[13]

The productive anticlines are of normal type, because of tangential compression.

Some authorities consider that compression, and consequent anticlinal uplift, occurred intermittently in response to successive phases of Alpine orogeny, providing here also very favorable conditions for anticlinal oil accumulation; however, the evidence on this point is as yet far from conclusive, and it is possible

[12] S. W. Tromp, *op. cit.*

[13] P. E. Kent, F. C. Slinger, and A. N. Thomas, *op. cit.*, 1951.

that effective folding occurred only during the culminating orogeny of late-Tertiary time.

Deformation was then intense, and the dominantly calcareous Mesozoic-Tertiary deposits were highly fractured, thus impairing the effectiveness of hard-rock cover.

There are strong indications in some fields that oil has migrated vertically through fractures and considerable thicknesses of rock, to accumulate finally in porous or broken limestones beneath the first plastic infrangible cover.

In the southwestern Iranian, Naft Khaneh, and Kirkuk pools, this cover consists mainly of lower Fars salt and anhydrite, which preserved these great oil fields. At Ain Zalah (northern Iraq), where oil of constant quality occurs sporadically in fractures through 4,000–5,000 feet of Middle Jurassic-Upper Cretaceous limestones, the cover is provided by plastic shales of the Eocene flysch facies. A similar facies of Upper Cretaceous-Eocene age seals the Raman Dagh (southeastern Turkey) oil pool.

There is, therefore, some support for the view that, given original petroliferous conditions somewhere in the section, the present occurrence of commercial oil pools depends mainly on the disposition of plastic cover; and that the reservoirs in the orogenic belt may contain mixtures of oil from several sources.

However, we are warned against any too general acceptance of this conclusion by an interesting example, in southwestern Iran, of Eocene marlstones which are a seal between distinct Tertiary and Cretaceous reservoirs in the same structure.

Moreover, some authorities discount the importance of any extensive vertical or lateral migration of oil, pointing to the adequacy of associated source and reservoir conditions at many horizons in the Zagros sections, to account for the indigenous origin of known accumulations.

In addition to these problems, many others await further investigation in the autochthone of the orogenic belt.

Dry holes have been drilled in geologically favorable locations, but the easy explanation that oil was never formed in the vicinity is not satisfying in the face of abundant residual oil traces.

It is possible that success or failure depends in some places on the coincidence of late anticlinal traps and original accumulations under stratigraphic and/or pre-orogenic structural controls.

If this is true, then good results may be obtained from a stratigraphic approach to oil-finding, involving studies of basinal morphology, sedimentation, and tectonics in relation to the areal distribution and migration of oil, and offering prospects for the discovery of surviving stratigraphic types of accumulation.

Even in the foreland, such considerations may account for some apparent anomalies in the occurrence of oil at specific stratigraphic positions.

Copyright 1970 by The American Association of Petroleum Geologists

The American Association of Petroleum Geologists Bulletin
V. 54, No. 12 (December, 1970), P. 2371-2394, 13 Figs., 3 Tables

Geology and Productivity of Persian Gulf Synclinorium[1]

MAURICE KAMEN-KAYE[2]
Cambridge, Massachusetts 02138

Abstract The writer attempts to reevaluate depositional history, diastrophism, structure, and oil productivity in the greater Persian Gulf area, and to offer additional reasons for the great petroleum richness of this area. Continuous subsidence and sedimentation through Phanerozoic time resulted in the deposition of maximum thicknesses of 25,000 ft, and possibly more than 30,000 ft under the present mountains and foothills of Iran. The zone of maximum thickness may best be regarded as one relatively thick prism within an immense compound prism of sedimentary rocks. The sedimentary prism also may be regarded as having been the site of persistent epicontinental conditions in a persistent platform environment.

The center of the prism was not greatly uplifted until near the end of Tertiary time, when the geometry of the sediments was profoundly modified. The effect of uplift in the area of study was to deform basement into a relatively simple syncline with a slightly steeper northeastern flank. Overlying sediments perhaps glided down the asymmetrical flank of the syncline to form high-frequency folds, that is, the present fold belt of Iran. The fold belt and largely undeformed platform together constitute a synclinorium, a feature which the writer regards as the most prominent characteristic of the region. The region thus is called the "Persian Gulf synclinorium" in this paper.

During the last two decades, a small amount of specific data relative to the synclinorium has been released. Nevertheless, an initial synthesis of the data into systemic isopach maps and a total-sedimentary-rock isopach map is attempted herein, with the aim of furthering a general understanding of the regional geology. Ordovician and especially Permian events appear to be important among the several marine transgressions. Thus the Permian Sea may have advanced southward from a Tethyan seaway into an ancestral Arabian Sea; this interpretation, which involves an ancestral Indian Ocean on the south, is as plausible as that of a supposed segment of Gondwanaland in the present Indian Ocean during Permian time.

Ultimate producible reserves of crude oil in the Persian Gulf synclinorium are believed to exceed 250 billion bbl, which may mean that as much as 500 billion bbl of in-place oil accumulated in major structures. The main causes previously suggested for prolific reserves were continuity of sedimentation, great total sedimentary volume, good caprocks, rich source materials, and long anticlines. The writer stresses the importance of long anticlines and the presence within them of great reservoir pore volume confined within closure. The writer further stresses the intercalation of source materials and reservoir, and the early release of oil into the reservoir where its accumulation protected initial porosity from the adverse effects of diagenesis. The important sources of the Persian Gulf synclinorium may not be those beds that now are obviously petroliferous.

Introduction

Lees (1950) was among the first to describe the fundamental depositional history and structure of the greater Persian Gulf area, and thus to outline the geologic background of its huge producible reserves. The authors who followed Lees gave valuable additional information, especially Dunnington (1967) and Stöcklin (1968), whose broad reviews brought to a close nearly two decades of publication of technical literature. However, despite many valuable and indispensable contributions during that period, important quantities of specific data were undisclosed and remained unavailable. This lack of availability possibly explains why such elementary illustrations as regional isopach and regional lithofacies maps for individual geologic systems seem not to have been published. In this study the writer undertakes this type of presentation. Perhaps also new for the area is an integration of interpreted and speculative isopach maps into a regional isopach map of total Phanerozoic sedimentary rocks. I hope that the results will stimulate the release of critically needed information, as well as the publication of more accurate isopach maps.

A maximum of about 30,000 ft of sedimentary rocks may be preserved in and below the foothills and mountains of Iran (calculations by other workers show as much as 40,000 ft). The maximum corresponds to part of the Zagros trough described by Stöcklin (1968). Although there can be no doubt of hypersubsidence in Iran, I suggest that the "trough" in reality is only one relatively thick prism within an immense compound prism of thick sedimentary rocks covering hundreds of thousands of square miles. If this interpretation is correct, I would regard the overall environment of deposition as that of a platform rather than that of a basin or

[1] Manuscript received, December 1, 1969; accepted, April 2, 1970.

[2] The writer is grateful to several explorationists who confirmed or added specific points in correspondence. The writer also is grateful to A. A. Meyerhoff and referees who offered valuable constructive criticism of the original manuscript.

geosyncline—despite the presence of widespread subsidence, linear Cretaceous troughs, and a Tertiary foredeep.

The regional and extra-regional platforms are interpreted to have existed through much of Phanerozoic time, and only in late Tertiary and Quaternary times were their geometry and architecture changed significantly. During those times marked uplift in Iran may have caused the basement to be deformed into a relatively simple syncline with a slightly steeper northeastern flank. On this asymmetric flank, high-frequency folding developed in the present fold belt. The development of the fold belt may be regarded as a kind of "overprinting" of the original platform, and the fold belt and the remaining sector of the platform together may be regarded as a synclinorium. The latter is a most noteworthy feature, sufficiently distinctive that I describe the whole region as the "Persian Gulf synclinorium."

Figure 1 shows that the Persian Gulf synclinorium can be defined by outcrops at the corners of an imaginary nearly equilateral triangle. The southwest corner is a salient of crystalline Precambrian rocks; the northern corner, folded Paleozoic sediments; and the southeastern corner, Infracambrian(?) sedimentary strata exposed in the core of an eroded arch. The synclinorium extends beyond the limits of the figure, but an idea of its magnitude is conveyed by the fact that the area of the synclinorium in Figure 1 alone is of the order of 450,000 sq mi.

This study concerns the depositional history, the diastrophism, the structure, and the productivity of the synclinorium. Depositional history includes persistent subsidence, thickness of sediment deposited, and lithofacies. Diastrophism includes relatively gentle regional movement developed during a time-span of geologic periods. Structure includes fold patterns of the platform and the fold belt, together with the assumed development of gliding phenomena in the latter. Productivity includes the physical production of wells, huge reserves, oil in place, and the factors which contributed significantly

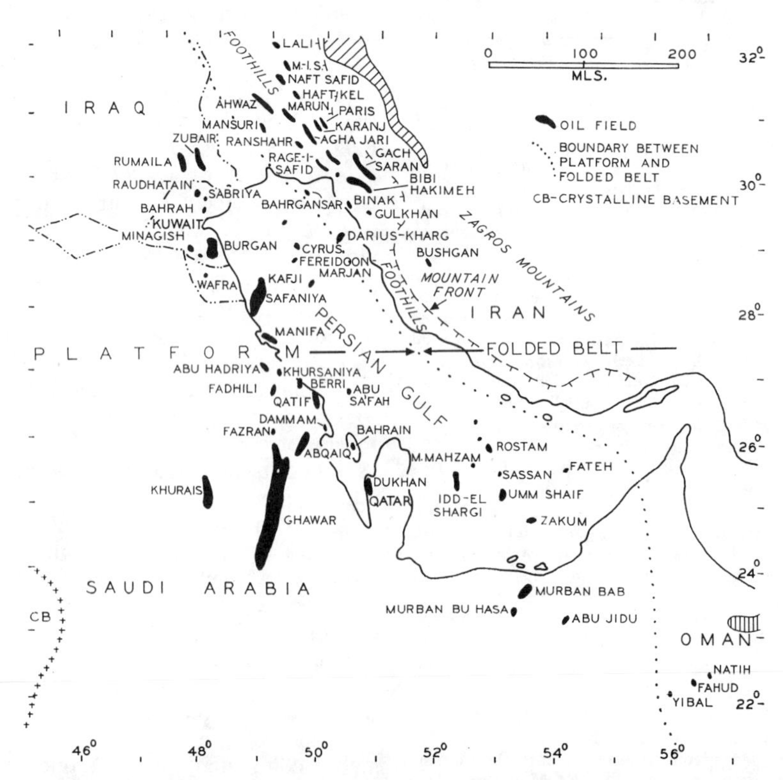

FIG. 1.—Location map of the Persian Gulf synclinorium. CB, crystalline basement, southwest (Saudi Arabia); oblique lines, area containing Paleozoic outcrops, north (Iran); vertical lines, Infracambrian(?) inlier, southeast (Oman). Control point identification is given in Table 3.

to these. Among these factors are source beds and reservoir rocks, and the relation between them is interpreted. The possible nature of petroleum sources for the Persian Gulf synclinorium also is examined.

Depositional History

General

Lees (1950, 1953a, b) differentiated between a pre-Permian and a post-Permian depositional history for the Middle East as a whole. He described a rather fragmentary sedimentary record during the Cambrian, Ordovician, Silurian, Devonian, and Carboniferous, and a regime of essentially continuous deposition during the long interval from the Permian to the Miocene.

In the pre-Permian section, Lees noted an absence of angular unconformities at gaps in the geologic record, and concluded that the early Paleozoic was a time of only mild tectonism. Stöcklin (1968) agreed with this view and pointed out that tectonic calm continued through succeeding periods until the Middle Triassic, at which time areas of central Iran east of the synclinorium became tectonically active.

Pre-Permian

The maximum pre-Permian column of the synclinorium, 10,000 ft thick, is in the mountains of Iran, and may include Infracambrian salt and terrigenous clastic sedimentary rocks in the lowest exposed beds. On the present platform toward the southwest, the column thins to zero where, in the subsurface, Permian oversteps pre-Permian onto basement. Terrigenous clastic lithofacies is prominent in the exposed column and also where wells have penetrated part of the column in the subsurface of the platform in Saudi Arabia and the Persian Gulf. Trilobites, graptolites, brachiopods, and other marine fossils are common in pre-Permian beds.

A pre-Devonian to Devonian sequence probably is present on the Arabian platform north of lat. 23°N to lat. 28°N (Powers *et al.*, 1966). This succession may have been deposited as the result of late pre-Permian or Permian uplift ancestral to the bulge now apparent at lat. 23°N. A similar uplift, possibly ancestral to the present NNE-SSW tectonic lines in southeastern coastal Arabia, may have taken place in Oman, where the evidence supplied by Tschopp (1967) suggests that the pre-Permian sequence was derived from the east, from a Cambrian or Ordovician nucleus. Uplifts such as the two described are foci of maximum pre-Permian beveling; it is reasonable to suppose that other foci of pre-Permian beveling exist on the present platform and that these also imparted patterns to pre-Permian paleogeology. Conditions appear to be different in the present fold belt of Iran. The paleogeologic patterns of the fold-belt area are likely to be pre-Carboniferous.

Variegated strata in pre-Permian lithofacies indicate episodes of regression. However, the common occurrence of a normal invertebrate fauna indicates that marine transgression was important in pre-Permian paleogeography. Ordovician graptolites discovered south of the Fahud oil field in Oman (Fig. 1; Morton, 1959) provide a basis for acceptance of the idea that, at least during Ordovician time, marine water may have advanced southward many hundreds of miles from a supposed northern Tethyan main seaway.

Permian

The thickness of Permian sediments deposited in the synclinorium was not great, and ranges from 700 to 1,500 ft in most localities (Fig. 2). However, thicknesses of 2,000 ft and more have been recorded in the fold belts of Iran and Oman. Lees (1953a, b) and later authors noted that, with the advent of the Permian transgression, carbonate and calcareous sediment deposition prevailed, in contrast to the predominance of terrigenous clastic deposition during pre-Permian time. Dolomite appears to constitute most of the Permian in the areas north, south, and east of the Dukhan oil field (Fig. 1). Limestone may be the principal carbonate farther west on the platform. A thin cap of red shale (part of the Sudair) forms an extensive cover above the carbonates on the platform.

The described Permian flora and fauna suggest that the main transgression did not begin until the latter half of the period. Depositional thickness patterns during transgression do not necessarily reflect diastrophism. Nevertheless, the isopach thinning in the Persian Gulf, if substantiated by additional data, may indicate that local movement took place there along a diffuse north-south axial trend. Some restriction of waters is indicated by the precipitation of anhydrite. Red shale deposition above the carbonate and anhydrite section suggests a broad uplift in the west and a change from open-marine to coastal-plain environment in latest Permian

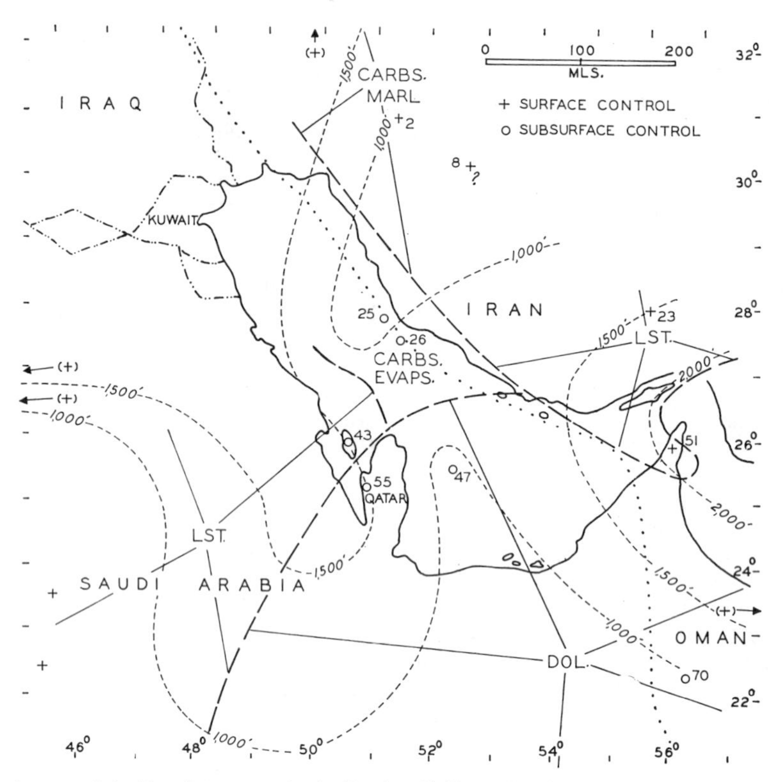

FIG. 2.—Isopach map of the Permian system in the Persian Gulf synclinorium and suggested lithofacies provinces. Contour interval 500 ft. Control point numbers are given in Table 3

time. Before this terminal uplift the average strandline of Permian waters generally was west of the synclinorium. However, Powers *et al.* (1966), basing their conclusions on drilling, showed an eastward swing of the average Permian strandline in south-central Arabia at about lat. 21°N. Farther east, in the present Oman coastal area, Tschopp (1967) reported that Permian marine limestone is present with terrigenous clastic strata at about lat. 20°N. On the basis of this evidence, an important southward trend component can be assumed to modify the general easterly trend of the average Permian strandline where it enters the present Arabian Sea (Fig. 3).

A Permian strandline also can be traced in India, on the basis of Permian freshwater beds with a diagnostic flora (the Lower Gondwana beds of Jacob, 1952), and two Permian marine transgressions onto the Indian shield (Sastry and Shah, 1969). The Indian strandline appears to turn sharply in the vicinity of Delhi and to lead down from a latitude near the Himalayas to one possibly as far south as lat. 22°N. It would be a simple paleogeographic solution to connect the Indian and Arabian ends by a strandline trending west-southwest across lat. 20°N. This solution is preferred by geologists who assume an unbroken Permian continent (Gondwanaland) that includes Antarctica. On the other hand it should be noted that a distance of at least 1,000 mi separates the two ends of the Permian strandlines shown in Figure 3. Across this distance there is space for a Permian Arabian Sea, and this sea might open in turn onto a Permian Indian Ocean, lying as today between India and Antarctica. The presence of marine Permian beds in Madagascar (Besairie, 1952) and in the Karampur well (lat. 29°58′38″N, long. 70°21′29″E) in the Indus basin of West Pakistan (Pakistan Geological Survey open files) significantly strengthens the idea of a Permian Indian Ocean. Nevertheless, in my opinion a choice between Permian ocean and Permian land should be kept open until there has been additional drilling in western India and Pakistan, and new ultra-deep drilling below the sea floor in the area of the

Arabian Sea. For the present even the old idea of land bridges should not be discarded.

Triassic

A complete Triassic sequence was not deposited on the western part of the platform (Powers *et al.*, 1966). However, the isopachs of Figure 4 only partly reflect this fact. The most noticeable effect shown by the isopachs is the presence of two thin areas which have the form of north and south "noses" at the center of the synclinorium. The southern "nose" projects northward to the Bahrein oil field (Fig. 1). The northern "nose" projects southwestward from the Iranian side. Although Triassic data are relatively sparse east of the "noses," it is evident that values increase considerably to exceed 4,500 ft in both Iran and Oman.

The Triassic lithofacies differs from that of the Permian in exhibiting a marked terrigenous clastic province which extends eastward on the present platform, approximately to the vicinity of the Ghawar oil field (Fig. 1). In other zones anhydrite beds are widespread, especially where dolomite is the main carbonate type.

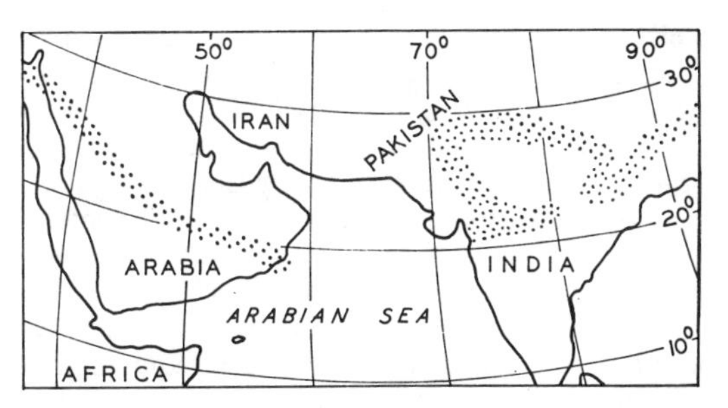

FIG. 3.—Presumed Permian strandlines in Arabia and India (stippled). Permian marine waters existed north of the strandlines. Conditions in much of present Arabian sea unproved to date. Indian peninsula from varied sources, including Sastry and Shah (1965) and Pakistan Geological Survey.

Only dolomite is found in some areas of the present platform and in certain areas of the present fold belt. However, despite the great geographic extent of the dolomite, petrographic evidence indicates that it is a replacement of limestone, and not a penecontemporaneously deposited dolomite.

The isopach "noses" indicate some major

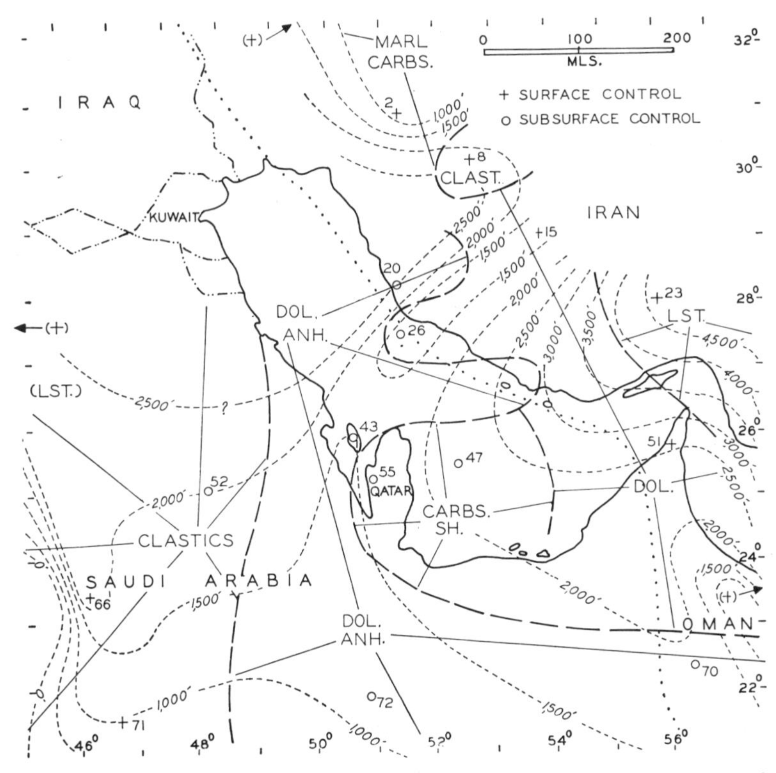

FIG. 4.—Isopach map of the Triassic system in the Persian Gulf synclinorium and suggested lithofacies provinces. Contour interval 500 ft. Control point numbers are given in Table 3.

uplift in the "thin" areas during Triassic time. The axis of uplift trends northward to Bahrein (Fig. 1) and then northeastward across the Persian Gulf into the present mountain area of Iran. The northward trend suggests a continuation of the tectonic movements that characterized the Permian. In contrast to the gentle Triassic uplift, Triassic subsidence appears to have been very marked farther east near the mouth of the present Persian Gulf and in the present Arabian Sea. West of the rise and higher on the present platform, the terrigenous clastic lithofacies contains continental strata, a fact which suggests that there was broad uplift farther west. During the Triassic, strandlines moved eastward and the area of the Triassic sea contracted. The presence of anhydrite may be interpreted to mean that marine waters were restricted as the sea contracted.

Jurassic

The Jurassic regional isopach pattern shows important changes from the Triassic. Despite a delay of deposition in the west until Middle Jurassic time, the thickest Jurassic sequences were deposited there. Thus a systemic isopach maximum of approximately 4,000 ft is present near the Khurais oil field (Fig. 1). The pattern suggests that the isopach maximum trends northward toward Kuwait. Post-Jurassic erosion affects the isopachs west of the maximum, but eastward from the maximum there is a consistent regional thinning. The result is that the thinnest Jurassic columns are in Iran. The southern increase in isopach values into Oman (Fig. 5) is based on uncertain data and may need substantial revision.

Except for terrigenous clastic strata in the southwest, the lithofacies of the rest of the Jurassic in the Persian Gulf synclinorium is predominantly carbonate. Subordinate lithologic types are shale, anhydrite, and halite. Anhydrite is ubiquitous, and extends from the Khurais-Burgan longitude on the west to the Persian Gulf coastal fringe on the east; anhydrite also is present far toward the southeast. Cycles of supposed Late Jurassic anhydrite in the big Saudi Arabian oil fields have been described by

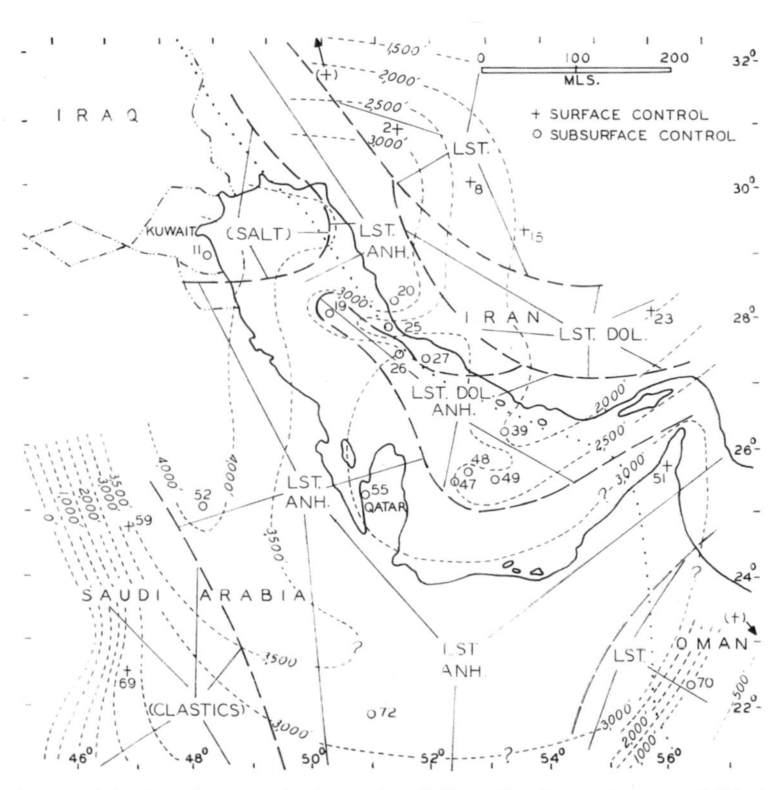

FIG. 5.—Isopach map of the Jurassic system in the Persian Gulf synclinorium and suggested lithofacies provinces. Contour interval 500 ft. Control point numbers are given in Table 3.

Steineke *et al.* (1958) and later authors. Older Jurassic anhydrite also has been described. In exposed Jurassic columns of the present foothills and mountains of Iran, anhydrite is sparse or absent, but Lower Jurassic anhydrite was reported to be a significant lithologic type in a deep boring in the M.I.S. (Masjid-i-Suleiman) foothill oil field (Fig. 1). Jurassic salt is present at the head of the Persian Gulf.

Dolomitized rocks are not widespread in Jurassic carbonate lithofacies. On the basis of available evidence, dolomite is confined to the Persian Gulf area and a small triangular area of onshore Iran. Limestone is by far the predominant carbonate. Current-washed calcarenite is a notable contributor to the limestone column, especially in the supposed Late Jurassic sequence, where it is interbedded cyclically with anhydrites. Calcarenite produces significant amounts of oil at Khurais, Ghawar, Abqaiq, and other coastal oil fields in the Arabian sector of the present platform (Fig. 1).

The isopach pattern does not lead to a clear interpretation of Jurassic diastrophism. Some incipient arching is suggested in the southern half of the Persian Gulf. The general tectonic pattern is significantly different from the northward trends suggested for the Triassic and possibly for the Permian. Jurassic subsidence is evident on the western side of the present platform, as indicated by the presence of a thick terrigenous clastic section, and by the presence farther north of salt. However, the fact that the section thins northeastward into the mountains of Iran should not be construed as evidence of uplift in that direction. On the contrary, the depositional environment of the northeast may well have been one of deeper epicontinental waters in which conditions occasionally were euxinic and where average rates of deposition may have been low. Major restrictions of marine waters took place in the Late Jurassic and resulted in the precipitation of anhydrite over wide areas of the present platform. The precipitation was cyclic, and the intercycles represent periods of shoaling and of clear water when currents washed carbonate sands clean and left them with porosity sufficient to provide for the great petroleum reservoirs of the present Arabian coastal area. Mina *et al.* (1967) showed that calcarenitic and other presumed Late Jurassic limestone porosity extended eastward into the Persian Gulf.

Cretaceous

Figure 6 reveals a considerably changed pattern of isopachs for the Cretaceous Period. The nucleus of the Cretaceous pattern is a thin ovoid area centered on the present Iranian shore of the Persian Gulf, where thicknesses as small as 1,200 ft are developed for a normal or nearly complete Cretaceous succession. A subsidiary thin area appears west of the ovoid area and trends east-northeast, thereby forming a large angle with the east-southeast trend of the ovoid area. North and east of the ovoid area, the Cretaceous isopach values increase considerably and have a northwestward trend which imparts to the isopach pattern a troughlike appearance. Maximum isopach values at the northwest end of the trend may be as great as 10,000 ft. South and southeast of the ovoid area the Cretaceous isopach values increase in two steps. The first involves a regular increase in values (widely spaced contours) to a maximum of about 7,000 ft. The second step is an abrupt eastward thickening to a maximum value possibly as great as 15,000 ft in the present Oman mountain belt. Northwest of the ovoid area the Cretaceous isopach values increase steadily to a maximum of at least 7,000–8,000 ft, *i.e.*, almost as great as the maximum developed in the linear troughlike trend on the northeast.

Sandstones in the middle Cretaceous mark an important development in the general lithofacies. They form an eastward bulge into the head of the Persian Gulf at least as far as the Cyrus oil field (Fig. 1), whereas farther south in the Gulf they bulge eastward toward and possibly beyond the Qatar Peninsula (Fig. 1). East of the areas of sandstone concentration, a lateral facies change to shale takes place. Shale is the principal clastic lithofacies and is widely distributed; in the Cretaceous of Iran and Oman, marl is present with the shale. Of the carbonate components, limestone is considerably more common than dolomite, and most of the known limestone lithologic types are present: *e.g.*, reefal, skeletal, oolitic, calcarenitic, argillaceous, and others.

The Cretaceous isopach pattern suggests a totally new diastrophic development in the history of the synclinorium. Perhaps the most outstanding feature is the already mentioned east-southeast-trending ovoid area. Because the sequence of this area contains many carbonates, I suggest that the ovoid area is a developing arch, not so much one of emergence as one of shoaling. If this interpretation is correct, the shoaling tendency persisted throughout Cretaceous time, because Mina *et al.* (1967) showed a

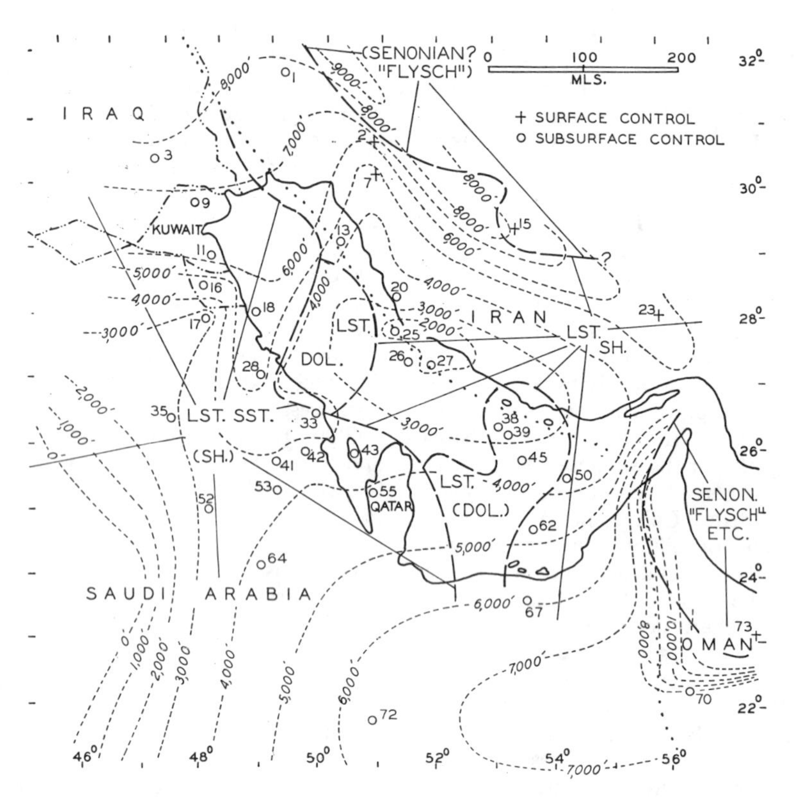

FIG. 6.—Isopach map of the Cretaceous system in the Persian Gulf synclinorium and suggested lithofacies provinces. Contour interval 1,000 ft. Control point numbers are given in Table 3.

Lower Cretaceous isopach "thin" in the same area. Similar reasoning suggests that there also was arching and/or shoaling along an east-northeast trend through northern Ghawar, Abqaiq, and Bahrein (Fig. 1), possibly ancestral to the arch now visible in outcrop at the Precambrian crystalline shield on the west. Both arches are evidence for a new tectonic instability during the Cretaceous. This instability also is manifested in a renewal of uplift in western areas of the platform and erosion of uplifted Paleozoic-early Mesozoic sandstones to supply sand for important middle Cretaceous reservoirs now present in the subsurface of Arabian coastal areas. These are the reservoirs at Burgan, the world's greatest onshore oil field (62 billion bbl), and Safaniya-Khafji (25 billion bbl), the world's greatest offshore oil field (Fig. 1; *see* Table 1). Other evidences of instability are numerous unconformities in the Cretaceous column, especially the well-documented Wasia-Aruma (middle-Upper Cretaceous) unconformity whose areal extent in the synclinorium is in the order of 200,000 sq mi. Subsidence relative to the main arch of the ovoid area was great in many sectors. Southern Iraq sank about 6,000 ft during Cretaceous time. The linear trough east of the foothill oil fields subsided as much as 8,000 ft at its northwestern end. However, maximum subsidence and maximum tectonic unrest occurred in Oman, where the relative downward movement was as great as 13,000–15,000 ft. Wilson (1969) interpreted the evidence in Oman to mean that ophiolites were poured out in Cretaceous time (on the sea floor?) and that large exotic blocks were dumped into local marine waters where the Oman Mountains now stand. However, although tectonic conditions definitely were unstable, it does not follow that Cretaceous diastrophism in Oman was necessarily of "orogenic" or "Alpine" intensity. Wilson believed that some of the Cretaceous clastic debris was provided by a seamount, so that not even simple emergence needed to be postulated to provide for some of the Cretaceous sediments in the unstable paleoenvironment of Oman. The writer (Kamen-Kaye, 1970) suggested that simi-

Table 1. Giant Oil Fields of Persian Gulf Region

Field	*Country*	*Onshore or Offshore*	*Geol. System (Main Production)*	*Reserves (Billions Bbl.)*
Burgan	Kuwait	Onshore	Cretaceous	62
Ghawar	Saudi Arabia	Onshore	Jurassic	45
Safaniya-Khafji	Saudi Arabia	Offshore	Cretaceous	25
Rumaila	Iraq	Onshore	Cretaceous	13.6
Agha Jari	Iran	Onshore	Oligo-Miocene	9.5
Abqaiq	Saudi Arabia	Onshore	Jurassic	9
Gach Saran	Iran	Onshore	Oligo-Miocene	8
Raudhatein	Kuwait	Onshore	Cretaceous	7.7
Ahwaz	Iran	Onshore	Oligo-Miocene	6
Manifa	Saudi Arabia	On- and offshore	Cretaceous	6
Marun	Iran	Onshore	Oligo-Miocene	6
Bibi Hakimeh	Iran	Onshore	Oligo-Miocene	4.5
Sabriya	Kuwait	Onshore	Cretaceous	4
Murban bu Hasa	Abu Dhabi	Onshore	Cretaceous	3
Paris (Faris)	Iran	Onshore	Oligo-Miocene	3
Qatif	Saudi Arabia	On- and offshore	Jurassic	3
Khursaniya	Saudi Arabia	On- and offshore	Jurassic	2.6
Abu Sa'fah	Saudi Arabia	On- and offshore	Jurassic	2.5
Wafra	Neutral Zone	Onshore	Cretaceous and Eocene	2.5
Dukhan	Qatar	Onshore	Jurassic	2.4
Minagish	Kuwait	Onshore	Cretaceous	2
Murban Bab	Abu Dhabi	Onshore	Cretaceous	2
Umm Shaif	Abu Dhabi	Offshore	Jurassic	2
Haft Kel	Iran	Onshore	Oligo-Miocene	1.9
Masjid-i-Suleiman (M.I.S.)	Iran	Onshore	Oligo-Miocene	1.9
Zubair	Iraq	Onshore	Cretaceous	1.9
Idd-el-Shargi	Qatar	Offshore	Jurassic	1.8
Khurais	Saudi Arabia	Onshore	Jurassic	1.5
Sassan	Iran	Offshore	Jurassic	1.5
Berri	Saudi Arabia	Onshore	Jurassic	1.4
Karanj	Iran	Onshore	Oligo-Miocene	1.3
Maydan-Mahzam	Qatar	Offshore	Jurassic	1.1
Dammam	Saudi Arabia	Onshore	Jurassic	1
Darius-Kharg	Iran	On- and offshore	Cretaceous	1
Fahud	Oman	Onshore	Cretaceous	1
Rostam	Iran	Offshore	Cretaceous	1
Total, billions of barrels				249.6

lar nonemergent conditions may have existed when Late Cretaceous (or Paleocene) "flysch" was dumped onto the sea floor of the linear trough in the present mountain areas of Iran ("Senonian flysch" of Fig. 6; after Falcon, 1958).

Tertiary

In the Iranian oil-field region, differences in Tertiary thicknesses as great as 10,000 ft are present between the anticlinal and synclinal axes of a single fold. Because of this, it is not feasible to present the Tertiary isopachs of the fold belt in their true aspect, especially with lack of data and small map scale. Nevertheless, the Tertiary isopachs may be represented as increasing in value from northeast to southwest across the foothill belt of the Iranian oil fields and as decreasing in value farther southwest up the southwestern flank of the present platform. This produces a regional synclinorial pattern. Thicknesses of sections in the cusp of the pattern presumably exceed 10,000 ft and could reach a maximum in the order of 15,000 ft (Fig. 7).

The lithofacies of Tertiary sediments is understood best from the sedimentary rocks of the present fold belt, where thicker sections were deposited and where a larger percentage was preserved than on the present platform. Carbonate dominates the lithofacies of lower Tertiary sediments; limestone is the main carbonate type present. An evaporite lithofacies with salt and anhydrite overlies the basal carbonates and occupies the lower Miocene interval. Upper Tertiary lithofacies is dominated by terrigenous clastic beds, particularly coarse clastic beds which are sandstone below and conglomerate above. This sequence ranges into the Quaternary. Similarly, on the present platform, limestone predominates in the lower Tertiary, but an evaporite sequence (anhydrite of the Rus Formation) appears in the Eocene. This Tertiary evaporite sequence is older than that in the fold belt.

The cusp of the Tertiary synclinorium is

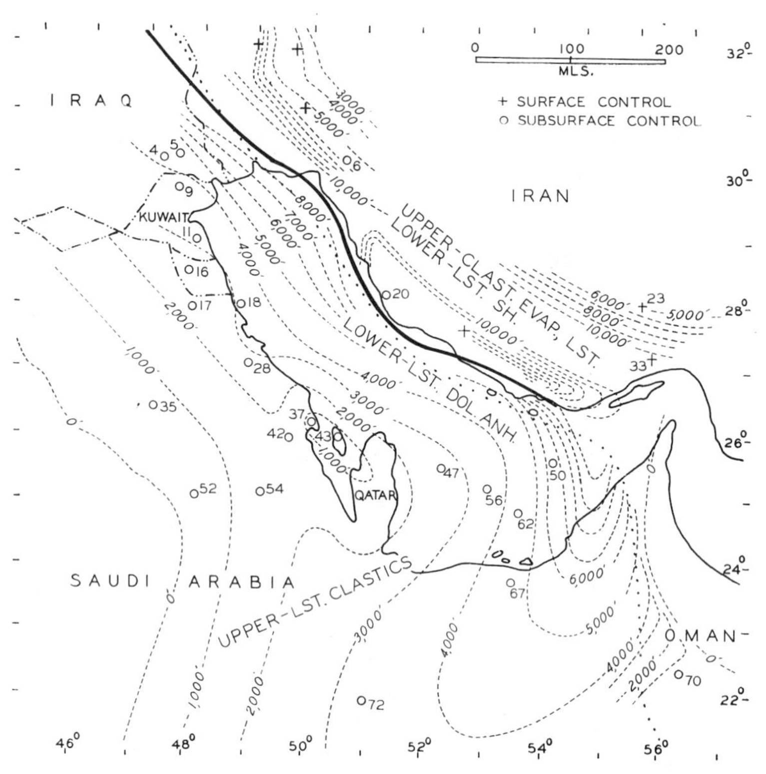

FIG. 7.—Isopach map of the Tertiary system in the Persian Gulf synclinorium and suggested lithofacies provinces. Contour interval 1,000 ft. Thick line along eastern coast of the Persian Gulf marks presumed main Tertiary foredeep. Control point numbers are given in Table 3.

marked on the platform side by a pronounced increase of isopach values across a short lateral distance. This thickening may be interpreted as an abrupt transition from platform to foredeep. The data are not adequate in some areas for defining the Tertiary foredeep closely, but a line may be drawn within comparatively narrow limits to define its position and trend (shown by the thick line of Fig. 7). The foredeep trends southeast across the Iraq-Iran border, then down the eastern shore of the present Persian Gulf. A Tertiary foredeep for Oman is not drawn in Figure 7, but additional data might reveal the presence of a foredeep adjacent to the zero isopach of the present mountain front. If a Tertiary foredeep is present in Oman, it and the foredeep of the Persian Gulf could be genetically related or even continuous. As mapped in Figure 7 the Persian Gulf foredeep coincides closely with a "geosynclinal axis" postulated by previous authors. However, the axis of previous authors appears to include the strata of more than one geologic system, whereas it is clear that the Tertiary isopach pattern is unique and that the pronounced foredeep of Figure 7 is strictly a Tertiary phenomenon.

The foredeep is the "classic" expression of subsidence in front of a rising mountain mass in an unstable tectonic environment. However, in this region, the period of maximum instability during all Phanerozic time was at the close of the Tertiary. The area of the present fold belt was covered by a widespread shallow sea as late as early Miocene (Asmari) time. Restriction began in post-early Miocene time, and the chemical deposition of evaporites increased. First, anhydrite was deposited; the deposition of salt followed. As the mountain front rose, Paleozoic and Precambrian(?) strata were exposed and then eroded, at times torrentially, to provide sand and gravel to the foredeep and adjacent areas. The rise of the mountain front was accompanied by tectonic strain, which was released by yielding in those formations containing anhydrite and salt. As other authors have noted, the area of the present platform remained largely unaffected by forces which deformed the fold belt. However, as Figure 7 shows, the Tertiary isopachs of the platform parallel those of the foredeep (or foredeeps). Because this parallelism is not characteristic in pre-Tertiary patterns (Figs. 2, 4–6), the parallelism may well be an expression of a slight response by the present platform area to stresses which contemporaneously were deforming the fold belt.

Total Sediment

Pre-Permian sedimentary rocks must be included to present an isopach of total sediment of the Phanerozoic time interval in the Persian Gulf synclinorium. However, very few pre-Permian control points—including the zero and maximum points—are available. Powers *et al.* (1966) suggested that the pre-Permian zero isopach should pass about 100 mi southwest of the Khurais oil field (Fig. 1). From this zero position the thickness of the pre-Permian would increase northeastward to a possible maximum in the order of 10,000 ft in the Zagros Mountains of Iran (northeast of the Bushgan oil field; Fig. 1). For additional control, the pre-Permian might be 7,000 ft thick east of Lali, 6,000 ft at Dukhan, and 4,000 ft at Fahud (Fig. 1). If these values are accepted for reconstructing a simple wedge-shaped prism, an approximate volume of pre-Permian rocks can be computed. Although such a method would, in most circumstances, be suspect, the subdued diastrophism of pre-Premian time suggests that this simple wedge model can be used, and therefore integrated with the isopachs of younger geologic systems. To do this, I added the values of pre-Permian, Permian, Triassic, Jurassic, Cretaceous, and Tertiary isopachs at the intersections of a one-degree latitude-longitude grid and constructed the total isopach map accordingly (Fig. 8).

The total isopach map includes erosional values at the edges of the synclinorium, but elsewhere represents the total thickness of the deposits. The predominant pattern is a northwest-southeast-trending maximum beneath the present foothills and mountains. The total thickness within the maximum trend is calculated to

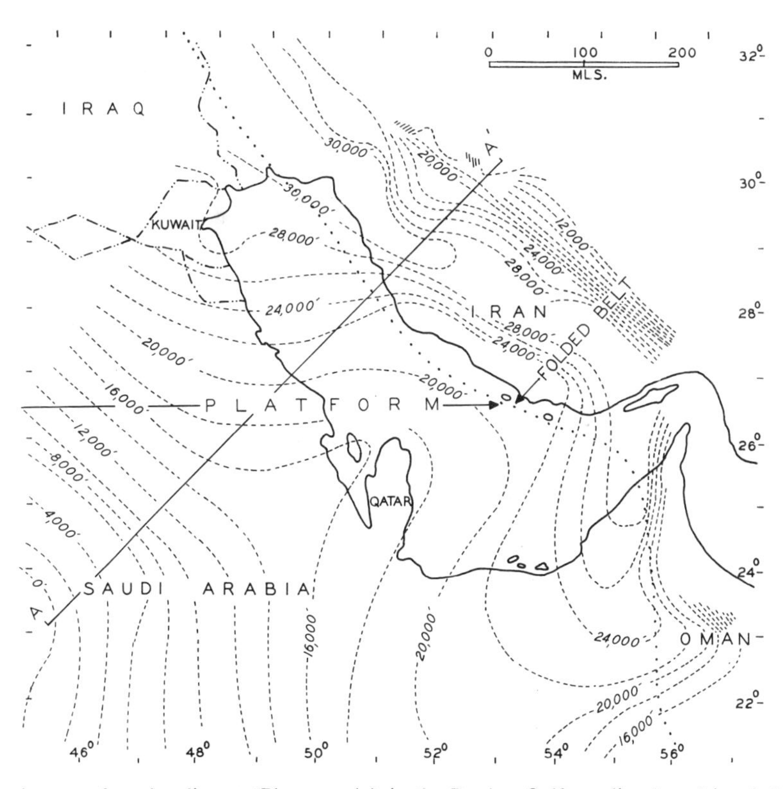

FIG. 8.—Isopach map of total sediment (Phanerozoic) in the Persian Gulf synclinorium. Line A-A′ refers to the cross section of Figure 12.

range from 28,000 to 33,000 ft. Although these values are unlikely to be very precise, they may be accepted as of the proper order of magnitude. The maximum trend itself is quite narrow.

The gross lithofacies of the fold belt may be studied in a lithologic column of total known outcropping sedimentary strata in "Mountain and Foothill Zone—Iran" by the British Petroleum Co. Ltd. (1956). The sequence in the column suggests that a threefold division of lithofacies is present: a middle unit (Permian-early Miocene) of carbonate; and upper (Miocene-Quaternary) and lower (Cambrian-Carboniferous) columns, each with terrigenous clastic sediments and evaporites. The lower column contains a notably thick and probably widespread Cambrian salt member below terrigenous clastic strata which characterize the Ordovician-Carboniferous section. The middle carbonate column is essentially limestone with minor shale and dolomite, and represents Permian to early Miocene time. The upper column, Miocene to Quaternary, contains a mixed lithofacies of terrigenous clastic beds, carbonates, and evaporites, in which the lithologic elements are shale, sandstone, conglomerate, limestone, anhydrite, and rock salt.

Lithofacies on the platform may be studied in a section described by Thralls and Hasson (1956). Pre-Permian sediments are eroded from a bulge in the central part of the area studied, but where present in other areas, the lithofacies proves to be entirely terrigenous clastic strata which are predominantly sandy. Carbonate rocks appear in the Permian and are predominantly limestone; in this respect, the section of the platform resembles that of the fold belt.

Drilling downdip from the belt of western outcrops probably has not yet revealed the total sedimentary column at any place on the platform. A few wells have shown the presence and nature of part of the lower Paleozoic, but most penetration has not been deeper than the Permian. The available data on outcrop and subsurface columns of the platform are synthesized in the regional cross section of Figure 9. The steep "dip" of the beds in the western part of the cross section is a gross distortion caused by an unusually large vertical scale exaggeration. Actually the regional structure is that of a platform overlain by a gently eastward-dipping sequence, from the Precambrian outcrop to the Tertiary foredeep in front of the present fold belt. Unconformities are numerous, and there may be five or more of regionally important unconformities above the Permian alone. Two are intersystemic: the Triassic-Jurassic and the Cretaceous-Tertiary.

Relatively thick pre-Permian strata almost certainly are present in the subsurface of the Persian Gulf and for some distance westward onshore. The known lithofacies is dominantly terrigenous clastic through the Ordovician, which likely is underlain by Cambrian deposits which include evaporites (with a major component of rock salt). Permian deposits are mainly

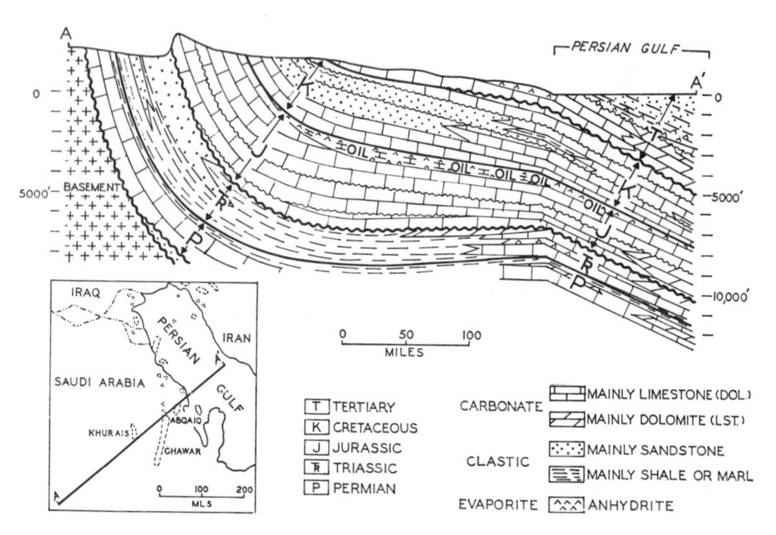

FIG. 9.—Cross section of the platform from Precambrian outcrop to the eastern coast of the Persian Gulf.

limestone, although dolomite makes a definite contribution to the column in some areas of the Persian Gulf. Triassic lithofacies is characterized by terrigenous clastic sedimentary rocks in the west, but limestone and subordinate dolomite and evaporite are characteristic in the east. Jurassic and Cretaceous lithofacies are largely carbonates, and limestone is the most important. However, the middle Cretaceous—a terrigenous clastic lens—is a major exception. Tertiary bed thicknesses are reduced considerably by erosion over most of the platform, and are thickest toward the eastern edge. There they constitute a column which includes carbonates, terrigenous clastics, and evaporites.

Cretaceous and Tertiary arching is reflected in the isopach pattern of total sediment, as well as in a bulging of the lines across the Persian Gulf tract of the synclinorium. However, the bulge is maintained some distance into the present fold belt, a condition which may be explained by the presence of a big frontal fold on the northeast shore of the Persian Gulf. Much Tertiary rock has been eroded from this frontal fold and the consequent reduction of total sediment causes an apparent anomaly in the isopach pattern. In addition to the arching, there are evidences of instability in the numerous unconformities known to be present. However, on the whole, the instability seems to have been moderate. Thus Fox (1956) assumed that the erosion of beds at the middle-Upper Cretaceous contact of the Burgan oil-field area in Kuwait (Fig. 1) took place under water. If he is correct, all of the major unconformities depicted in Figure 9 may not be an expression of emergence and conditions may have been relatively quiet when gaps in deposition developed, especially where the gaps were between carbonates.

A casual glance at the small width of the belt of maximum total sediment might lead one to conclude that maximum deposition had taken place in a narrow trough. Actually, the narrowness of the "trough" is a result of the incomplete isopach thickness on the east side. This incompleteness is a result of large-scale erosion of the rising mountain front. Stöcklin (1968) indicated a trough in the same general area as the "trough" of total sediment, but it is probable that Stöcklin's trough was much broader and extended farther east. In fact the presumed broadness of Stöcklin's trough is in keeping with my concept that the trough in reality is only the thickest part of a supraregional compound prism.

Diastrophism

Except for the Pliocene-Pleistocene episode of mountain uplift and correspondingly intense movements in the present fold belt, the history of the Persian Gulf synclinorium is one of definitely subdued diastrophism. Lack of information on the pre-Permian precludes a system-by-system analysis of diastrophism during that large time interval. However, the extensive development of pre-Permain terrigenous clastic beds suggests that pre-Permian rivers at times flowed from moderately raised lands. Broad spread of the clastics may have been effected by currents working across wide and persistent continental shelves. By Permian time the lands generally had been peneplaned, and conditions were propitious for a broad transgression of marine waters during the Late Permian. However, the Permian transgression, though important, was short lived. Before the end of Permian time there was uplift in the west and deposition of coastal or littoral redbeds across considerable areas in Arabia and the Persian Gulf. The uplift was stronger during Triassic time, a conclusion supported by the increase in grain size in the same areas. Also to judge from Permian and Triassic isopach patterns, shoaling began in marine waters on incipient north-south trends across the Persian Gulf. Farther east Triassic movements were stronger, as shown by the pronounced subsidence and increased thicknesses.

By Jurassic time subsidence was widespread. In western areas subsidence was accompanied by increasing deposition, particularly the precipitation of evaporites near the close of the period. In eastern areas subsidence was accompanied by deepening water, local development of euxinic environments, and by a smaller amount of deposition. Subsidence continued to be pronounced during the Cretaceous; deposition equaled subsidence in many areas. However, Cretaceous diastrophism increased notably compared with earlier periods. A resumption of uplift took place in the west, where exhumation of Paleozoic and Mesozoic sandstones supplied material for important sand development in the middle Cretaceous of eastern Saudia Arabia, Kuwait, and western Persian Gulf. Upward movement also took place on the eastern shore of the Persian Gulf, where shoal water may have persisted throughout Cretaceous time.

In contrast with these western events, subsidence was unusually pronounced in Iraq, Iran, and Oman. A narrow trough developed in Iran

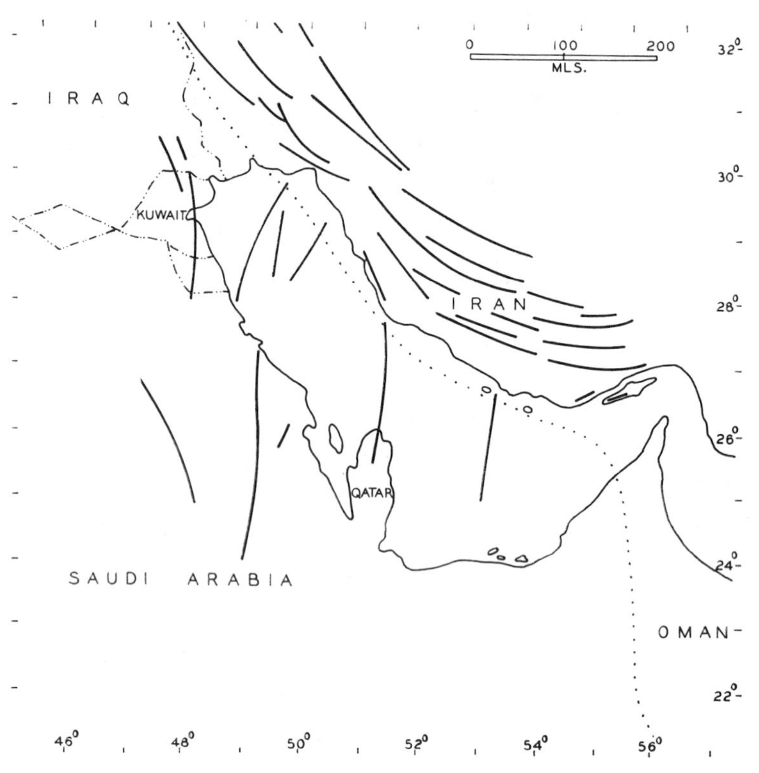

FIG. 10.—Tectonic map of the Persian Gulf synclinorium. Thick lines are anticlinal axes or crests of uplifts.

that was filled during Late Cretaceous and early Tertiary (Paleocene) times with "flysch." In Oman ophiolites were poured out, possibly on the sea floor, and total subsidence may have been as great as 15,000 ft. During Tertiary time subsidence diminished considerably in most areas, but near the end of the period increased to form depressions at the foot of rising mountain fronts. Elsewhere Tertiary diastrophism generally was restrained. Arching of the Ghawar-Bahrein area, begun in the Cretaceous, probably continued into and possibly through Tertiary time.

STRUCTURE

The boundary between the fold belt and platform is shown on Figure 10 by a dotted line. Generally the line divides a northeastern province of high-frequency, parallel, northwest-southeast striking folds from a southwestern province of uplifts with an overall northward component. It is clear that the trends in one structural province make a large angle with the trends in the other.

Mina *et al.* (1967) supplied new data on axial trends of the platform and reported that, in the area of the Darius oil field (Fig. 1), the fold belt structural trend had been superimposed on the platform trend. A similar superposition of structural trends may be present at the Bahrgansar oil field at the head of the Persian Gulf (Fig. 1). A third superposition of structural trends is possible in the eastern Persian Gulf near the small island of Lavan, north of the Sassan oil field (Fig. 1). These superpositions are those of a younger fold system on an older one. In this region, the younger fold system, that of the fold belt, is late Tertiary or younger. The older, northward-trending system of the platform is middle Mesozoic or older. Not surprisingly, the phenomenon of superposition is now seen in detail only in the boundary area between the two main tectonic provinces. However, Henson (1951) believed that the present geology of the fold belt could be interpreted to mean that older north-trending structural trends probably had preceded the now dominant northwest-southeast trend of parallel folds. The direction of the earlier trends, in my opinion, suggests that the platform extended well into the fold belt. This view, which was assumed by Henson, would tend to support the interpretation here that the platform and its epicontinental seas occupied all of the Persian Gulf synclinorium through long periods of time.

The northward structural trends suggest the relative independence of the platform from tectonic pressure generated in the fold belt. However, on a truly regional scale the independence of the platform is seen to be less than absolute. A good example of regional-scale structure is shown on Figure 11, a map of the top of the Cretaceous. The contours are molded into a broadly northwest-southeast trend across much of the Persian Gulf, *i.e.*, a trend at least subparallel with the trend of the fold belt. A similar subparallelism is seen in the isopach pattern of the Tertiary (Fig. 7). The northwest-southeast orientation of axes in the Zubair and Rumaila oil fields (Fig. 1) also may be significant, because these axes parallel those of the mountain belt.

The structure of the whole synclinorium is presented schematically in Figure 12. Regional structural features discussed in preceding paragraphs can be seen. These include the eastward tilt of sediments on the platform, the Tertiary foredeep facing the frontal fold, and the frontal fold itself. Distortion of the vertical scale prevents the inclusion of the high-frequency folds in the Tertiary trough and in the Zagros Mountains east of the frontal fold. The basement is shown to have a smooth form, and to be structurally simple, *i.e.*, a partly compound syncline, in contrast to the synclinorium of the overlying sediments. The smoothness of the basement outline is not related to the vertical scale and is meant to imply that *décollement* took place in the sedimentary cover. The obvious surface of *décollement* would be at or near the top of the Cambrian salt. The possibility of *décollement* also could be deduced from the irregular tectonic style of the fold belt, particularly as displayed in foothill and mountain cross sections presented by the British Petroleum Company Ltd. (1956). These cross sections also revealed the presence of some box-shaped folds like

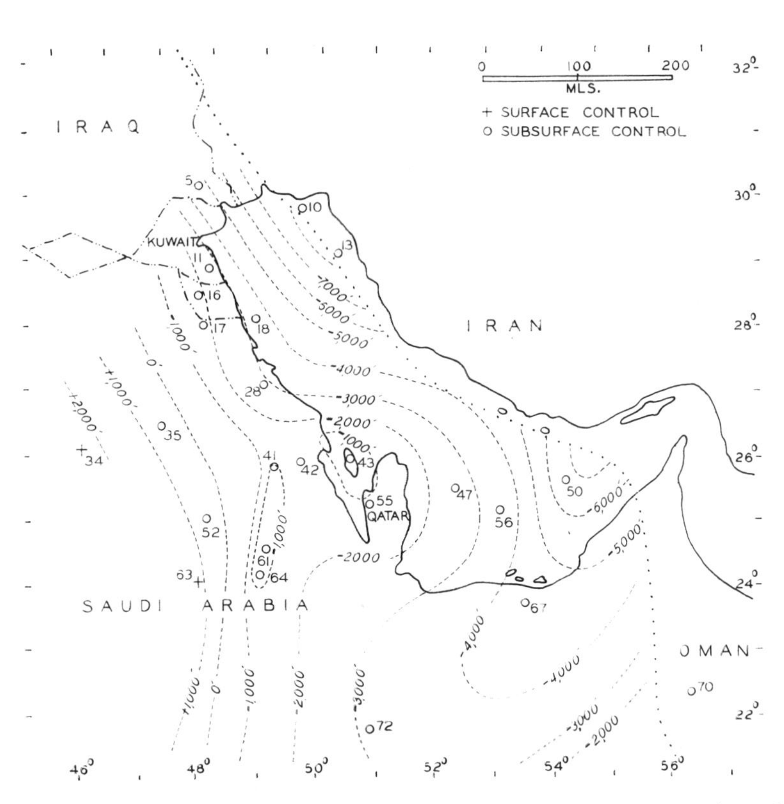

FIG. 11.—Platform structure contours; structural datum is top of Cretaceous. Contour interval 1,000 ft. Elevation datum is sea level.

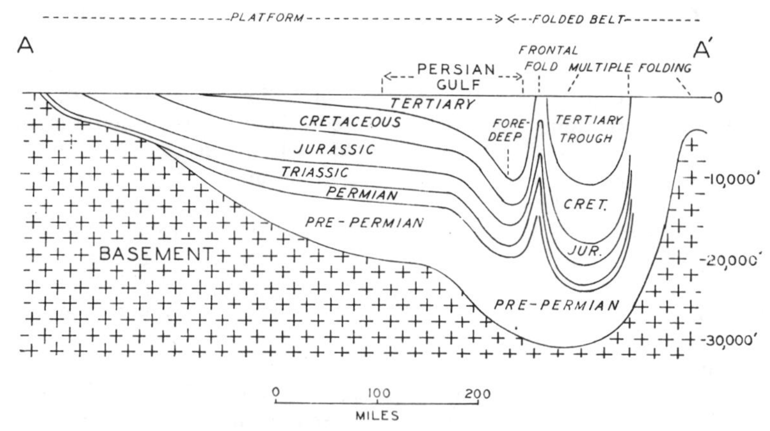

FIG. 12.—Cross section of the Persian Gulf synclinorium. Location is line A-A' of Figure 8.

those of the "classic" Jura type. Lees (1950) arrived at the conclusion of *décollement* long before the writer:

It would seem as if the thick skin of sedimentary rocks became detached from its rigid basement and folded into these long simple wrinkles, at least along the mountain front. Perhaps the Cambrian salt series has supplied the lubrication permitting this freedom of movement just as the Jura mountains owe their present form in part to the lubricating quality of the Triassic salt.

De Sitter (1956) attributed the Jura lubrication and *décollement* to Triassic anhydrite (without mention of salt).

The structure of the fold belt has been attributed by several authors to orogeny, especially "Alpine orogenesis." Insofar as orogeny is the folding of mountain belts by tangential compression, the term would seem to be misapplied in part to the fold belt if *décollement* of the gliding type is understood. The drawing of Figure 12 may be viewed in this respect. Basement is assumed to have been uplifted approximately 25,000 ft, and the sedimentary rocks above basement thus would have attained great height. From this great height the sediments could have glided downward and come to rest in compressional folds. The factor governing such a development would be relatively simple uplift rather than orogeny. Uplift would have supplied gravitational energy to the sediments for their eventual folding (and, thus, appearance of tangential compression). However, this energy would be distinct from the energy of "tangential stress" in classic "orogenesis."

Although the main gliding of *décollement* is presumed to take place near the base of the sedimentary column on a surface of Cambrian salt, an independent disharmonic movement of beds took place much higher in the column, specifically in the lower Miocene (Fars) beds. In this higher example, salt also is an obvious agent of lubrication. One of the effects of the disharmony is to displace westward the surface axes of the foothill oil fields. Cross sections by Lees (1953) and later authors show this same westward movement, and also show the advance of incompetent salt-bearing beds across oil-bearing limestone folds below them. The westward advance is the same as that assumed for the *décollement* gliding of the deep beds. Faults are prominent in the disharmonic movement of the upper beds, and the writer speculates that, locally, the faulting may be rotational on salt. In such cases a cross section of the fault surface would appear as a segment of a circle, concave upward.

Regarding platform structure, several authors from Lees (1953) until very recently have described and discussed repeated structural growth as an important characteristic of productive anticlines and domes. Fox (1956) noted in the great Burgan oil field of Kuwait (Fig. 1) evidence for structural uplift after middle Cretaceous (Cenomanian) time. This uplift was sufficiently strong that it apparently resulted in submarine erosion of beds on the Burgan fold, and the uplift was followed by almost continuous growth of the structure into post-Eocene time. Because drilling was limited almost exclusively to Cretaceous beds, nothing could be deduced about pre-Cretaceous growth. However, such growth may be anticipated if Fox's idea is accepted that buried Cambrian salt is the main agent of structural growth (although Jurassic salt also is present at Burgan).

In the big Ghawar oil field (Fig. 1), the Arabian American Oil Company Staff (1959) assumed that growth of the Ghawar structure began in Early Cretaceous time. As at Burgan the growth was accelerated in post-middle Cretaceous time and was accompanied by truncation of beds; movements continued at least until middle Eocene time. On the basis of these assumptions, Ghawar's Jurassic oil was considered to have occupied a stratigraphic trap and to have moved only when post-Jurassic structural growth took place. However, the Jurassic calcarenite reservoirs in the supposed stratigraphic trap seem to coincide approximately with the present elongate oil-field position. This suggests to the writer an alternate explanation —that the calcarenites were deposited on a long north-south fold which already was growing in Jurassic time. Lack of deep drilling to Lower Jurassic or pre-Jurassic beds precludes the possibility of confirming a pre-middle Cretaceous time of structural growth. If, contrary to gravimetric evidence, Cambrian salt should have caused structural growth at Ghawar, the earliest time of structural growth could be notably older than the Jurassic.

The Dukhan oil field on the Qatar peninsula (Fig. 1) is on a platform fold, and also has Jurassic calcarenite reservoirs. The geometry of the fold was examined in detail by Sugden (1962) who, in addition, made a mathematic analysis of growth. Sugden concluded that the folding could have commenced as long ago as the Late Jurassic. Deep drilling has been done in the general region but its density is insufficient to permit interpretation of the earliest times when structural growth could have commenced.

I believe that it is justifiable to suggest that in many, and perhaps in all areas of the platform, the structural growth of individual oil-field and other folds began at least as early as Jurassic time. Structural growth earlier than Jurassic time remains to be proved by deeper drilling, but could well have taken place if deeply buried Cambrian salt proves to be the agent of growth. If the salt were sufficiently mobile to move with little overburden, the beginning of structural growth in platform folds could date to the Paleozoic.

Productivity

General

Tremendous quantities of oil have accumulated in the Persian Gulf synclinorium. The total amount of oil in place could be in the order of 500 billion bbl. Estimates of recoverable reserves made in recent years converge on a figure of about 200 billion bbl. Later available figures are still higher. A graph by Dunnington (1967) may be read to mean that known reserves were discovered in two phases: (1) a sharp rise in the period 1950–1957; (2) a less sharp but still marked rise in the period 1957–1966. Additional reserves continued to be discovered but the rate of discovery has yet to be appraised.

The huge total of reserves of crude oil is reflected in the high productivity of individual wells. Spectacular wells capable of producing 20,000 bbl/day or more have captured popular attention. However, even if such wells are disregarded in a broad survey, the general level of well productivity in the Persian Gulf synclinorium is extraordinary. Probably more than 50 percent of the wells in the area are capable of producing 5,000 bbl/day for several years. Many wells are capable of producing 10,000 bbl/day and a few may be capable of producing 15,000 bbl/day for several years.

Reserves

Burke and Gardner (1969) issued a significant list showing concensus figures for possible reserves of the world's largest oil fields. The figures themselves, with reference only to the Persian Gulf synclinorium, and with added descriptive notations, appear in Table 1.

Van Dyke (1969) reported a major oil field which straddled the Saudi Arabian-Iran boundary in the Persian Gulf, and which "industry sources" believed to contain 10 billion bbl of recoverable reserves (Fereidoon-Marjan). Thus *the total producible reserves of oil in the Persian Gulf synclinorium may exceed 260 billion bbl.*

Burgan, Kuwait (Fig. 1), is a huge accumulation of oil with ultimate producible reserves believed to be in excess of 60 billion bbl (one unpublished company source gives the minimum reserve as 66 billion bbl). In such terms Burgan is about 600 times greater than a minimum "giant" oil field, which Halbouty (1968) defined as one with minimum ultimate producible reserves of 100 million bbl. The Burgan structure is a domoid anticline of considerable proportions, but its area is far from exceptional for an oil field in the synclinorium, perhaps about 140,000 acres to judge from available diagrams. If this acreage figure is approximately correct, the structure may have perhaps 400,000 bbl/acre of ultimately producible oil. The great wealth of the field is enhanced if the

Table 2. Distribution of Persian Gulf Oil by Age

Age	*Billion Bbl*
Tertiary	44.6
Cretaceous	133.0
Jurassic	71.0

shallowness of the producing reservoirs is considered. The average depth of production at Burgan is 4,800 ft (Oil and Gas Journal, 1968). The combination of huge reserves, areal compactness, and shallow depth easily make Burgan the most remarkable oil field in the world. The magnitude of the reserves comes even more sharply into focus when one remembers that they are half of the 120 billion bbl of reserves commonly attributed to the whole United States.

Ghawar, in Saudi Arabia (Fig. 1), is second (Table 1) with a reported reserve figure of 45 billion bbl of oil, though an unpublished Saudi Arabian–financed study has placed the recoverable reserve at 65–75 billion bbl. In Ghawar, the spectacular feature is a length of at least 160 mi, which leads to an acreage calculation of as much as 600,000. Using the 45-billion-bbl figure, I calculate a productivity of the order of 70,000 bbl/acre. However, the productivity could be considerably greater if recent claims concerning the reserves of Ghawar can be substantiated. Should these claims prove to be true, the reserves of Burgan and Ghawar together might exceed the reserves of the whole United States, even if the major newly discovered reserves of Alaska are included.

Distribution

The distribution of oil fields in the Persian Gulf may be considered first in time (age), and second in time and position (age and geography). With respect to time it is instructive to group the oil fields by geologic system and to assign each system a total of reserves. Use of the data from Table 1 leads to the figures for ultimate producible reserves in Table 2.

Dunnington (1967), supported by others working in the area, contended that much of the oil now in Tertiary reservoirs ascended through fractures from underlying Cretaceous limestones. Apart from the fact that presumed contiguous source beds may be present in the Tertiary, the final resting place of the oil also should be a factor influencing a decision to assign it statistically to the Tertiary.

Regarding Jurassic oil, a different type of ar-

Table 3. Control Points for Map Figures

Control Point	*Name*	*Country*	*Location*
1	M.I.S.	Iran	N31°55′ E49°15′
2	Tang-i-Dina	Iran	N31°05′ E51°20′
3	Ratawi	Iraq	N30°30′ E46°50′
4	Rumaila	Iraq	N30°15′ E47°25′
5	Zubair	Iraq	N30°20′ E47°40′
6	Gach Saran	Iran	N30°10′ E50°50′
7	Tang-i-Gurguda	Iran	N30°15′ E50°55′
8	Banish/Bizan	Iran	N30°15′ E52°40′
9	Rhaudhatain	Kuwait	N29°50′ E47°40′
10	Bahrgansar	Iran	N29°55′ E49°40′
11	Burgan	Kuwait	N29°00′ E47°55′
12	Cyrus	Iran	N29°05′ E49°30′
13	Darius	Iran	N29°15′ E50°15′
14	Kuh-i-Beriz/T.i-Ashk	Iran	N29°00′ E52°15′
15	Tang-i-Pohreh/Katak	Iran	N29°30′ E53°15′
16	Wafra	Neutral Zone	N28°35′ E47°55′
17	Jauf	Saudi Arabia	N28°05′ E47°55′
18	Safaniya	Saudi Arabia	N28°15′ E48°50′
19	F.P.C. A-1	Iran	N28°05′ E50°05′
20	Kuh-i-Mund	Iran	N28°15′ E51°15′
21	—	Iran	N28°30′ E53°30′
22	Bizdan S.E.	Iran	N28°40′ E54°30′
23	Kuh Gakun	Iran	N28°05′ E55°55′
24	F.P.C. B-1	Iran	N27°50′ E50°20′
25	Dopco C-1	Iran	N27°55′ E51°10′
26	Iminoco D-1	Iran	N27°35′ E51°30′
27	Lapco LE-1	Iran	N27°30′ E51°50′
28	Abu Hadriya	Saudi Arabia	N27°15′ E49°00′
29	Fadhili	Saudi Arabia	N26°55′ E49°10′
30	Khursaniya	Saudi Arabia	N27°10′ E49°15′
31	Iropco F-1	Iran	N27°20′ E51°00′
32	Dopco G-1	Iran	N27°15′ E51°20′
33	Tang-i-Shah Katar	Iran	N27°20′ E56°05′
34	—	Saudi Arabia	N26°10′ E45°55′
35	Ma'agala	Saudi Arabia	N26°25′ E47°20′
36	Qatif	Saudi Arabia	N26°40′ E49°55′
37	Dammam	Saudi Arabia	N26°20′ E50°10′
38	Lavan	Iran	N26°50′ E53°10′
39	Iminoco O-1	Iran	N26°25′ E53°20′
40	Iminoco T-1	Iran	N26°35′ E53°35′
41	Ghawar	Saudi Arabia	N25°50′ E49°10′
42	Abqaiq	Saudi Arabia	N26°05′ E49°40′
43	Bahrain	Bahrain	N26°00′ E50°35′
44	Rostam	Iran	N25°50′ E52°50′
45	Pegupco U-1	Iran	N25°55′ E53°30′
46	Ghawar	Saudi Arabia	N25°45′ E49°20′
47	Idd-el-Shargi	Qatar	N25°30′ E52°20′
48	M. Mahzam	Qatar	N25°40′ E52°30′
49	Sassan	Iran	N25°35′ E53°10′
50	Fateh	Dubai	N25°35′ E54°25′
51	J. Qamar	Oman	N25°45′ E51°05′
52	Khurais	Saudi Arabia	N25°05′ E48°05′
53	Ghawar	Saudi Arabia	N25°25′ E49°10′
54	Ghawar	Saudi Arabia	N25°10′ E49°10′
55	Dukhan	Qatar	N25°15′ E50°50′
56	Umm Shaif	Abu Dhabi	N25°15′ E53°10′
57	Juweiza	Oman	N25°15′ E55°45′
58	—	Saudi Arabia	N24°25′ E46°15′
59	—	Saudi Arabia	N24°50′ E46°40′
60	—	Saudi Arabia	N24°35′ E46°50′
61	Ghawar	Saudi Arabia	N24°40′ E49°05′
62	Zakum	Abu Dhabi	N24°50′ E53°45′
63	—	Saudi Arabia	N24°05′ E47°55′
64	Ghawar	Saudi Arabia	N24°10′ E48°55′
65	—	Saudi Arabia	N23°35′ E45°30′
66	—	Saudi Arabia	N22°30′ E46°00′
67	Murban	Abu Dhabi	N23°50′ E53°45′
68	—	Saudi Arabia	N22°30′ E45°20′
69	—	Saudi Arabia	N22°35′ E46°50′
70	Fahud	Oman	N22°20′ E56°30′
71	—	Saudi Arabia	N21°35′ E46°50′
72	ST-1	Saudi Arabia	N22°00′ E50°55′
73	Wadi Mi-Aidin	Oman	N23°00′ E57°40′

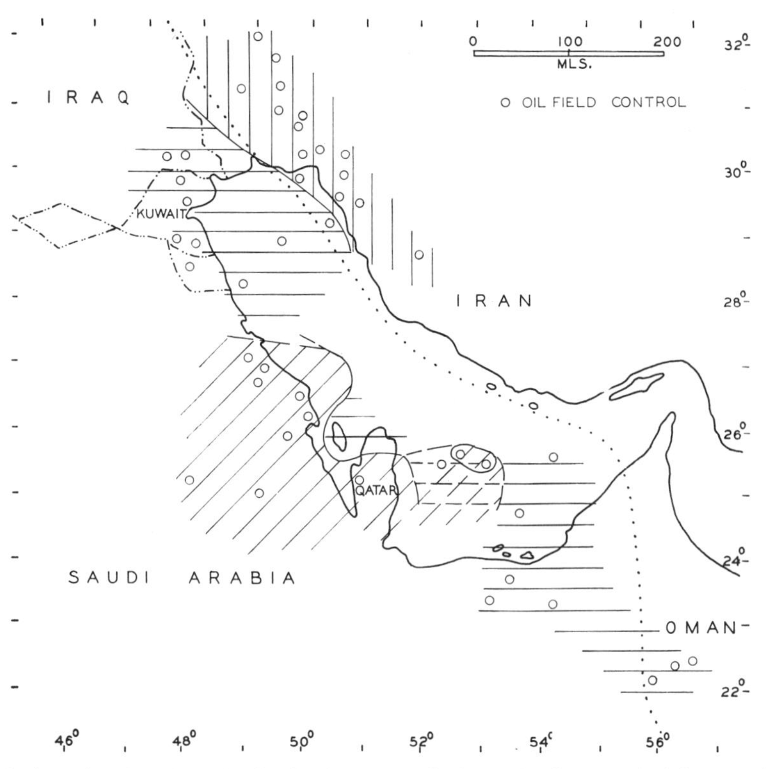

FIG. 13.—Distribution of major oil production in the Persian Gulf synclinorium. Vertical lines: major production from the Tertiary. Horizontal lines: Cretaceous. Oblique lines: Jurassic.

gument was raised by Banner and Wood (1964). They disputed the assigned Late Jurassic age of the Arab platform reservoirs. Dunnington (1967) strongly opposed the contention of Banner and Wood, insisting that Arab-formation oil was indeed Jurassic. Nevertheless, the possibility of a future revision of Late Jurassic to Early Cretaceous age should not be ignored completely. If such a revision eventually were to be made only a negligible amount of reserves would remain in the Jurassic, whereas more than 60 billion bbl of oil would be added to the Cretaceous. Such an addition would raise Cretaceous reserves to a mammoth 190 billion bbl of oil or nearly 80 percent of the total for the Persian Gulf synclinorium.

Figure 13 shows distribution of the oil fields by age and geography. Three ages of production show in patterns which appear to form tiers. The southwestern tier (diagonal lines) contains Jurassic oil fields in the Saudi Arabian sector; the middle tier (horizontal lines) contains Cretaceous oil fields in the Persian Gulf sector; and the northern tier (vertical lines) contains Tertiary oil fields in the Iranian sector. From the distribution of these tiers, one may infer that a narrow swath of Cretaceous production may be found to occupy almost the whole length of the Persian Gulf. Oil fields of Cretaceous age might then be found to extend from Zubair in the northwest to Fahud in the southeast, a total distance of nearly 800 mi.

Causes of Productivity

General

In view of the great petroleum abundance and productivity of the Middle East, of which the Persian Gulf synclinorium is a major part, many petroleum explorationists have reflected on the causes of such abundance. However, relatively few have published their views on the subject. Among the few are three geologists of the British Petroleum Company Ltd., and their comments form almost a tradition over two decades. Lees (1950) introduced the subject with the following statement.

> The exceptional richness of this great oil field province is the consequence of an unusually favourable conjunction of all the factors which control oil accumulation—original richness of source rocks, good res-

ervoir rocks, large anticlines with extensive drainage areas and excellent cover rocks preventing the escape of oil to the surface.

Lees also stated:

The giant size of the anticlines, the excellence of the reservoir rocks and the competence of the cover rocks would avail little were it not for the original oil richness, and this and other factors in favourable conjunction are responsible for the great importance of the Middle East oil province.

Law (1957) made additions to the above concepts:

The Persian Gulf area is remarkable for the volume of its oil reserves, the size of individual oil fields, and the productivity of individual wells. The abundance of oil is due not to any single factor but to the coincidence of all the conditions favorable to its generation, entrapment and preservation. These conditions include a great volume of marine sediments, many of them favorable to the generation of oil; structural and stratigraphic traps which were present during initial migration of the oil; numerous large anticlines, excellent reservoir and cap rocks; and lack of important unconformities in the oil-bearing sediments.

Falcon (*in* Dunnington, 1967) summarized his views as follows:

1. A long history of quiet sedimentation, including abundant source beds in a marine sub-tropical environment with restricted water circulation.
2. Thick reservoirs of sandstone and limestone, covering wide areas.
3. Excellent quickly deposited cap rocks.
4. Salt at depth to assist the building of growth structures on the foreland and of rapidly formed large folds in the foothill belt.

Caprocks

All three authors emphasized the importance of caprocks or cover rocks. However, Lees himself noted that at Burgan, Kuwait (Fig. 1), the oil was contained and preserved by as little as 100 ft of shale over the crest of the structure. This shows that one of the largest—if not the largest—oil fields in the world can be contained under a relatively insignificant cap. Moreover, the Burgan caprock is not an evaporite; many geologists have attributed the oil abundance in the Persian Gulf synclinorium to the presence of an evaporite caprock. The shale cap at Burgan has no unusual properties, and its role in that oil field suggests that the importance of evaporite caprocks as contributors to unusually great productivity may have been overstated.

Sediment Volume

Falcon and Law both stressed the considerable time of sedimentation and great volume of sediments as important factors in the generation of outstanding productivity. I believe that it is pertinent to compare the sedimentary volume of the Middle East prism with that of the Gulf of Mexico basin. The sedimentary volume of the basin is much greater than that of the prism, yet it is unlikely that the Gulf of Mexico basin could exceed or even rival the Middle East prism in ultimately recoverable reserves of crude oil, even if another 100 billion bbl of oil were assigned to undrilled underwater areas of the Gulf of Mexico basin. This comparison, and another on a volume of sediment per unit area basis, suggest the possibility that regional sedimentary volume, although important, need not be an overriding factor in the generation of enormous oil reserves.

The sedimentary volumes of both areas are the result of unusually great subsidence (hypersubsidence) of the basement. Thus, a significant vertical parameter, subsidence, might be an important element in a theory of widespread vertical migration of oil from great depths through connecting fissure systems. Although important vertical migration is theoretically possible and actually has been demonstrated to have taken place through limited vertical intervals in several Iranian oil fields, that mechanism is unlikely to be among the more important factors that determine great productivity.

Reservoir Volume

Although the total volume of sediments present in the Persian Gulf synclinorium may not be a deciding factor in the great oil productivity, the reservoir volume may be of the greatest importance. Lees (1950) and Law (1957) stressed the great lengths of the anticlines in the region. In the fold belt only a fraction of the total length of anticlines is occupied by oil, but this fraction is itself sufficiently long to act as a "multiplier" of reservoir volume. The best illustration of the multiplying effect of length to big reservoir volume is the major Ghawar (Saudi Arabia) anticline on the platform (Fig. 1). The entire 160 mi of this anticline is underlain by reservoir beds. The thickness of the reservoirs, with an average of possibly 200 ft, is not exceptional. Yet the multiplying effect of this thickness and a record-breaking length is a major factor causing the accumulation of the mammoth producible reserves at Ghawar (45 billion bbl or more). The Burgan oil field, however, is huge for a different reason. The high "multiplier" in this case is not so much the length of the structure as the reservoir thickness. The average thickness may be as great as

800 ft—hence the 62 billion bbl in only 140,000 acres.

Porosity Volume

The great volume of reservoirs, although a fundamental requirement, would be of negligible importance for productivity if permeability and porosity were not sufficient to permit the free movement of oil into the reservoirs. The thick sandstones at Burgan have retained almost all of their original porosity and thereby provide the space required to hold record or near-record volumes of oil. Concerning Middle East carbonate reservoirs, Illing (*in* Dunnington, 1967) observed that dolomite reservoirs are a minor factor, and that much primary porosity seemed to have been preserved in limestone. Examples noted were the porous and permeable pelletoid calcarenites of the Jurassic Arab Formation of the platform. Ghawar contains this type of porosity and permeability, and for this oil field alone one might speculate that the volume of producing reservoir is of the order of 20 million acre-ft. Illing observed further that in the fold belt even the calcareous mudstones, which comprise much of the Tertiary Asmari oil-field reservoirs, retained a significant amount of porosity and permeability. However, the consensus of opinion regarding the fold belt has been that strong deformation of Tertiary and pre-Tertiary limestones opened a system of fissures to provide much secondary porosity and permeability. Thus, regardless of the lithofacies or structural province, a great volume of reservoir beds has been translated into a great aggregate volume of pore space or of induced porosity. In actual figures it seems feasible to assign to the Persian Gulf synclinorium as a whole a volume of oil-field pore space in excess of 100 million acre-ft. Therefore, the total pore space available must be a most significant factor among those which contribute importantly to the tremendous productivity of the region.

Types of Source Beds

Large total pore volume is not an exclusive property of the Persian Gulf synclinorium, an area which excels only because the reservoir pore space is filled with oil. The fact that so much reservoir space *is* filled with oil means that prolific source materials were available. Excluding the reservoirs which actually contain the oil, the most petroliferous rocks of the area are the polybituminous marls and marly limestones of the fold belt, which Lees interpreted to be possible source beds for Iranian oil. The prefix "poly" refers to the fact that at least two hydrocarbon materials are present in the formations, one kerogen and the other soluble bitumen. The proportion of soluble bitumen is relatively small, but the area of polybituminous formations is so large that great quantities of petroleum must have collected in them. However, the evidence indicates to the writer that large quantities of potential oil were never expelled. Although there can be no question that the polybituminous formations are petroleum rocks, one definitely must question whether they were the sources of the oil, because they do not appear to have transferred major volumes of oil to closely associated reservoirs.

Close Relations of Source Rock and Reservoir

In Jurassic beds of the platform, Hedberg (1964) noted that in some places the Arab Formation oil accumulations were sandwiched between anhydrite beds, a situation which strongly supported the idea that some Arab oil had originated in carbonate muds closely associated with the reservoirs. Accumulation was thus related to intercalation. In the main Arab D member, the productive zone at Ghawar, the intercalation relations are less clear and less well documented. Nevertheless, I would interpret Arab D subsurface geology in the following terms.

1. The probable source beds now contain little petroleum or bitumen, as is true of the polybituminous formations of Iran, but consist of a carbonate mudstone lithofacies which today is not the obvious source of the oil.
2. The mudstone source is closely intercalated with the limestone, and the carbonate mudstones are part of a relatively large body of muddy carbonate lithofacies.

Intercalation in a dominantly limestone lithofacies also is a characteristic of reservoirs in the fold belt. Thus Thomas (1950) noted a close association of the reservoirs with carbonate mudstones in the Tertiary limestone columns of many foothill oil fields.

In the middle Cretaceous clastic lithofacies of Burgan, the tremendous accumulation also invites interpretation in terms of close relations between source beds and reservoirs. Shale lenses are known to be present in the reservoir sandstones of the Cretaceous Burgan Formation, but it is not clear whether some of these actually are not lenses, but lateral gradations with thick shale bodies. The major reservoir sandstone is underlain by a thin but continuous shale, and this shale also could communicate with a thicker shale body on the east. Intercalations, or space relations which potentially amount to intercalations, are definite possibilities in the sandstone and shale of the Burgan producing column. However, despite the close relation of sandstone and shale, the shale appears to be essentially unpetroliferous. In fact the shale may be only marginal marine, because Owen and Nasr (1958) reported an absence of Foraminifera and the presence of plant remains and resin lenticles. Yet the shale now present in the producing column of Burgan is the obvious source for the oil. The conditions at Burgan suggest that, as seems to be true for the carbonate lithofacies of Ghawar and allied oil fields, the source is an inconspicuous one, particularly in its lack of petroliferous character.

Early Migration

If the compaction of shale or of other muddy rocks is assumed to furnish energy for the migration of oil into the reservoir, the movement should take place almost immediately after—if not partly during—deposition of the reservoir. Such early movement could well have taken place on the platform because structures grew almost constantly after Jurassic time. Dunnington (1967) believed that early movement of oil into a reservoir was an important factor in preserving original porosity, because the pores filled with oil would be protected almost completely from diagenetic processes. Dunnington was writing of the fold belt and of the movement of platform oil into stratigraphic traps; regardless, the same reasoning should apply to the interpretation presented here.

Significant Factors in Oil Accumulation

Ideas developed in the preceding sections lead me to believe that the close association of source rocks and reservoir is perhaps the most significant factor in the development of great productivity. The same ideas also explain the unpetroliferous nature of the intercalated source beds. The source beds suggested here should be assumed (as Thomas, 1950, and earlier authors assumed) to have been of such a nature that even small percentages of organic matter or of protohydrocarbons would be sufficient to supply as much petroleum as the reservoir could hold. Where generation and accumulation of oil are assured by the presence of such source beds, by the intercalation of source and reservoir, and by early movement into open pores, the "multiplier" of enormous pore volume enters the equation to provide conditions in which productivity of mammoth proportions is possible.

Nature of Source Beds

The real nature of petroleum sources continues to be elusive. Clearly the source is something more than a simple fine-grained lithofacies. Otherwise, wherever intercalated fine- and coarse-grained rocks are present in the sedimentary column, significant accumulations of oil should be found. This is not the case. Where major oil accumulations are present, only a few —commonly only one—zones in a complete sedimentary column contain oil. Other potential reservoirs are barren. Moreover, oil tends to be present in relatively segregated areas. Restriction of accumulation to one (or a few) zones in the total section suggests special paleoecologic conditions for the sources, and the restriction of accumulations to selected areas suggests special paleogeographic conditions for the source materials. In terms of Figure 13, the special paleoecologic conditions might have developed during Jurassic time mainly in a special paleogeographic situation peculiar to the southwestern part of the platform; in the Cretaceous along a nearly linear belt on the eastern side of the platform; and in the Tertiary along the present fold belt. Paleogeologic conditions, on the other hand, may not have played an important role in the generation of large quantitites of oil. As now distributed (*see* isopach patterns of Figs. 5–7), the major accumulations of oil do not appear to bear close relations to the diastrophic history of individual geologic periods.

Conclusions

1. The geologic history of the present Persian Gulf synclinorium is characterized by several important inundations by marine water. The inundations were not just in embayments of the main Eurasian Tethyan seaway on the north, because the extent of the waters was that of broad epicontinental seas. The spread of the earlier inundations definitely was not as extensive as that of later ones, but the Ordovician sea may have advanced as far south as an ancestral Arabian Sea. The same conclusion may be drawn for the widespread Late Permian incursion. It is suggested that Permian correlations leave open the choice between a Permian ancestral Arabian Sea (Indian Ocean) and a Permian Gondwanaland, even though Permian marine sections are present in Madagascar,

Tanzania, Burma, Thailand, Indonesia, and Western Australia.

2. Although the general regime of subsidence and deposition was interrupted significantly in Triassic time east of the area of the present synclinorium, the overall paleoenvironment during Phanerozoic time was that of a platform extending across many hundreds of thousands of square miles. The geometry of the resulting sediments formed a compound prism, even though subsidiary basins, linear troughs, and a foredeep developed at different times. Geosynclinal conditions do not appear to have developed to any important extent.

3. The fundamental sediment prism was modified by strong uplift in late Tertiary-Quaternary time, and the basement may have assumed the shape of a fairly simple regional but asymmetric syncline. Sediments are presumed to have glided down the asymmetric flank of the syncline on a bed of Cambrian salt, and thereby to have produced the high-frequency folds and other structures of the fold belt. The rest of the original platform remained almost unaffected and formed the western side of the present Persian Gulf synclinorium, a feature which dominates the whole region in which the big oil fields now are present.

4. Possibly more than 500 billion bbl of oil accumulated in the oil fields of the synclinorium; of this amount more than 250 billion bbl may be ultimately producible reserves. Of the 250 billion bbl, two fields alone—Burgan in Kuwait (62 billion) and Ghawar in Saudi Arabia (45 billion or more)—are believed to be capable of producing 100 billion bbl or more (some estimates suggest 130–140 billion bbl).

5. Explorationists have mentioned numerous possible reasons for the unsurpassed productivity of the area. A large volume of sediments deposited under little disturbed tectonic conditions has been considered to be important. However, much larger volumes of sediment were deposited in the Gulf of Mexico basin, and they show little prospect of even approaching the Persian Gulf synclinorium in total productivity. Thick cover rocks, and their capacity for acting as a seal for large quantities of oil, also have been mentioned as an important factor, yet the world's reported greatest oil field, Burgan, exhibits minimal cover at the crest of the structure.

6. The *present* petroleum richness of certain fine-grained rocks has led to their consideration as the original source beds, but their actual contribution of oil to reservoirs is at best uncertain and quite possibly negligible. Occurrence and distribution of oil in the area suggest instead that "source rock" is relatively scarce—possibly unidentifiable, because the original petroleum-forming materials are now gone. Thus the original beds in which the source materials were present—although undoubtedly a fine-grained lithofacies, are lacking in petroleum manifestations and apparently are impossible to distinguish from other types of barren fine-grained lithofacies.

7. The writer therefore suggests that factors of major importance in the generation of great productivity are:

a. Tremendous aggregate pore volume of reservoirs, confined in closure, and preserved largely in long anticlines;

b. Closure association of the beds which contained the original source materials (these beds now are not obviously petroliferous) and the reservoir rocks, mainly by intercalation; and

c. Early release of oil from the source beds and early movement of oil into the reservoir rock, with the consequent preservation of large volumes of original porosity before the adverse effects of diagenesis could set in.

Selected References

Arabian American Oil Company Staff, 1959, Ghawar oil field, Saudi Arabia: Am. Assoc. Petroleum Geologists Bull., v. 43, no. 2, p. 434–454.

Baker, N. E., and F. R. S. Henson, 1952, Geological conditions of oil occurrence in Middle East fields: Am. Assoc. Petroleum Geologists Bull., v. 36, no. 10, p. 1885–1901.

Banner, F. T., and G. V. Wood, 1964, Lower Cretaceous-Upper Jurassic stratigraphy of Umm Shaif field, Abu Dhabi marine areas, Trucial Coast, Arabia: Am. Assoc. Petroleum Geologists Bull., v. 48, no. 2, p. 191–206.

Besairie, H., 1952, Les formations du Karroo à Madagascar, *in* Symposium sur les séries de Gondwana: 19th Internat. Geol. Cong., Algiers, p. 181–186.

Beydoun, Z. R., 1966, Geology of the Arabian peninsula—Eastern Aden Protectorate and part of Dhufar: U.S. Geol. Survey Prof. Paper 560-H, 49 p.

British Petroleum Company Ltd., Staff, 1956, Geological maps and sections of south-west Persia: 20th Internat. Geol. Cong., Mexico, portfolio volume.

Burke, R. J., and F. J. Gardner, 1969, The world's monster oil fields and how they rank: Oil and Gas Jour., v. 67, no. 2, p. 43–49.

Dunnington, H. V., 1967, Stratigraphical distribution of oilfields in the Iraq-Iran-Arabia basin: Inst. Petroleum Jour., v. 53, no. 520, p. 129–153.

Elder, S., 1963, Umm Shaif oilfield. History of exploration and development: Inst. Petroleum Technology Jour., v. 49, no. 478, p. 308–314.

Falcon, N. L., 1958, Position of oil fields in southwest Iran with respect to relevant sedimentary basins, *in* Habitat of oil: Am. Assoc. Petroleum Geologists, p. 1252–1278.

Fox, A. F., 1956, Oil occurrence in Kuwait, *in* Symposium sobre yacimientos de petróleo y gas: 20th Internat. Geol. Cong., México, v. 2, p. 131–158.

——— and R. C. C. Brown, 1968, The geology and characteristics of the Zakum oil field, Abu Dhabi: 2d AIME Regional Tech. Symposium Proc., Dhahran, Saudi Arabia, sec. A, p. 101–116.

Hajash, G. M., 1967, The Abu Sheikdom—the onshore oilfields history of exploration and development: 7th World Petroleum Cong. Proc., Mexico, v. 2, p. 129–139.

Halbouty, M. T., 1968, Giant oil and gas fields in United States: Am. Assoc. Petroleum Geologists Bull., v. 52, no. 7, p. 1115–1151.

Hedberg, H. D., 1964, Geologic aspects of origin of petroleum: Am. Assoc. Petroleum Geologists Bull., v. 48, no. 11, p. 1755–1803.

Henson, F. R. S., 1951, Observations on the geology and petroleum occurrences of the Middle East: 3d World Petroleum Cong. Proc., The Hague, sec. 1, p. 118–140.

Jacob, K., 1952, A brief summary of the stratigraphy and palaeontology of the Gondwana system with notes on the structure of the Gondwana basins and the probable directions of movements of the Late Carboniferous ice sheets, *in* Symposium sur les séries de Gondwana: 19th Internat. Geol. Cong., Algiers, p. 153–174.

James, G. A., and J. G. Wynd, 1965, Stratigraphic nomenclature for Iranian Oil Consortium Agreement area: Am. Assoc. Petroleum Geologists Bull., v. 49, no. 12, p. 2182–2245.

Kamen-Kaye, M., 1970, Late Cretaceous eugeosynclinal sedimentation, gravity tectonics and ophiolitic emplacement in Oman Mountains, south Arabia: discussion: Am. Assoc. Petroleum Geologists Bull., v. 54, no. 3, p. 536–537.

Kuwait Oil Co. Ltd., Staff, 1953, Kuwait, *in* The science of petroleum: Oxford Univ. Press, v. 6, pt. 1, p. 99–100.

Law, J., 1957, Reasons for Persian Gulf oil abundance: Am. Assoc. Petroleum Geologists Bull., v. 41, no. 1, p. 51–69.

Lees, G. M., 1950, Some structural and stratigraphic aspects of the oilfields of the Middle East: 18th Internat. Geol. Cong., London, pt. 6, p. 35–44.

——— 1953a, The Middle East, *in* The science of petroleum: Oxford Univ. Press, v. 6, pt. 1, p. 67–72.

——— 1953b, Persia, *in* The science of petroleum: Oxford Univ. Press, v. 6, pt. 1, p. 73–82.

Mina, P., M. T. Razaghnia, and Y. Paran, 1967, Geological and geophysical studies and exploratory drilling of the Iranian continental shelf—Persian Gulf: 7th World Petroleum Cong. Proc., Mexico, v. 2, p. 870–903.

Morton, D. M., 1959, The geology of Oman: 5th World Petroleum Cong. Proc., New York, sec. 1, paper 17, p. 227–290.

Oil and Gas Journal, 1968, World production: v. 66, no. 53, p. 109–136.

Owen, R. M. S., and S. N. Nasr, 1958, Stratigraphy of the Kuwait-Basra area, *in* Habitat of Oil: Am. Assoc. Petroleum Geologists, p. 1252–1278.

Pakistan Geological Survey, 1949: Quetta, unpub. open file logs and reports.

Powers, R. W., L. F. Ramirez, C. D. Redmond, and E. L. Elberg, Jr., 1966, Geology of the Arabian peninsula: sedimentary geology of Saudi Arabia: U.S. Geol. Survey Prof. Paper 560-D, 147 p.

Qatar Petroleum Co., Ltd., Staff, 1956, Qatar—occurrence of oil and gas, *in* Symposium sobre yacimientos de petróleo y gas: 20th Internat. Geol. Cong., México, v. 2, p. 161–169.

Sastry, M. V. A., and S. C. Shah, 1969 (1964), Permian marine transgression in peninsular India, *in* Gondwanas: 22d Internat. Geol. Cong., New Delhi, p. 139–150.

Sitter, L. U. de, 1956, Structural geology, 1st ed.: New York, McGraw-Hill, 552 p.

Steineke, M., R. A. Bramkamp, and N. J. Sander, 1958, Stratigraphic relations of Arabian Jurassic oil, *in* Habitat of oil: Am. Assoc. Petroleum Geologists, p. 1294–1329.

Stöcklin, J., 1968, Structural history and tectonics of Iran: review: Am. Assoc. Petroleum Geologists Bull., v. 52, no. 7, p. 1229–1258.

Sugden, W., 1962, Structural analysis, and geometrical prediction for change of form with depth, of some Arabian plains-type folds: Am. Assoc. Petroleum Geologists Bull., v. 46, no. 12, p. 2213–2228.

Thomas, A. N., 1950, The Asmari Limestone of southwest Iran: 18th Internat. Geol. Cong., London, pt. 6, p. 35–44.

Thralls, W. H., and R. C. Hasson, 1956, Geology and oil resources of eastern Saudi Arabia, *in* Symposium sobre yacimientos de petróleo y gas: 20th Internat. Geol. Cong., México, v. 2, p. 14–32.

Tschopp, R. H., 1967, The general geology of Oman: 7th World Petroleum Cong. Proc., Mexico, v. 2, p. 230–242.

Van Dyke, L. F., 1969, Four groups to quit Iranian search: Oil and Gas Jour., v. 67, no. 35, p. 76–79.

Wilson, H. H., 1969, Late Cretaceous eugeosynclinal sedimentation gravity tectonics, and ophiolite emplacement in Oman Mountains, southeast Arabia: Am. Assoc. Petroleum Geologists Bull., v. 53, no. 3, p. 626–671.

BULLETIN OF THE AMERICAN ASSOCIATION OF PETROLEUM GEOLOGIST
VOL. 41, NO. 1 (JANUARY, 1957), PP. 51-69, 5 FIGS.

REASONS FOR PERSIAN GULF OIL ABUNDANCE[1]

J. LAW[2]
Calgary, Alberta, Canada

ABSTRACT

The Persian Gulf area is remarkable for the volume of its oil reserves, the size of individual oil fields, and the productivity of individual wells. The abundance of oil is due not to any single factor but to the coincidence of all the conditions favorable to its generation, entrapment, and preservation. These conditions include a great volume of marine sediments, many of them favorable to the generation of oil; structural and stratigraphic traps which were present during initial migration of the oil; numerous large anticlines; excellent reservoir and cap rocks; and lack of important unconformities in the oil-bearing sediments. Variations in oil gravity may be due to varying amounts of terrigenous source material.

INTRODUCTION

The Persian Gulf sedimentary basin (Fig. 1) embraces eastern and possibly western Iraq, southwestern Iran, the Persian Gulf and adjacent sheikdoms, eastern Saudi Arabia, and the states along the coast of the Arabian Sea.

Topographically the area can be divided into (1) a southwestern highland area covering southwestern Saudi Arabia, Yemen, and the Aden Protectorate, (2) a low central area, largely desert, covering most of Iraq and Saudi Arabia, the extreme southwestern corner of Iran and the Persian Gulf, and (3) a northeastern mountain and foothill system comprising the Kurdish ranges of northern Iraq, the Zagros ranges of Iran, and the Oman ranges.

Only part of the basin has been explored but its proved reserves at the end of 1955 were already 126 billion barrels, representing two-thirds of the world total (*Oil and Gas Journal*, December 26, 1955). During 1955 a daily average of more than 3,000,000 barrels of oil was produced from fewer than 600 wells, an average of more than 5,000 barrels per well; many individual wells are capable of sustained production of more than 20,000 barrels per day.

During 1955 production was obtained from 21 oil fields. Six more oil fields and one gas field were shut in or awaiting completion of discovery wells. More than half the production came from three fields: Burgan in Kuwait, Ghawar in Saudi Arabia, and Kirkuk in Iraq. Each of the first two fields has proved reserves of more than 20 billion barrels and several of the other fields in the area have reserves of more than one billion barrels.

About two-thirds of the oil being produced is of 34°–38° API gravity, and roughly one-third is of 32° API.

The purpose of the present paper is to show how the abundance and high gravity of the oil in the present oil-field area have been caused by the geological conditions.

[1] Manuscript received, May 28, 1956.

[2] The California Standard Company, formerly with Anglo-Iranian Oil Company Limited. The writer thanks A. Allison, S. Elder, A. J. Goodman, R. M. S. Owen, T. P. Storey, and A. N. Thomas for their helpful criticism.

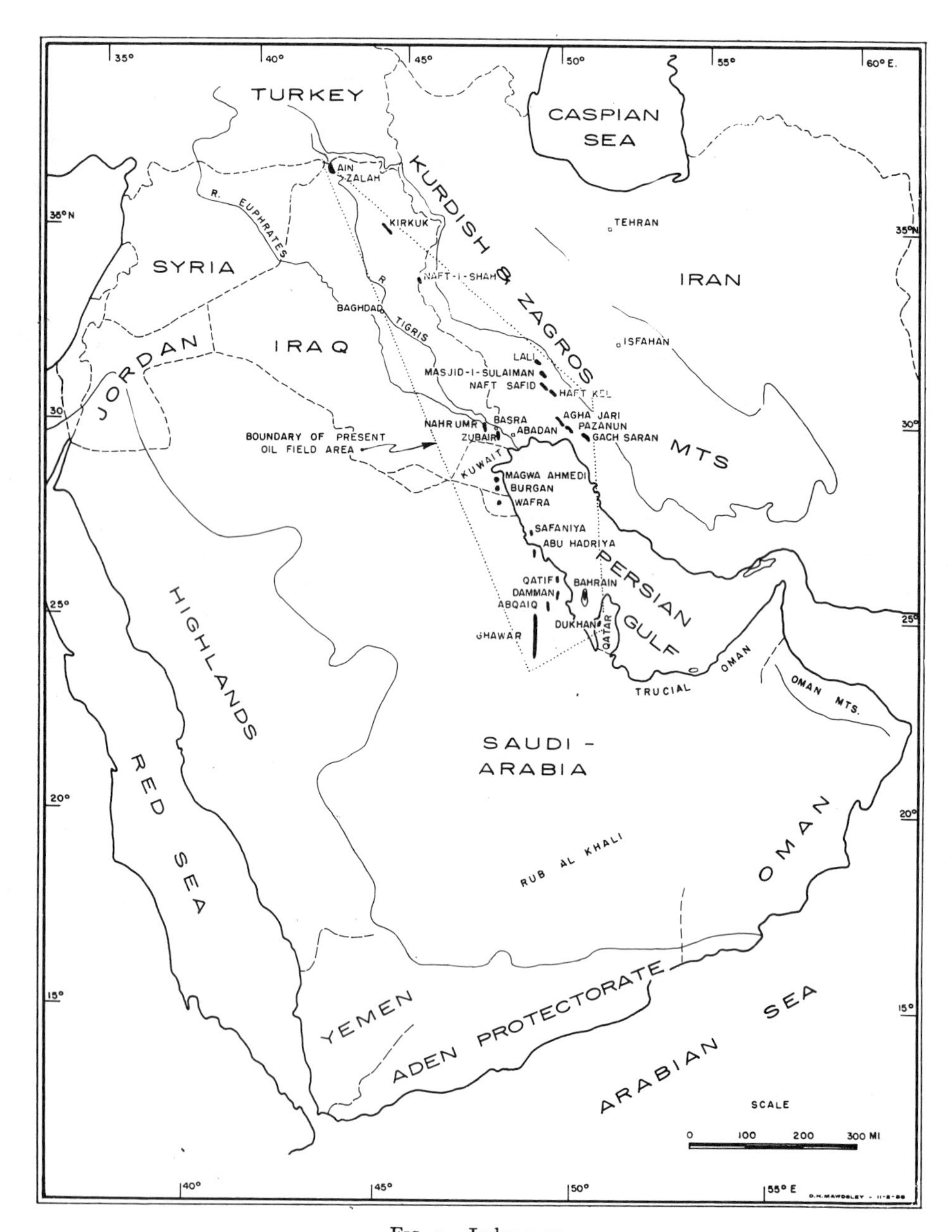

FIG. 1.—Index map.

GENERAL DESCRIPTION OF AREA

The following summary is mainly drawn from the papers listed as references. The reader is referred to these publications for details.

STRATIGRAPHY

Control is largely limited to outcrop areas fringing the basin and to oil-field development areas. Furthermore, the regional framework of sedimentation is still open to debate (Henson, 1951, and discussion). The outline given here is therefore tentative but provides a background for the ensuing discussion.

The Persian Gulf basin is bounded on the northeastern side by the outcrop of Paleozoic and Early Mesozoic rocks in the folded Kurdish and Zagros mountains (Fig. 1). Northeast of these ranges younger rocks reappear. On the southwestern side the basin is limited by a shield which occupies roughly the position of the highlands of western Arabia (Fig. 2). The southeast side also is partly defined by a shield area situated off the southeastern coast of Arabia (Henson, 1951), but thick sediments are continuous from the upper Persian Gulf area southeastward through Oman (Lees, 1928) and southern Iran. According to Baker (1953) the basin extends northwestward into Egypt, Cyprus, and Turkey.

Most of the sediments in the basin including all those from which oil is produced are of post-Carboniferous age (Fig. 4). However, the Cambrian system is several thousand feet thick and includes salt which appears as piercement domes in southern Iran. The Ordovician to Carboniferous periods are poorly represented and need not be considered here. Most of the post-Carboniferous sediments were deposited in one of two tectonic environments (Fig. 2).

1. A cratonic area which may have occupied the entire basin during the Permian period extended throughout all but the most northeasterly part of the basin during Triassic to lower Miocene time. This cratonic area covered that part of the Arabian peninsula lying outside the shield areas, together with southern Iraq and part of southwestern Iran. Henson (1951) divided it into a stable shelf, fringing the shields, with thin continental and shallow-water marine sediments, and, east of this, an unstable shelf with intracratonic geosynclines and much thicker marine sediments. All the oil fields now producing in the Persian Gulf area are in these latter sediments.

2. An orthogeosynclinal belt bordered the northeastern side of the basin in northeastern Iraq and southwestern Iran. An orthogeosyncline probably formed along the northeastern margin of the present basin during the Early Mesozoic. It migrated southwestward during Late Mesozoic and Tertiary time, and its sediments overlie shelf deposits of the first environment in most of northeastern Iraq and southwestern Iran (Fig. 3). At least the northeastern part of this area experienced folding during the Upper Cretaceous (Lees, 1950; Kent *et al.*, 1951) and the whole area was folded by a major orogeny in the Pliocene. Subsequent elevation formed the Zagros and Kurdish mountains and foothills.

Permian rocks have been reported in Iraq, Iran, and several parts of Arabia.

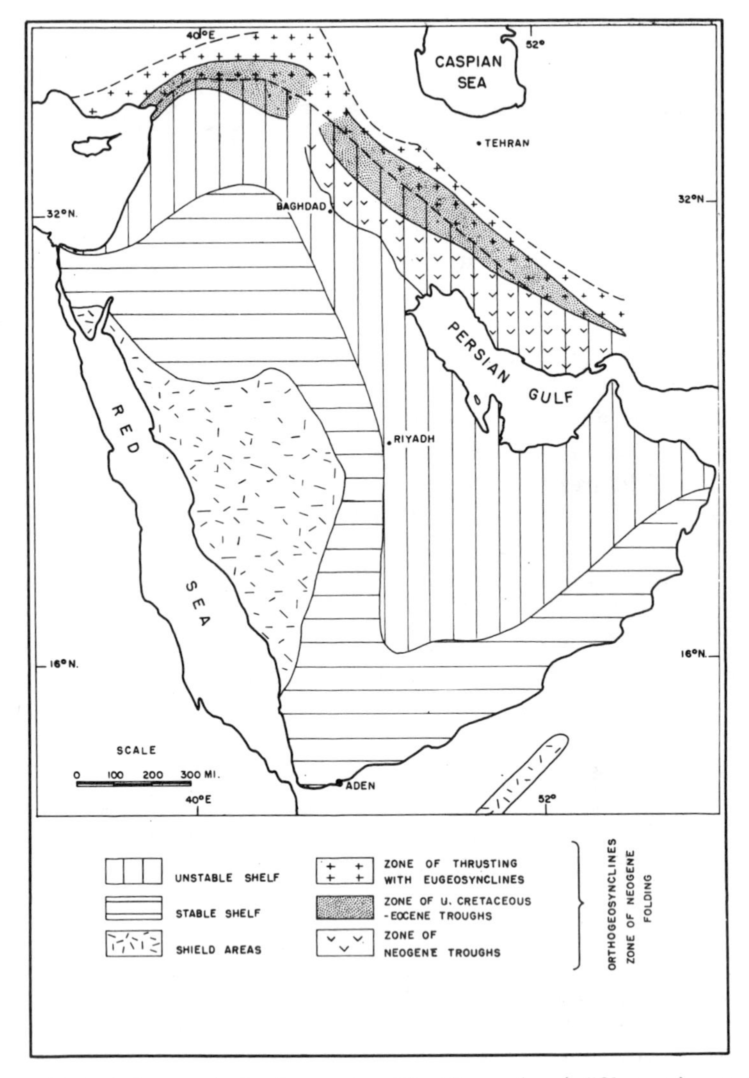

FIG. 2.—Tectonic framework of sedimentation. After Henson (1951), "Observations on the Geology and Petroleum Occurrences in the Middle East," Pl. III, generalized, and modified in Zagros Mountains area. *Proc. 3d World Petroleum Congress*. By permission of E. J. Brill, Leiden, Netherlands.

They consist of limestones with maximum thickness of 3,000 feet, with some red and green shales on the Arabian side of the basin.

Along the northeastern border of the basin Triassic, Jurassic, and Lower Cretaceous rocks are radiolarian cherts, basic lavas and intrusions, shales, and limestones. Though no thickness figures have been published, the lithologic character of the rocks and their distribution parallel with subsequent tectonic strike lines suggest that they were deposited in a eugeosyncline.

In northern Iraq and in Iran, Triassic deposits on the shelf, southwest of

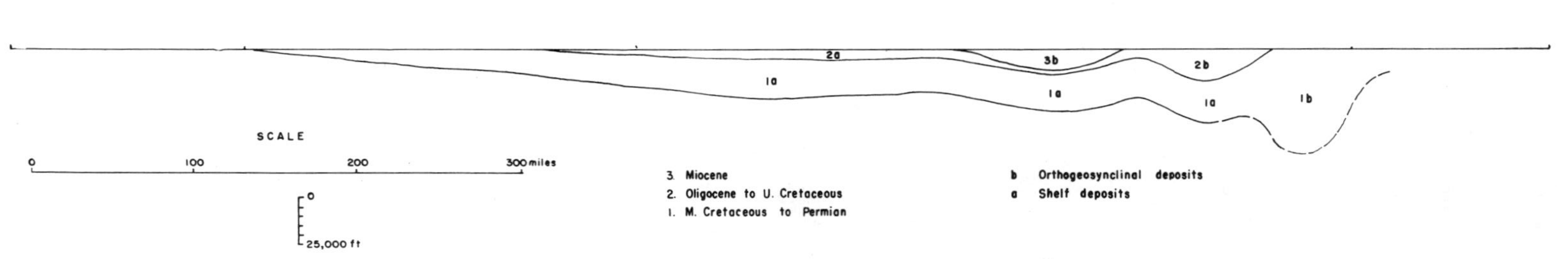

FIG. 3.—Diagrammatic section across Persian Gulf basin before Pliocene orogeny.

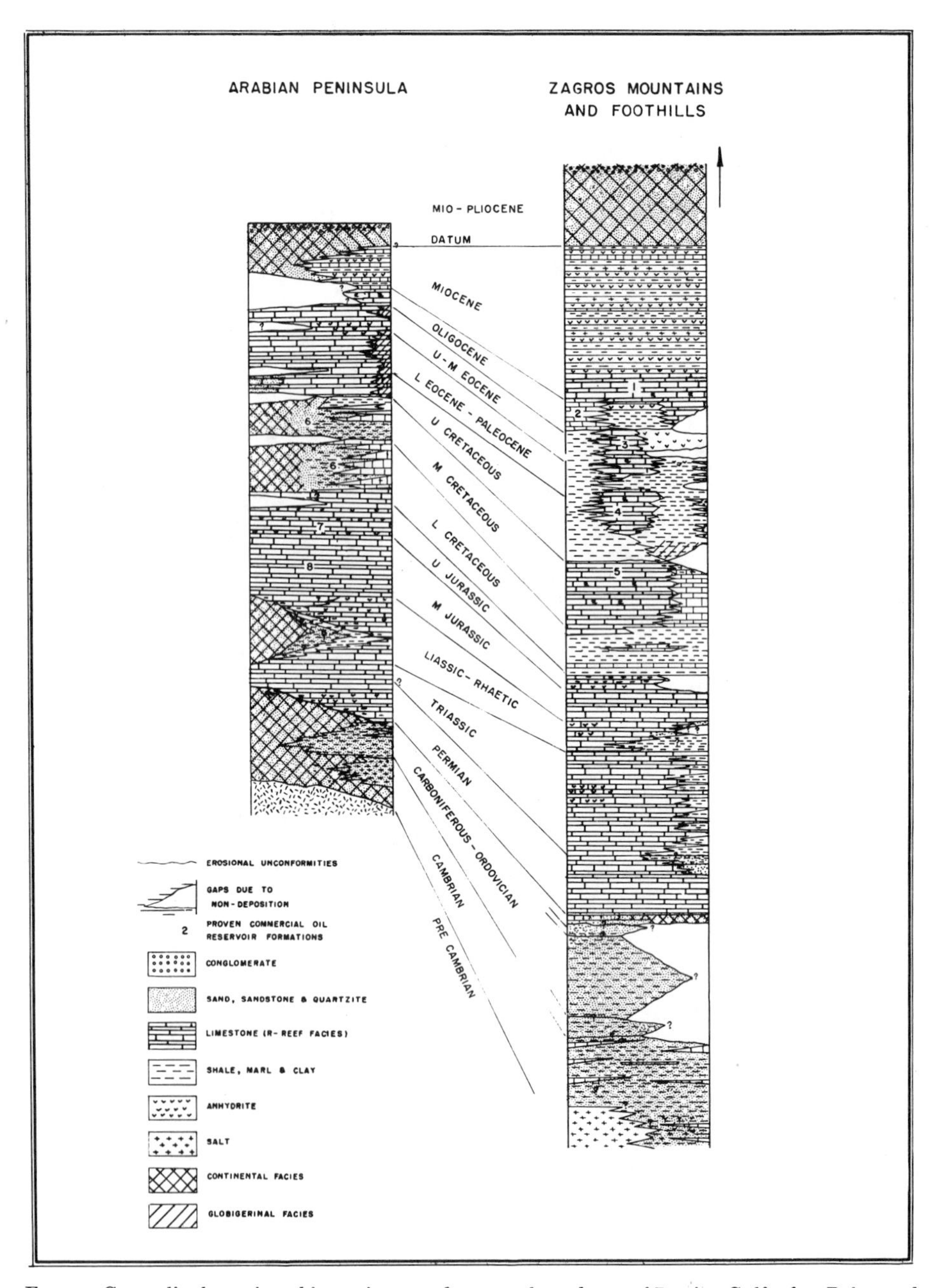

FIG. 4.—Generalized stratigraphic sections southwest and northeast of Persian Gulf, after Baker and Henson (1952), northeastern section modified after Kent *et al.* (1951).

the orthogeosynclinal belt, are limestones, dolomites, and anhydrites; southwest of the Persian Gulf, rocks of this age are mainly sandstones and shales. Jurassic rocks deposited on the shelf consist of several thousand feet of carbonates and anhydrites with subordinate shales. Most of the Arabian fields produce from Jurassic carbonates under anhydrite or limestone cap rocks. Over most of the shelf Lower and Middle Cretaceous sediments consist of thick carbonates, partly of reef or bank type, with some shales and marlstones. Middle Cretaceous carbonates are productive in several of the foothill fields of Iraq and Iran. Near the head of the Persian Gulf the Lower and Middle Cretaceous epochs are largely represented by sands and shales which form the producing fields in this area.

The increase in the proportion of terrigenous clastics, and Cretaceous vulcanicity in Oman and Syria, may have been associated with orogenic (Lees, Ion, Falcon, and Lehner, discussion of Henson, 1951), taphrogenic (Henson, 1951), or geanticlinal movements, which took place in the possibly eugeosynclinal belt during the Cretaceous. During the Upper Cretaceous and Eocene this belt was elevated and eroded, giving rise to more than 10,000 feet of flysch deposits. These were laid down in a northwest-southeast-trending orthogeosyncline (miogeosyncline?) which developed within the area now occupied by the Zagros Mountains.

Shelf conditions continued in the rest of the basin. There the Upper Cretaceous rocks are much thinner and consist largely of thin-bedded limestones and marls. Eocene beds are also much thinner and are limestones, including reef types, and marls. The Kirkuk field of Iraq produces in part from Eocene reef and fore-reef limestones.

Post-Eocene sediments are mostly confined to the area north of the Persian Gulf. Oligocene deposits are reef carbonates and marls. The lowest Miocene beds consist of limestones with subordinate anhydrites and shales. The upper part of the Kirkuk reservoir consists of Oligocene and lower Miocene reef limestones. The Asmari limestone, which is the main producing formation in Iran, comprises lower Miocene fine-grained foraminiferal limestones together, in some fields, with Oligocene reef limestones.

During later Miocene time 5,000 feet or more of red and gray marls, anhydrites, and salts—the lower and middle Fars formations—were deposited in an orthogeosyncline lying between the Persian Gulf on the southwest and the Kurdish and Zagros mountains on the northeast. These beds are overlain by 3,000 feet of brown continental sandstones and marls of latest Miocene age.

Folding movements which took place north of the Gulf during the Pliocene were accompanied by the deposition in this area of 10,000–20,000 feet of brown continental conglomerates, sandstones, and siltstones.

STRUCTURE

Within the present oil-field area there are two main structural provinces.

1. In northeastern Iraq and southern Iran, below violently contorted salt-bearing beds, there are very large, simple folds striking northwest-southeast

with steep in some places vertical, southwestern flanks, broad tops, and moderately dipping northeastern flanks (Fig. 5a). There are normal faults and large reverse and thrust faults in the incompetent beds, but only a few small normal and reverse faults in the underlying limestones. The size and simplicity of the folds in the latter beds may be due to the lubricating action of underlying Cambrian salt (Lees, 1950) and to the competence of the thick carbonate beds involved. This area lies within the belt of orthogeosynclinal development and in-

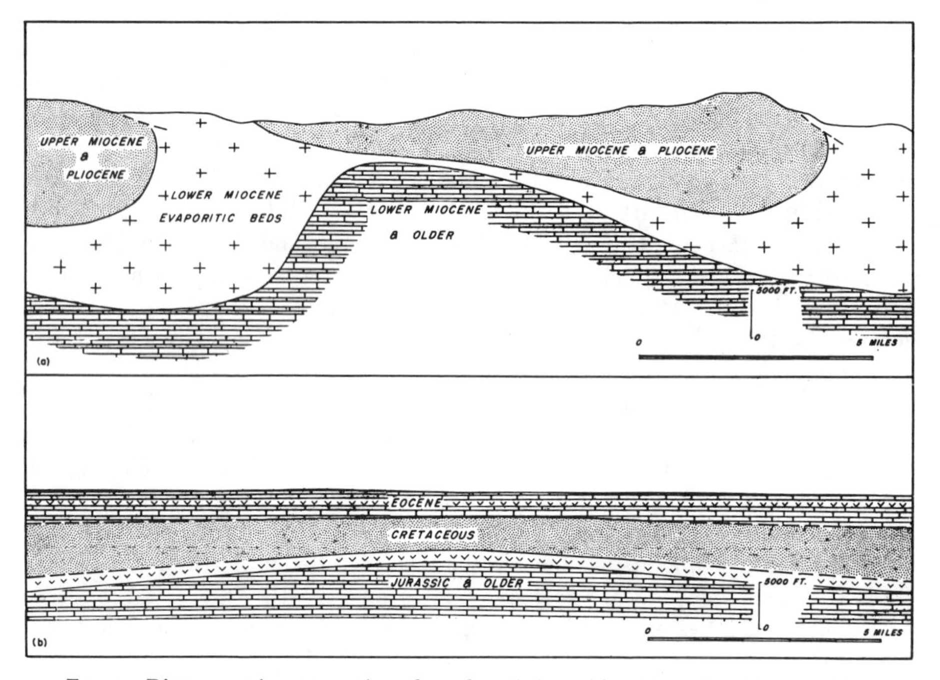

FIG. 5.—Diagrammatic cross sections through anticlines: (a) in Foothills; (b) on Arabian side of basin.

cludes the foothills of the Kurdish and Zagros mountains. The main folding took place in the Pliocene, though Upper Cretaceous movements caused folding and thrusting in the northeastern part of the Zagros Mountains (Lees, 1950) and gentle folding in the front ranges (Kent *et al.*, 1951).

2. South of the Persian Gulf and in southern Iraq, the anticlines, though large, have dips of only a few degrees (Fig. 5b). These folds strike north-south and some or all of them developed slowly from Upper Jurassic time onward (Henson, 1951; Kuwait Oil Company, 1953, Fig. 2; Daniel, 1954). Some of these folds are oval, others are markedly elongate. They may have been caused by the orogenic forces which culminated in the Cretaceous folding in Oman, or by deep-seated salt plugs (Lees, 1950). Salt movement, if it occurred, may have been controlled in some places by faulting (Henson, 1951). Normal faults of small throw have been encountered on some anticlines.

OIL FIELDS

Particulars of the known fields are shown in Table I. Producing reservoirs include both sands and carbonates. They range in age from Miocene to Middle Jurassic and in depth from 600 to 11,000 feet.

The limits of most Persian Gulf fields have not yet been defined by drilling and in some wells the oil-water level has not been reached. In addition many anticlines have not been tested and no attempt has so far been made to discover stratigraphic traps. The chance that undiscovered fields exist, even in the present oil-field area, is therefore excellent. The eventual increase in reserves by extensions and new discoveries is likely to be substantial and could amount to several times the present proved reserves, which are 126 billion barrels.

For the 600,000 cubic miles of probably oil-producing sediments in the present oil-field area, the proved reserves represent 210,000 barrels of recoverable oil per cubic mile.

According to Weeks (1950), the total amount of recoverable oil (oil produced plus proved reserves) in different petroleum provinces of the United States, ranges from 6,000 to 200,000 barrels per cubic mile of sediments, and the total ultimate production from United States sediments is likely to average about 50,000 barrels per cubic mile. Thus the sediments in the present oil-field area of the

TABLE I. OIL AND GAS FIELDS IN PERSIAN GULF AREA

Field	*Area*	*Age of Producing Reservoir(s)*	*Reservoir Rock*	*Cap Rock*	*API Degrees Gravity of Oil*	*Minimum Depth in Feet*
Naft-i-Shah	Iran foothills	Miocene	Carbonate	Salt, anhydrite, marl	42	2,400
Lali	Iran foothills	Miocene	Carbonate	Salt, anhydrite, marl	36	5,000
		Cretaceous	Carbonate	Marl	35	8,500
Masjid-i-sulaiman	Iran foothills	Miocene	Carbonate	Salt, anhydrite, marl	38	600
Naft Safid	Iran foothills	Miocene	Carbonate	Salt, anhydrite, marl	35	3,000
Haft Kel	Iran foothills	Miocene	Carbonate	Salt, anhydrite, marl	38	2,000
Agha Jari	Iran foothills	Oligo-Miocene	Carbonate	Salt, anhydrite, marl	34	4,500
Pazanun	Iran foothills	Oligo-Miocene	Carbonate	Salt, anhydrite, marl	Gas	5,500
Gach Saran	Iran foothills	Oligo-Miocene	Carbonate	Salt, anhydrite, marl	32	2,500
Kirkuk	Iraq foothills	Miocene to Eocene	Carbonate	Salt, anhydrite, marl	35	900
Ain Zalah	Iraq foothills	Upper and Middle Cretaceous	Carbonate	Shale	33	5,000
Bai Hassan	Iraq foothills	Tertiary	Carbonate	?	?	4,200
Butmah	Iraq foothills	Cretaceous	Carbonate	?	34	3,700
Jambur	Iraq foothills	Tertiary	Carbonate	?	?	5,800
West Tigris	Iraq	Tertiary, Cretaceous, Jurassic	Carbonate	?	16–35	700
Rumaila	S. Iraq	Cretaceous	Sand	Shale	33	11,000
Zubair	S. Iraq	Lower Cretaceous	Sand	Shale	33	10,500
Magwa-Ahmadi	Kuwait	Middle and Lower Cretaceous	Sand	Shale	32	3,500
Burgan	Kuwait	Middle and Lower Cretaceous	Sand	Shale	32	3,500
Wafra	Kuwait-Saudi Arabia Neutral Zone	Middle and Lower Cretaceous	Sand	Shale	24	3,500
Bahrein	Bahrein	Middle Cretaceous	Carbonate	Shale	33	2,500
Dukhan	Qatar	Upper Jurassic	Carbonate	Anhydrite	35	5,000
Safaniya	Saudi Arabia	Cretaceous	Sand	Shale	27	5,500
Abu Hadriya	Saudi Arabia	Middle Jurassic	Carbonate	Limestone	36	10,100
Fadhili	Saudi Arabia	Middle Jurassic	Carbonate	Limestone	38	9,800
Qatif	Saudi Arabia	Upper Jurassic	Carbonate	Anhydrite	35	6,900
Dammam	Saudi Arabia	Upper Jurassic	Carbonate	Anhydrite	35	5,000
Abqaiq	Saudi Arabia	Upper Jurassic	Carbonate	Anhydrite	38	5,800
Ghawar	Saudi Arabia	Upper Jurassic	Carbonate	Anhydrite	35	5,900

Data from Lees (1953), Baker (1953), Kerr (1953), *Oil and Gas Journal* (December 26, 1955).

Persian Gulf have already been shown to have as much oil per cubic mile as sediments in the richest petroleum provinces of the United States, and indications are that the Persian Gulf sediments will eventually be proved considerably richer than these United States sediments.

CAUSES OF LARGE OIL RESERVES

VOLUME OF MARINE SEDIMENTS

With the exception of the upper Miocene and Pliocene continental beds, most of which lie north of the Persian Gulf, the sediments present in the Persian Gulf basin are almost exclusively marine. Oil-producing formations are at present limited to the Jurassic, Cretaceous, and Tertiary systems, but some of the oil now in these rocks may have been derived from Triassic or Permian formations. Neither the evaporitic part of the Miocene Fars formations nor the pre-Permian rocks are likely to have generated oil in significant quantity. The volume of probably oil-bearing sediments depends, therefore, on the thickness from the base of the Fars formations to the base of the Permian system.

In a generalized section in the Iranian foothills, Lees (1950, Fig. 2) showed the thickness from the base of the Fars formations to the base of the Permian as 14,500 feet. That this figure is of the right order has been confirmed by more recent work on the Zagros Mountains front by Kent *et al.* (1951, Fig. 2). In the Basra area the sedimentary thickness of the Eocene to Permian interval is known to be about 17,000 feet (Owen and Nasr). In the Arabian fields area, the depth to the base of the Permian was estimated by Kerr (1953) as 17,000 feet, and Lees (1928) indicated that in Oman the thickness to the same horizon was about 14,000 feet. The average thickness of potentially oil-bearing sediments in the present oil-field area should therefore be about 3 miles.

The fields now producing are within an area of 200,000 square miles (Fig. 1) around the head of the Persian Gulf. Thus, there are approximately 600,000 cubic miles of potentially oil-bearing sedimentary rocks in the present oil-field area. The volume of possible oil-bearing sediments in the whole basin is much greater, perhaps more than twice this figure. By comparison, the volume of possibly petroliferous sedimentary rocks in all the basins of the United States has been estimated as 2,000,000 cubic miles (Weeks, 1950).

Thus the Persian Gulf basin is one of the largest in the world and its great reserves of oil are due, in part, to the large volume of marine sediments which it contains.

RICHNESS OF SOURCE ROCKS

The source of oil both in the Middle East and elsewhere is as yet incompletely understood. However, it is generally agreed that: (1) most of the world's oil was formed in marine sediments; (2) the best source rocks were fine-grained; (3) stagnant reducing conditions were the most favorable for oil generation; and (4) certain sedimentary environments have been particularly effective in the generation and accumulation of oil.

The Persian Gulf basin has many rocks which fulfill all these requirements. Most of the sediments present are marine. Shales and marls, some of them bituminous, occur at many horizons and are likely source rocks of oil. The most notable of these rocks are the Lower Cretaceous shales, on both sides of the Gulf, and the Upper Cretaceous to Miocene marls in the foothills. In addition to these largely terrigenous muds there are abundant fine-grained limestones which have almost certainly acted as source rocks (Thomas, 1950; Henson, 1951; Kent *et al.*, 1951).

Stagnant conditions could readily develop in the relatively deep water in front of barrier reefs and between isolated reefs and banks. Reefs occurred extensively in the Jurassic, Cretaceous, and Tertiary periods at least. Stagnant conditions could also develop in front of deltas such as may be represented by the Cretaceous sands near the head of the Persian Gulf (Owen and Nasr.) Both these environments are well known as prolific generators of oil in North America, the first in West Texas and Alberta, the second on the Gulf coast.

Weeks (1952) has drawn attention to the importance of silled basins in producing stagnant reducing conditions, and of the Persian Gulf basin he remarked: "This basin was silled to closed throughout the greater part of its long history. Only a small change in basin closure was necessary to pass from a silled basin with huge volumes of associated oil to a more completely closed basin with salt or anhydrite deposition."

Except during the Cambrian and Miocene it is perhaps more probable that some silled depressions were present rather than that the whole basin was silled. This would be more in accordance with the existence of reefs near the middle of the basin and with Henson's observation that the unstable shelf "is divided up into a pattern of major paleogeographic basins and swells." Conditions in these smaller basins might well have been more conducive to the generation of oil than those in a single large basin, insofar as the smaller basins could have become more frequently and effectively silled, and the oil once formed would have to migrate a shorter distance in order to reach reservoir rocks fringing each basin. Whether silled basins existed or not, there is no doubt that source and reservoir rocks were in close association throughout the huge area of the unstable shelf and not merely along one hinge line or facies trend.

Notwithstanding the foregoing, it is possible that some oil is indigenous to the reservoir rocks in which it is now found. Organic material has been reported (Van Tuyl and Parker, 1941, pp. 43–44; Levorsen, 1954, p. 488) to occur within the shells and skeletons of several marine organisms, including corals; such material would be preserved from oxidation. Since many of the Persian Gulf area reservoir rocks are coral reefs and other organic limestones, this source may have made an important contribution to the oil formed.

TYPE OF HYDROCARBONS

In the present oil-field area of the Persian Gulf basin the proportion of oil to gas is high. Many of the oil fields have either no gas cap or a small one. Very few of the large hydrocarbon accumulations are gas pools only. Of these the the largest is Pazanun, in Iran, and even this field may be proved to contain oil, as the water level has not yet been reached.

This relative scarcity of gas may be due to its differential entrapment outside the present oil-field area and flushing of oil updip into that area. The fact that most of the currently producing oil fields lie at depths of less than 5,000 feet supports this idea. If it is correct, oil from a large part of the basin will have been concentrated in the present oil-field area and the largely untested region outside it will be found less prolific of oil than the area already developed.

However, it is equally probable, throughout the basin, that the proportion of liquid hydrocarbons originally generated was unusually high.

As with gas, very heavy oil or tar does not form a high proportion of the total hydrocarbons in the Persian Gulf area (though large reserves of sulphurous 16° API gravity oil have been found near Baghdad (Baker, 1953)). But several other oil provinces have very large accumulations of tar oil which represent a greater part of the total hydrocarbons. Since this oil is not included in reserve estimates, the volume of hydrocarbons in these provinces appears lower than it actually is.

It is possible, therefore, that Persian Gulf sediments, though unquestionably very rich in hydrocarbons, do not contain more than those of all other areas, but rather that the large reserves figure is due to the high proportion of hydrocarbons in the light-oil range. For example, western Canada has hundreds of billions of barrels of very heavy oil in the Athabasca "tar" sands; the gas reserves of the region are about 20 trillion cubic feet; and the proved reserves of recoverable "light" oil are 3 billion barrels. The total hydrocarbon reserves in place in western Canada are of the same order as those in the Persian Gulf basin but there is no comparison between the reserves of light oil.

RESERVOIR ROCKS

Reservoir rocks of Persian Gulf oil fields have in many places very high porosity and permeability. They include rocks such as deltaic sands, reef limestones, and oölites, which are well known for their excellent reservoir qualities. Many of the limestones have been dolomitized, and the sands show very little cementation. Reservoir formations usually have a stratigraphic thickness of several hundred feet, and, in some strongly folded fields, pay zones have a vertical thickness of several thousand feet.

On the Arabian side of the basin many reservoir rocks are exceptionally good. For example, in Saudi Arabia, where the main reservoir is oölitic partly dolomitized limestone, drilling without mud returns is common practice; the formation is sufficiently porous and permeable to take up drill cuttings (McConnell, 1951). The producing reservoir at Burgan in Kuwait consists of up to 800 feet of soft clean well sorted medium- to coarse-grained sands which have water saturations as low as 5 per cent. Such conditions not only provide very large reservoirs but allow a high proportion of the oil in place to be recovered.

In the Iranian fields, the permeability of the Asmari limestone, which is the main producing formation, is usually less than 0.001 millidarcy (Gibson, 1948), and at Ain Zalah in the Iraq foothills the permeability of the First Pay limestone is too small to measure (Daniel, 1954). Such permeability would usually be too low to allow commercial production; yet in all these foothills fields, individual wells can produce at a rate of more than 20,000 barrels of oil per day. Such high production rates are due to fractures which provide intercommunication throughout the reservoir (Gibson, 1948; Daniel, 1954). The fractures were caused mainly by the Pliocene orogenic movements, though some jointing may have occurred during consolidation of the sediments.

Despite the low permeability, recovery factors in Asmari limestone reservoirs have been estimated between 11% and 25%, and, considering only the porous oil-bearing limestones, the recovery factors were estimated at 20%–23% (Lees, 1953a). Fracturing of reservoirs in the folded belt allows wells to be spaced as much as 2 miles apart. In fields with thin pay zones, fracturing might present a serious disadvantage by allowing coning of edge or bottom water, but this difficulty rarely arises in the Persian Gulf area where pay zones in producing fields are thick.

DISTRIBUTION AND SIZE OF ANTICLINES

Because of the presence of more than one fold system, anticlines occur throughout the greater part of the basin. The structures are of immense size. Lengths of more than 20 miles are common both north and south of the Gulf, and anticlines are known which exceed 100 miles in length. North of the Gulf many folds have amplitudes exceeding 10,000 feet. Many of the anticlines are filled to the spill-point with hydrocarbons. Thus most of the fields are 10–20 miles long and some are much longer. For example, the Kirkuk field in the Iraq foothills has continuous production for nearly 60 miles and the indicated length of the Ghawar field in Saudi Arabia is more than 100 miles. The size of the anticlines not only allows very large single accumulations to form, but makes their discovery by surface geology, seismic work, or structure test holes relatively easy.

Some anticlines received oil from far beyond their closed areas. For example, Lees (1953b) pointed out that "the late Tertiary movements . . . must have had an important influence on oil migration by depressing a broad tract of territory now occupied by the Persian Gulf and the sunk-land of Iraq." In other words, the fields in Kuwait, and near the Hasa coast of Saudi Arabia lie along a hinge line.

TIME OF TRAP FORMATION

South of the Persian Gulf and the River Tigris many of the anticlines show evidence of growth beginning during or soon after deposition of the reservoir rocks (Fig. 5b). In this area, therefore, anticlines were formed early enough to trap oil during its migration.

In the foothills of Iraq and Iran the evidence is not so clear. Oil may have accumulated in anticlines formed during the Upper Cretaceous movements, but most reservoirs in the area are of Tertiary age and in them there is no evidence of penecontemporaneous folding. Such may have occurred but if so its effect on younger formations has been obscured by salt flowage in the lower Fars beds which overlie the Tertiary reservoirs. In any case these reservoirs were covered by 5,000–10,000 feet of overburden prior to the beginning of the main (Pliocene) folding movements, and the oil must have migrated into reservoir rocks before this time. Therefore if anticlines did not form soon after deposition of the reservoir rocks, the Tertiary oils must either have been derived from pre-existing Mesozoic structural or stratigraphic traps by vertical migration through fractures caused by the Pliocene orogeny (Lees, 1950), or they must have been stored in Tertiary stratigraphic traps until anticlines were formed to hold them (Henson, 1951). In the reefs, the latter mode of accumulation was envisaged by Henson (1950a, p. 237) thus:

> . . . oil was generated in the righly organic basinal depths in front of these reefs. . . . As oil was pressed out by compaction, it found a ready outlet along the bedding through porous, fore-reef carrier beds, closely interfingering with the source rocks, which led it up to the reef-reservoir. There much of it no doubt escaped, but some would be trapped in irregular pockets and porous zones within the dense matrix, or would be covered by cemented, fore-reef limestones deposited in the higher, shallower shoals in front of the reef. A primary stratigraphic accumulation of oil would be formed. Later, when folding and faulting occurred, fractures would permit this oil to spread through the reef complex . . . secondary concentration would follow in any accessible, structural trap, and an oil field would result.

Such a hypothesis would apply to the reef or partly reef Tertiary reservoirs at Kirkuk in Iraq, Agha Jari, Pazanun, and Gach Saran, in Iran, and the Middle Cretaceous reservoir at Lali, Iran.

As regards the upper Asmari limestone, small stratigraphic traps were probably widely distributed throughout the reservoir because most of the rock has extremely low permeability, and the more permeable beds are so restricted in extent that they can not usually be correlated between wells; thus any oil formed in the upper Asmari or migrating into it from below would remain in the first porous lens it encountered until released by fracturing.

This method of entrapment is dependent on numerous local variations in permeability to form stratigraphic traps and on hardness to allow fracturing during folding. Carbonates possess these characteristics in greater degree than most sands or sandstones so that the predominantly calcareous sediments of the Persian Gulf area may have been responsible for the retention and accumulation of much oil which would either have escaped from sand reservoirs or would have remained in small uneconomic traps.

To recapitulate, south of the Persian Gulf, where the high permeability of the reservoir rocks would have permitted oil to migrate out of the area, folds were available during or soon after the deposition of the reservoir rocks to trap the migrating oil. North of the Persian Gulf also, traps were present during initial migration of the oil, though it is not certain whether these were anticlines which began to form before the Pliocene movements or whether the oil was held in stratigraphic traps until anticlines were formed to hold it.

CAP ROCKS

The cap rocks over producing fields are listed in Table I. South of the Persian Gulf and River Tigris, the cover rocks are anhydrites, limestones, or shales, and are less than 100 feet thick in some fields. In the Dukhan field of Qatar, where the cap rock anhydrite is only 60 feet thick, there is a fault with a throw of 70 feet, but the seal has been preserved "through diagenesis or plasticity of anhydrite under load" (Henson, 1951).

Cap rocks on the Arabian side of the basin are adequate only because of the gentleness of the folding in this area. But in the foothills, where the rocks are strongly folded and fractured and where such poor cover would be ineffective, either salt or thick marl is present to seal the reservoirs. Lees (1950) has pointed out that if it were not for the presence of salt, oil pools which have minimum cover of less than 1,000 feet—such as those at Kirkuk and Masjid-i-Sulaiman—might not have been preserved. Even with salt cover, there are numerous seepages from the foothills fields.

Thus the distribution of the various types of cap rock is exactly what is required by the tectonics in different parts of the area.

TECTONIC FRAMEWORK OF BASIN

The Persian Gulf basin is composite laterally (Levorsen, 1954). The northeastern flank was structurally controlled by the elevation of the Zagros Mountains in which early Mesozoic and Paleozoic rocks are exposed. In this area much oil has been lost by erosion. Rocks which are oil-bearing farther southwest, here crop out in eroded anticlines, and pebbles of bitumen derived from older oil fields have been found in Pliocene conglomerates (Kent *et al.*, 1951). This destruction of oil fields was limited to a relatively narrow belt formed by the mountains along the northeastern side of the basin. Few oil fields have been exposed in the foothills.

The broader southwestern flank of the basin was controlled by the subsidence which accompanied deposition of the sediments, and most of the oil fields on this side of the basin have been preserved. If this flank of the basin had been of structural origin, that is, if the basin had been folded into a very large syncline, oil fields might have been exposed here also.

Some authors have pointed out the value of unconformities in providing permeable reservoir rocks and stratigraphic traps, and have suggested that unconformities are an asset to a petroliferous basin. This is true only where rocks with primary permeability are absent or where alternative traps are not available to trap the oil during its migration. Except under these conditions, unconformities, by permitting escape of oil at old land surfaces, are a liability.

Throughout the Persian Gulf basin, Mesozoic and Tertiary unconformities though numerous were all of short duration (Fig. 4). According to Gussow (1955) 2,000 feet or more of overburden are probably required to cause flush migration of oil. Except in the Zagros and Kurdish mountains, erosion has seldom been great enough to remove this thickness of sediments and expose oil fields. With more pronounced erosional breaks, cap rocks on some reservoirs might have been breached. For example, the oil field at Burgan in Kuwait has a minimum cap-rock thickness of only 83 feet beneath an unconformity; slight additional erosion would have destroyed this big trap.

Thus the almost continuous subsidence of the basin since the Carboniferous and the consequently limited amount of erosion have allowed the preservation of most of the oil fields.

SCARCITY OF LARGE FAULTS

Faults occur in several of the Persian Gulf fields. In the foothills of the Zagros Mountains there are reverse faults, with throws of several thousand feet, in the salt-bearing beds which overlie the reservoir rocks. However, in the reservoir beds and cap rocks themselves, faults of this magnitude are absent. It is true that faults do occur in these beds: normal faults are present at Kirkuk in Iraq (Daniel, 1954) and there are both normal and reverse faults in the Iranian reservoir rocks; there are several normal faults in the Burgan field in Kuwait and the fault in the Dukhan field has already been discussed. Nevertheless, few if any faults cutting reservoir and cap rocks have been of sufficient size to breach the seals on oil fields and allow the oil to escape.

RARITY OF IGNEOUS ACTIVITY AND METAMORPHISM

Igneous rocks are absent from the present oil-field area and are uncommon in the rest of the basin, except in the possibly eugeosynclinal belt and in the Cretaceous system in Oman and Syria. Metamorphic rocks also are absent from the present oil-field area and very rare outside it. Thus little or no oil has been destroyed by cracking under high temperature or pressure.

OIL GRAVITY

Most of the oils in the Persian Gulf are of high or, less commonly, medium gravity API (Table I). The only notable exceptions are in the low-gravity oil fields near the River Tigris in Iraq (Baker, 1953).

No theory has yet been advanced which will explain all variations in oil gravity. However, some of the factors which influence gravity are known empirically (for example, papers by Thom, Reger, Barton, Bartram, and Taff in *Problems of Petroleum Geology*, 1934) though their relative importance differs from one region to another. These factors are the following.

1. Depositional environment of source rocks: most light oils were formed in marine sediments, many heavy oils were formed in brackish or continental sediments.
2. Migration history.

3. Depth of burial: in any one area the deeper oils tend to be lighter.
4. Age: older oils tend to be lighter.
5. Orogenic pressure: the gravity of oil usually increases with the degree of orogenic pressure to which the enclosing rocks have been subjected. Above a critical pressure only gas occurs and metamorphic rocks are usually barren of hydrocarbons.
6. Inspissation: some oils have become heavier due to escape of light fractions through imperfect cap rocks or at unconformities.
7. Contact with circulating meteoric waters: oils which have come into contact with meteoric waters are usually heavier than those at the same stratigraphic horizon which have not.

The effects of migration are uncertain, and may differ from place to place, but the influence of the other factors is fairly well established.

It follows, therefore, that the high gravity of most of the Persian Gulf oils is due, partly, to the following factors.

1. The almost exclusively marine nature of the oil-bearing sediments. A few may be of brackish-water origin. None is continental.
2. The presence at some time of several thousand feet of overburden on the currently producing fields. The younger (Tertiary) oils of the foothills were buried to greater depths than most of the older (Mesozoic) oils farther south, though much of the load on foothills fields has since been removed by erosion.
3. The occurrence of gentle to moderate folding without metamorphism.
4. The effective separation of the oil fields from the outcrop of their producing formation—by distance on the southwest side of the Persian Gulf, and by deep synclines and depressions in the foothills. This has prevented degradation of the oil by meteoric water.

It is not so easy to explain variations in gravity in the area. These can be accounted for either by present (Table I) or even by past depths of burial. Nor is there a correlation between age and gravity. The Tertiary and Jurassic oils are of similar gravity, and most of them are lighter than most Cretaceous oils, which vary among themselves by as much as 14° API.

The factor causing the greatest differences in gravity appears to have been the nature of the source and reservoir rocks. On this criterion, most of the oils belong in one of the three following categories.

1. Oils in carbonate reservoirs and probably largely of carbonate origin. These range in gravity from 35° to 42° API and include all the Jurassic carbonate oils of the Arabian peninsula and the oils in those Iranian reservoirs (Naft-i-Shah, Lali, Masjid-i-Sulaiman, Naft Safid, Haft Kel) where only the non-reef part of the Asmari limestone is present (Thomas, 1950).
2. Oils in carbonate reservoirs, mostly reef carbonates, and probably generated mainly in marls and argillaceous limestones. These oils range in gravity from 31° to 35° API and include those in the Middle Cretaceous limestones at Lali in Iran, the Eocene to Miocene limestones at Kirkuk in Iraq, and the Oligo-Miocene (Asmari) limestones at Gach Saran and Agha Jari in Iran.
3. Oils in sand reservoirs and probably generated in shales. The gravity range here is 24° to 33° API and this group includes nearly all the oils in Cretaceous sand reservoirs near the head of the Persian Gulf.

There are some exceptions to this classification, of which the most important are in the third group. At Zubair, for example, the main (Third Pay) producing reservoir contains oil of 33° API gravity, but the Fourth Pay has 42° API gravity oil, and an Upper Cretaceous limestone contains tar oil. In the Burgan reservoir of Kuwait where the average gravity is 31.5° API, gravities as high as 36° API have occasionally been recorded and are believed to be due to gravity separation.[3]

Nonetheless, the relationship between oil gravity and the amount of terrigenous material is too constant to be fortuitous. Whether it indicates differences in

[3] R. M. S. Owen, personal communication.

source material, in environment of origin, or in the history of the oils after generation is not known with certainty, and more than one of these factors may have had some effect. But in the Burgan field, the shales interbedded with the reservoir sands contain abundant plant remains and resin lenticles but no foraminifera (Owen and Nasr), whereas limestones, in many places in the Persian Gulf basin, contain abundant remains of foraminifera, corals, algae, and other marine organisms. This suggests that gravity differences were caused by differences in the amount of land-plant detritus or of precipitated humic material (Snider, 1934) or both in the source rocks. That is to say, oil formed from such material is heavier and more asphaltic than oil formed from marine organisms. A similar relationship has been reported by Taff (Van Tuyl and Parker, 1941) in California, and by Kugler (*ibid.*) in Trinidad.

CONCLUSIONS

The huge oil reserves in the Persian Gulf area are the result not of any single factor but of the coincidence of all the conditions favorable for the formation, accumulation, and preservation of hydrocarbons. The most important of these conditions were the following.

1. A very large volume of marine sediments including some which were deposited in stagnant reducing conditions.
2. Excellent reservoir rocks and consequent high recovery factors.
3. The association of reservoir rocks and source rocks throughout a very extensive shelf area. Reefs are common and thick deltaic sands also occur. These rocks are well known as prolific producers of oil in North America.
4. Immense anticlines, caused at least partly by the competence of thick limestone beds in conjunction with the plasticity of underlying salt. Due to the presence of more than one fold system these anticlines occur in a large part of the basin.
5. The availability of structural and stratigraphic traps during migration of the oil. The predominance of carbonate rocks in the basin may have been of critical importance in retaining oil in stratigraphic traps until anticlines were formed.
6. The occurrence of fractures which have breached many small stratigraphic traps in areas of low or variable permeability and allowed accumulation in highly productive anticlinal traps. Here also the abundance of carbonates has been important since they fracture more readily than most fine-grained sandstones and siltstones.
7. Efficient cap rocks, in particular the presence in the foothills of salt which seals strongly fractured reservoirs from which oil might otherwise have escaped.
8. The absence, outside the mountains, of prolonged erosional intervals so that loss of oil by seepage at unconformities has been minimized.
9. The absence, in reservoir rocks, of large faults which might have broken the seals on oil fields.
10. Rarity of igneous activity or metamorphism which might have destroyed some of the oil.

In addition, a high proportion of the total hydrocarbons is in the form of light oil. The reason for this is not certain. It may be original. On the other hand, there may have been differential entrapment of gas outside the present oil-field area. The high gravity API of most of the oil is due to the marine origin of the source rocks, to considerable depth of burial either past or present, to moderate but not excessive orogenic stress, and to the absence of circulating meteoric water. Variations in gravity are commonly related to the amount of terrigenous material in the source and reservoir rocks. This may be due to differences in the quantity of land-plant source material in different formations.

REFERENCES

Baker, N. E., and Henson, F. R. S., 1952, "Geological Conditions of Oil Occurrence in Middle East Fields," *Bull. Amer. Assoc. Petrol. Geol.*, Vol. 36, No. 10.

———, 1953, "Iraq, Qatar, Cyprus, Lebanon, Syria, Israel, Jordan, Trucial Coast, Muscat, Oman, Dhofar and the Hadramaut," *Science of Petroleum*, Vol. VI, Pt. 1.

Daniel, E. J., 1954, "Fractured Reservoirs of Middle East," *Bull. Amer. Assoc. Petrol. Geol.*, Vol. 38, No. 5.

Gibson, H. S., 1948, "Oil Production in Southwestern Iran," *World Oil* (June 14).

Gussow, W. C., 1955, "Time of Migration of Oil and Gas," *Bull. Amer. Assoc. Petrol. Geol.*, Vol. 39, No. 5.

Henson, F. R. S., 1950a, "Cretaceous and Tertiary Reef Formations and Associated Sediments in Middle East," *ibid.*, Vol. 34, No. 2.

———, 1950b, "The Stratigraphy of the Main Producing Limestone of the Kirkuk Oilfield," *Int. Geol. Congr.*, 1948, Pt. VI.

———, 1951, "Observations on the Geology and Petroleum Occurrences in the Middle East," *Proc. Third World Petrol. Congr.*, Sec. 1. E. J. Brill, Leiden, Netherlands.

Kent, P. E., Slinger, F. C. P., and Thomas, A. N., 1951, "Stratigraphical Exploration Surveys in South-West Persia," *ibid.*, Sec. 1.

Kerr, R. C., 1953, "The Arabian Peninsula," *Science of Petroleum*, Vol. VI, Pt. 1.

Knebel, G. M., and Rodriguez-Eraso, G., 1956, "Habitat of Some Oil," *Bull. Amer. Assoc. Petrol. Geol.*, Vol. 40, No. 4, pp. 547–61.

Kuwait Oil Company Ltd., 1953, "Kuwait," *Science of Petroleum*, Vol. VI, Pt. I.

Lees, G. M., 1928, "The Geology and Tectonics of Oman and Parts of South-Eastern Arabia," *Quar. Jour. Geol. Soc. London*, Vol. 84, Pt. 4.

———, 1950, "Some Structural and Stratigraphical Aspects of the Oilfields of the Middle East," *Int. Geol. Congr.*, 1948.

———, 1953a, "Persia," *Science of Petroleum*, Vol. VI, Pt. 1.

———, 1953b, "The Middle East," *ibid.*

Levorsen, A. I., 1954, *The Geology of Petroleum*. W. H. Freeman, San Francisco.

McConnell, P. C., 1951, "Drilling and Producing Techniques . . . in Abqaïq Field," *Oil and Gas Jour.* (December 20).

Owen, R. M. S., and Nasr, S. N., "Stratigraphy of the Kuwait-Basra Area," forthcoming publication.

Problems of Petroleum Geology, 1934, Amer. Assoc. Petrol. Geol.

Snider, L. C., 1934, "Current Ideas Regarding Source Beds for Petroleum," *ibid.*, pp. 51–66.

Steineke, Max, and Bramkamp, R. A., 1952, "Mesozoic Rocks of Eastern Saudi Arabia" (abst.), *Bull. Amer. Assoc. Petrol. Geol.*, Vol. 36, No. 5, p. 909.

Thomas, A. N., 1950, "The Asmari Limestone of South-West Iran," *Int. Geol. Congr.*, 1948, Pt. VI.

Van Tuyl, F. M., and Parker, Ben H., 1941, "The Time of Origin and Accumulation of Petroleum," *Quar. Colorado School Mines*, Vol. 36, No. 2.

Weeks, L. G., 1950, "Concerning Estimates of Potential World Oil Reserves," *Bull. Amer. Assoc. Petrol. Geol.*, Vol. 34, No. 4.

———, 1952, "Factors of Sedimentary Basin Development That Control Oil Occurrence," *ibid.*, Vol. 36, No. 11.

GENERATION, MIGRATION, ACCUMULATION, AND DISSIPATION OF OIL IN NORTHERN IRAQ[1]

H. V. DUNNINGTON[2]
Kirkuk, Iraq

ABSTRACT

Most of the known oil accumulations of Northern Iraq probably originated by upward migration from earlier, deeper accumulations which were initially housed in stratigraphic or long-established structural traps, and which are now largely depleted. The earlier concentrations had their source in basinal sediments, into which the porous, primary-reservoir limestones pass at modest distances east of the present fields.

Development of the region favored lateral migration from different basinal areas of Upper Jurassic and Lower-Middle Cretaceous time into different areas of primary accumulation. Important factors affecting primary accumulation included—1. early emergence and porosity improvement of the reservoir limestones, followed by burial under seal-capable sediments; 2. the timely imposition of heavy and increasing depositional loads on the source sediments, and the progressive marginward advance of such loads; 3. progressive steepening of gradients trending upward from source to accumulation area; 4. limitation of the reservoir formations on the up-dip margin by truncation or by porosity trap conditions. In late Tertiary time, large-scale folding caused adjustments within the primary reservoirs, and associated fracturing permitted eventual escape to higher limestone reservoirs, or to dissipation at surface.

The sulfurous, noncommercial crudes of Miocene and Upper Cretaceous reservoirs in the Qaiyarah area are thought to stem from basinal radiolarian Upper Jurassic sediments, which lie down dip, a few tens of miles east of these fields. Upper Cretaceous oils of Ain Zalah and Butmah drained upward from primary accumulations in Middle Cretaceous limestones, which were filled from basinal sediments of Lower Cretaceous age situated in a localized trough a few miles northeast of these structures. The huge Kirkuk accumulation, now housed in Eocene-Oligocene limestones, ascended from a precedent accumulation in porous Middle-Lower Cretaceous limestones, which drew its oil from globigerinal-radiolarian shales and limestones of the contemporaneous basin, a short distance east of the present field limits.

Eocene-Oligocene globigerinal sediments, considered by some the obvious source material for Kirkuk oil, seemingly provided little or no part of the present accumulation. The reservoir formation may have been filled from these sources, to lose its oil by surface dissipation during the erosional episode preceding Lower Fars deposition. Upper Cretaceous basinal sediments probably contributed nothing to known oil field accumulations, though they may have subscribed to the spectacular impregnations of some exposed, Upper Cretaceous reef-type limestones. Neither Miocene nor pre-Upper Jurassic sediments have played any discernible role in providing oil to any producing field. Indigenous oils are thought to be negligible in the limestone-reservoir formations considered.

INTRODUCTION

The exceptionally rich oil accumulations of the Middle East have attracted much comment in recent years. In this paper the sedimentary and tectonic history of the North Iraq sector of the Persian Gulf-Iraq-Iran geosyncline is reviewed briefly, and inquiry is pursued into the probable origins and modes of accumulation

[1] Read before the Association at New York, March 30, 1955. Manuscript received, April 5, 1955. The opinions expressed in this paper are those of the writer, and not to be construed as necessarily reflecting the views of the geological department of the Iraq Petroleum and associated companies.

[2] Divisional Paleontologist, Iraq Petroleum Company Limited, Kirkuk, Iraq. The writer is indebted to the management and to the Chief Geologist of Iraq Petroleum and associated companies for opportunity to prepare and permission to publish this paper. Acknowledgements are gratefully extended to many colleagues in the past and present employ of these companies for their contributions to the stratigraphic synthesis. Particular thanks are due to R. C. van Bellen whose interpretations of Tertiary sections have been freely drawn upon, to R. G. S. Hudson who has been responsible for stratigraphic determinations of most of the Mesozoic macrofaunas, and to R. Wetzel, whose field studies in Kurdistan and elsewhere have provided many of the thickness controls and sample collections on which the map constructions have been based.

of the oils found in the Kirkuk, Butmah, and Ain Zalah; Bai Hassan and Jambur; and Qaiyarah, Najmah, Jawan, and Qasab fields. The first three named fields, which are producing at present, yielded a total output of 194 million barrels of oil in 1954, mostly from the Kirkuk structure. Before development drilling was suspended in 1939, the four connected fields of the Qaiyarah area had been proved to contain, within their two superimposed limestone reservoirs, very large reserves of heavy, sulfurous, unmerchantable oil. Jambur and Bai Hassan are newly confirmed fields, under development, but not yet in production. It is too early for their importance to be fully assessed, though it is clear already that they will prove to be rather small fields in comparison with the nearby Kirkuk.

The region considered is that part of Iraq which lies north of Latitude 33° North. The area embraced is approximately 90,000 square miles.

Several recent papers treat of the Middle East as a whole and touch upon sedimentary history, tectonics, oil geology, or oil-field development in Northern Iraq. History of exploration has been dealt with briefly, by Lees (1950a) and by Barber (1948). Later information on exploration and development appears in the annual reviews of "developments in foreign fields" featured in successive July numbers of the Association *Bulletin,* and in papers by Baker (1953), Wellings (1954), and Daniel (1954). The general geological background is covered in publications by De Boeckh, Lees, and Richardson (1929), Lees and Richardson (1940), Lees (1950a, b, 1951, 1952, 1953), and Henson (1950a, b, 1951a, b). Some of these contributions discuss oil-origin and -migration problems of the region.

Rock unit nomenclature for Northern Iraq is being prepared for publication in the near future. References to rock units by name have been excluded from this paper, except in the cases of the Euphrates limestone, Lower Fars and Upper Fars-Bakhtiari units, for which time-stratigraphic treatment would have been inappropriate.

The isopach and facies maps are based upon all available collections of surface and well samples, and on unpublished records of unsampled measured sections. Stratigraphic limits have been determined, and facies variants differentiated, principally upon the evidence provided by thin sections.[3] Age determinations of surface samples of Paleozoic to Jurassic age have been based largely on extensive macrofossil collections.

Stratigraphic interpretations summarized in map form represent the combined findings of a large number of geologists and paleontologists, including C. André, R. C. van Bellen, E. J. Daniel, G. F. Elliott, T. F. Grimsdale, F. R. S. Henson, J. M. Hudson, R. G. S. Hudson, A. Keller, G. M. Lees, J. McGinty, D. M. Morton, K. al Naqib, J. Robinson, A. H. Smout, R. Wetzel, the writer, and others.

THE TECTONIC FRAMEWORK

As introduction to the fundamental architecture of the North Iraq basin it suffices to make the simple division into "nappe zone," "folded zone," and "unfolded

[3] All rock samples from wells and outcrops are studied in thin section as part of routine examinations carried out in geological laboratories of the Iraq Petroleum and associated companies. Over 300,000 thin sections of rock samples from North Iraq are on file at Kirkuk.

area" which is illustrated in Figure 1. These units were differentiated during the late-Tertiary (mainly Pliocene) Alpine orogeny, which raised the high Zagros Mountains in the area adjacent to and beyond the northeastern frontiers of Iraq, and cast up the large, elongate, anticlinal folds of the foothills and frontal ranges. All the known oil fields of Northern Iraq are contained in anticlinal traps within the "folded zone."

An elaborate analysis of the tectonic construction of the Middle East in general, including Northern Iraq, has been made by Henson (1951b), who stresses the important role played by pre-Miocene vertical movements in the "unfolded area" and in the "folded zone." Henson considers that folding due to late-Tertiary compression was superimposed on and molded against an earlier, deep-seated, block-faulted framework, inherited from the pre-Miocene history of the region. Other accounts, by De Boeckh, Lees, and Richardson (1929), Lees and Richardson (1940), and Lees (1950a, b, 1951, 1953), have accepted uncomplicated compressional folding, with a broad, "normally folded zone" intervening between the "nappe front" and the unfolded "foreland."

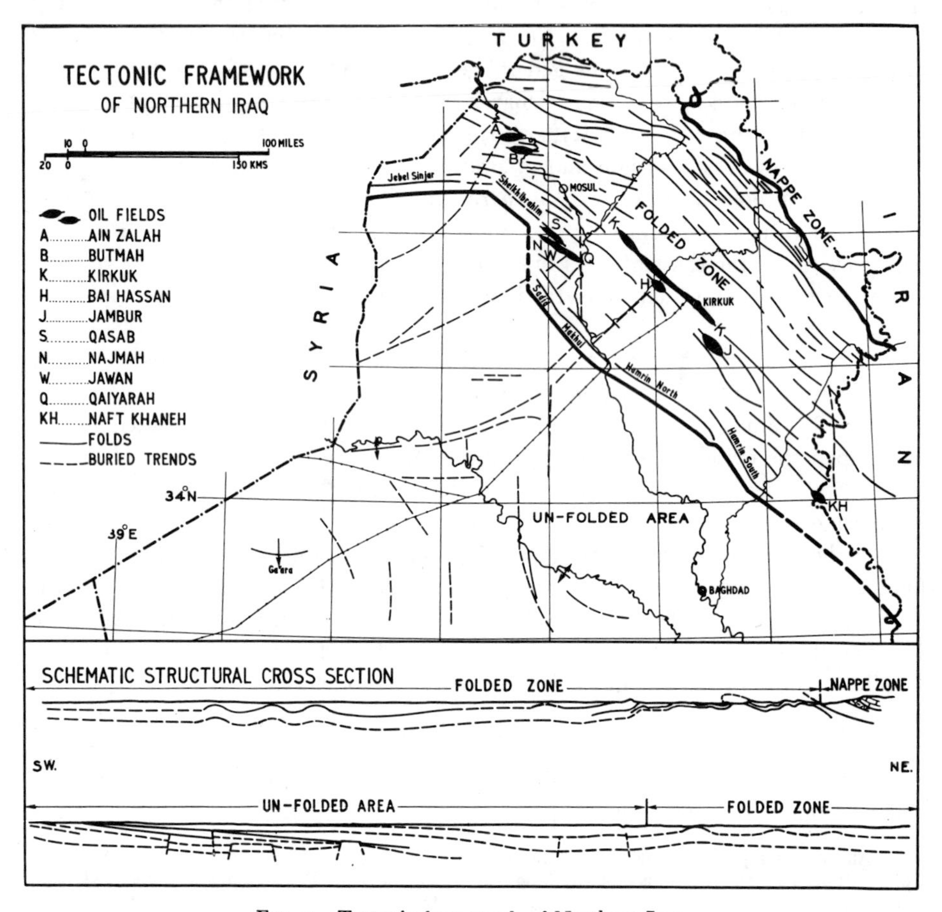

FIG. 1.—Tectonic framework of Northern Iraq.

Tripartite division into unfolded, folded, and nappe zones, on the dominantly northwest-southeast trend of the Zagros, is appropriate for post-Miocene time, and also is harmonious with existing concepts of a Cretaceous-Pliocene geosyncline developing on the same Zagros trend, with its axis migrating more or less progressively southwestward, until the paroxysm of orogeny in Pliocene times. Conditions in Jurassic and pre-Jurassic time were somewhat different. The basinal trend ran more nearly north-south than northwest-southeast in the known Jurassic stages, and the Jurassic-Cretaceous transition was probably a time of tectonic disturbance, with widespread uplift and tilting of fault-bounded blocks on the western margins of the basin. Several deduced faults of this age are directed north-south, across the Zagros trend. The axes of thickest basinal deposition for the Jurassic stages underlie thin developments of shallow-water Lower Cretaceous sediments.

Although the "unfolded area" has been referred to by De Boeckh, Lees, and others as a "foreland," it is now emphasized that a very thick sedimentary sequence underlies much of it (Lees, 1953). The outcrops of the Arabo-Nubian massif or "shield" lie far to the south and southeast of the limits of the region considered.

Gravimetric data and slight structural manifestations in the flat-lying Tertiary cover provide evidence of important buried structural elements throughout the "unfolded area." Lateral variations in sedimentary thicknesses over small distances are deducible, and it is concluded that most of these variations are controlled by pre-Miocene faulting, of various dates, the relief due to faulting having been erased at surface due to subsequent erosion and deposition. The tectonic pattern suggested by the evidence is a complex one, resulting from interplay of vertical movements, on fault trends which are oriented predominantly north-south, east-west, northwest-southeast, and northeast-southwest, the importance of individual trends showing marked local variations. Some of the prominent detected structural features are continuous for some tens or even hundreds of miles. Some pass beyond the limits of the region considered, into Syria and Southern Iraq. A few transgress directly across the limit of the unfolded area into the folded zone (Fig. 1).

Within the unfolded area, a single large broad dome on an east-west alignment occurs in the southwest of the region, in the Ga'ara Depression north of the Wadi Hauran. This dome, which exposes Lower Triassic sediments in its culmination, is on the flank of a very gentle broad uplift which must reflect basement arching. A second prominent east-west fold, narrow but gentle, is known at Anah, on the Euphrates, in a fault trough setting which argues strongly for fault origin of the anticline itself. Other anticlinal features within the "unfolded area" are small and indefinite, with very low dips and dubious or no closures; most of them bear no directional relationship to the Zagros-trend folds of the folded zone, or to the deduced directions of late-Tertiary tangential pressures.

The boundary between the folded and unfolded zones is abrupt. The southernmost and westernmost anticlines of the folded zone, Jebels Sinjar, Sheikh Ibrahim, Sadid, Makhul, and Hamrin North and Hamrin South, are surprisingly large folds, tens of miles long, and rising in some cases several thousands of feet out of the adjacent synclines. The boundary follows a more or less arcuate line, from northwest-southeast in the southeast to east-west in the northwest, but it is offset sharply in the area west of Qaiyarah. Here its course is picked out by the north-south alignment of the pitching ends of five rather small and feeble anticlines. The line plunges below alluvium in the plains east of Baghdad (or else is offset northward, near Naft Khaneh, to the foot of the exposed Iraq-Iran frontier folds).

The folded zone has an average width of about 100 miles, and the alignments of the individual folds and of the zone itself follow the Zagros swing from east-west in the north to northwest-southeast in the south. There is an over-all tendency for anticlines to increase in amplitude and tightness as the "nappe front" is approached, the most northeasterly folds being themselves thrust and locally overturned. But there are many departures from the superficially natural expectation that the folds should diminish in strength with increasing distance from the thrust front. Thus Hamrin North, Makhul, and Sinjar are much more powerful and much more elevated folds than the anticlines which occur immediately to the north and northeast, whereas flat, low domes appear, between steep, narrow, anticlinal neighbors, in the middle of the zone. Other oddities include the existence of great elevational differences between adjacent anticlines of otherwise similar dimensions, the pitching down of successive anticlines along linear features, and the appearance of marked asymmetries in some folds, and of occasional opposed asymmetries in some pairs of adjacent folds. These and other anomalies provide evidence for the opinion of Henson (1951b) that the late-Tertiary folding in the "folded zone" was intimately affected by a pre-existing complex of faults, which left residual features to buttress or deflect folding, and which predisposed the region to react in irregular and complex fashion to simple tangential pressures.

Where directional characters can be read into the complications, one or other of the four main tectonic trends discerned in the "unfolded area" is generally found to be involved. As some few important tectonic features on these trends pass directly from folded into unfolded terrain it is argued that the "folded zone" was probably traversed, prior to folding, by a tectonic network comparable with that which is now detectable in the unfolded area.

STRATIGRAPHIC DEVELOPMENT

PRE-JURASSIC

The Precambrian basement is not exposed anywhere in Northern Iraq, and deep boring has not yet penetrated the base of the Triassic.

Paleozoic rocks crop out in the extreme north of the region close to the Turkish frontier, where Cambro-Orodovician shales and quartzites are overlain by a thin, incomplete Lower Carboniferous marine succession, which is covered by thick Upper Permian limestones. There are no discernible angular discordances at the unconformities. Permo-Carboniferous limestones are recorded from Harbol, in southeastern Turkey (Tasman, 1949), and limestones of similar age, overlying bituminous Devonian shales and sandstones, have been described from Hazro, northeast of Diyarbekir (Tolun, 1949). According to Ternek (1953), Upper Permian limestones crop out extensively in the area between Lake Van and the Turco-Iraq frontier north of Amadia—schists, phyllites, and quartzites underlying the limestones are also dated as Permian by this author, but it seems probable, from the nature of the succession known in Iraq, that these clastics are at least as old as Lower Carboniferous.

Graptolite-bearing Lower Paleozoic clastics have been encountered in some deep wells in Syria, where they underlie Permo-Carboniferous limestones and marine shales, siltstones, and quartzites. Thick lower Paleozoic sections of marine clastics occur in Northern Saudi Arabia, and Cambrian limestones, shales, and sands are known in Jordan.

In the Bakhtiari Province of Iran the thick, exposed, Cambrian succession includes bedded salt in its lower parts, and the intrusive salt of the spectacular domes of the southern Persian Gulf, Laristan, etc., is of Cambrian age (Lees, 1953b, etc.). Lees (1952) and others (e.g., O'Brien, 1950) have suggested tentatively that Cambrian rocks, including important salt components, may underlie much of the "normally folded zone" of the Zagros. Such salt could have permitted *décollement*-type folding over an unfolded basement (O'Brien), or it could have facilitated the development of markedly different fold patterns in the underlying basement and in the overlying sedimentary cover (Lees, 1952). The absence from Northern Iraq of any apparent salt dome, despite the great thickness of post-Cambrian sediments, argues against the presence of any thick salt series at depth.

Though the positions, dimensions, and directional trends of pre-Triassic basins are unknown, it is probable that thick deposits of marine Paleozoic sediments underlie most of the region.

Triassic sediments are known from many exposures in Kurdistan, from deep well sections on the Qalian, Atshan, and Butmah anticlines and on Syrian anticlines, and from exposures in the Ga'ara depression in the southwest of the region. The upper part of the Triassic is calcareous or dolomitic; the lower part is dominantly clastic, comprising sandstones and quartzites in the southwest, and shales and marls with subordinate limestones in Kurdistan. Only the calcareous upper division has been reached in wells in Iraq. Subsurface successions include frequent intercalations of anhydrites and calcareous shales, which are absent or inconspicuous at outcrop. The total thickness of the Triassic section at exposure exceeds 5,500 feet. The same order of total thickness is estimated for the Quaiyarah area, and some 4,000 feet is present in central Syria. The Ga'ara sections are much thinner, and show sand intercalations in the limestone sequence, and continental features in the lower clastic division. The Triassic basin occupied all Northern Iraq and extended far into Persia, Turkey, and Syria; its geography remains largely unknown, except that one approach to the margin lay in the southwest of the region considered.

The occasional evaporites of the well sections indicate intermittent barred-basin conditions, and the entire sedimentary suite suggests slow subsidence over the whole basinal area. Oil was probably generated from time to time within this environment, and indications of indigenous oils have been noted in some wells, and at outcrop. But the faunas of the time were sparse, and planktonic organisms were rare, so that these rocks may offer only limited potentialities as oil source rocks.

JURASSIC

Liassic.—In Kurdistan the Liassic rocks comprise an upper dolomitic limestone unit, about 600 feet thick, and a lower limestone-shale-anhydrite sequence which is extensively slump bedded. The Liassic of the Ga'ara is partly obscured, but includes evaporitic limestones and dolomites with some dispersed sand and intercalated shales.

The wells show much thicker sections than those of Kurdistan, and a more mixed assemblage of rock types, including bedded anhydrites, argillaceous limestones, shales, dolomites, and oölitic and pseudo-oölitic limestones. The bipartition of the Liassic of the mountain sections is not matched by any simple lithological division of the subsurface successions.

Generation of oil probably occurred in the semi-barred basinal environment indicated by the sediments, but total organic content of the formations was probably never great.

The Liassic basin was more localized than that of the Triassic. The thickest sedimentation and the deepest-water facies, indicating the center of the basin, is found in the wells to the west of Mosul. Facies correlations suggest a trend approximately northwest-southeast in this area, but the basin axis may have swung westward into Syria to cross the Syria-Iraq frontier at about the latitude of Jebel Sinjar.

Middle Jurassic.—The Middle Jurassic is the earliest time interval for which any serious paleogeographic reconstruction has been attempted in map form (Fig. 2). Isopachs in the northern part of the area are fairly well controlled by surface

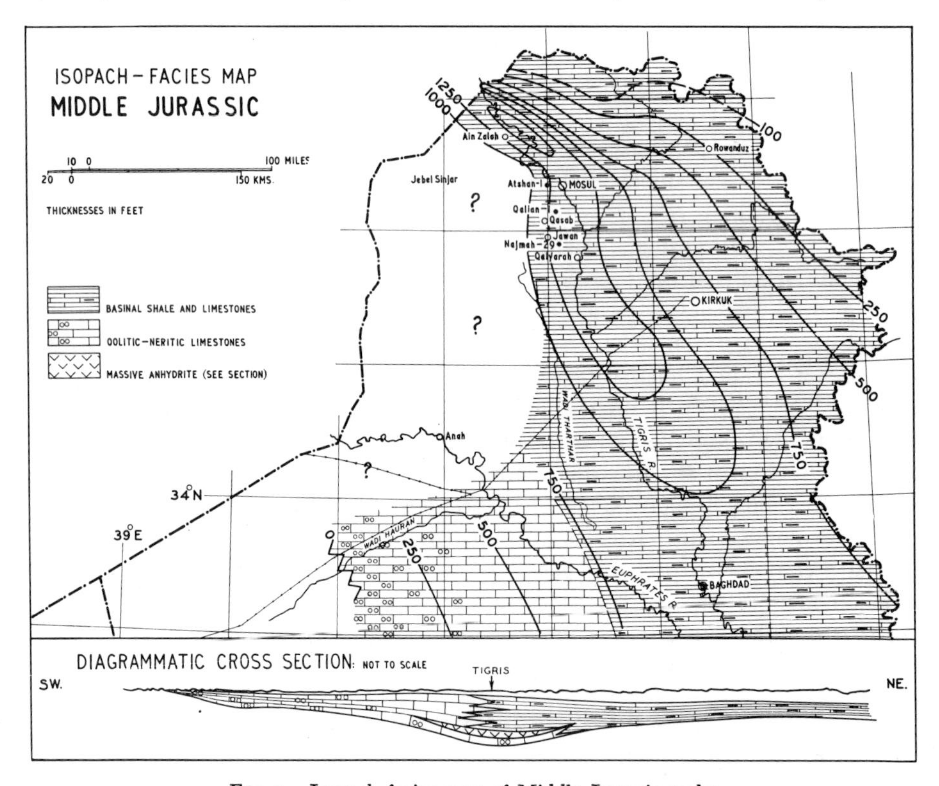

FIG. 2.—Isopach-facies map of Middle Jurassic rocks.

and subsurface reference sections; the two "trough" features on the line of the Tigris, and the swings of contours in the mountain area are almost unavoidable. There are certainly other irregularities in actual thickness, not suggested by available controls, which may be revealed by eventual drilling.The construction of the southern and western parts of the map hangs entirely upon the small group of exposures in the Wadi Hauran.

The thicknesses and facies of Middle Jurassic sediments remain unknown throughout the large area which lies between the Wadi Hauran, in the southwest, and the Atshan—Ain Zalah wells, close to the Tigris, in the north.

No attempt is made in this or other figured isopach maps to differentiate between contours which are adequately supported by observation and those which are in large measure conjectural.

Isopachs are intended to show existing thicknesses. Original thicknesses were much reduced by early Neocomian erosion in the area west of the Tigris and north of Atshan. The original area of thickest deposition probably lay between the Tigris, north of Mosul, and Jebel Sinjar. The original thickness trend may have been east-west, rather than northwest-southeast in this area.

Similarly, the succession has been planed off by pre-Albian erosion in the southwestern part of the area; the zero-thickness line reflects erosional convergence. The original basin margin lay to the west of Muhaiwir. But the sediments here are neritic limestones, locally oölitic and sandy, indicating approach to the continental limit.

Slight erosional unconformity without visible discordance is found between Middle and Upper Jurassic rocks in the Qalian and Najmah wells, close to the area of maximum depositional thickness, yet in the mountain zone, where the sediments are very thin, deposition was continuous, and the facies are more "basinal" than are found elsewhere in the region. Sedimentation was probably rapid, in the Tigris area, on a rapidly subsiding but always relatively shallow, shelving floor, while slow deposition continued in the deeper mountain zone, which was more remote from the basin margins, and which subsided slowly.

Fauna and lithological facies are generally similar in the Tigris area and Kurdistan. *Posidonia* shales and limestones, and *Radiolaria*-rich sediments occur in both areas, as does super-abundant debris of a problematical organism of planktonic habit.[4]

But the initial sediments of the Middle Jurassic in the zone of thick sedimentation are pellety, oölitic limestones with a rich benthonic fauna, including numerous gastropods, rare *Haurania* spp., and ubiquitous pellets of the encrusting foraminifera *Nubecularia*. A discontinuous, bedded-anhydrite unit intervenes in some wells between this shallow-water limestone and the main mass of the radiolarian sediments above. The pellety limestone and anhydrite are not developed in Kurdistan.

Rocks of this time interval are markedly bituminous both in the mountain zone and in the thick-sedimentation belt. Semi-euxinic depositional conditions are presumed, and source potentialities are rated high. Indigenous oil has been noted in the upper part of the sequence in some wells, but no oil has been produced from these

[4] Identified with "filaments d'Algues?," Pl. VIII, 2; Pl. IX, 1, in J. Cuvillier and V. Sacal (1951).

formations, which are of very low permeability. In the northwestern part of the region, ample opportunity for dissipation of indigenous oil was afforded by long exposure following Berriasian uplift and preceding Aptian transgression.

Upper Jurassic (Callovian-Kimmeridgian).—Figure 3 illustrates the distribution of the principal facies and total thickness isopachs of the Upper Jurassic. The conditions of the Middle Jurassic were largely repeated in this time interval. Again the mountain zone shows very thin, ammonitiferous, radiolaria-rich, euxinic shales and limestones, with a thin terminal anhydritic unit which is dated as Kimmeridgian and tentatively equated with the Hith anyhdrite formation which caps the "Arab zone" fields of the Hasa and Qatar. Again the thickest sedimentation is found in the Tigris area, where very thick oölitic and pseudo-oölitic and chemical limestones culminate in a massive-bedded anhydrite, some 600 feet thick. The western limits of existing sediments of this age were imposed by erosion of early Neocomian date, which bared the uplifted area northwest of Mosul down to the Middle Jurassic. The planing down of a tilted fault block embracing the Najmah, Qalian, and Atshan wells is illustrated; the existence and positioning of the bounding faults is conjectural.

Upper Jurassic sediments are not exposed in the southwestern part of the region, where the youngest sediments seen, below transgressive Albian sandstones, are of Middle Jurassic age.

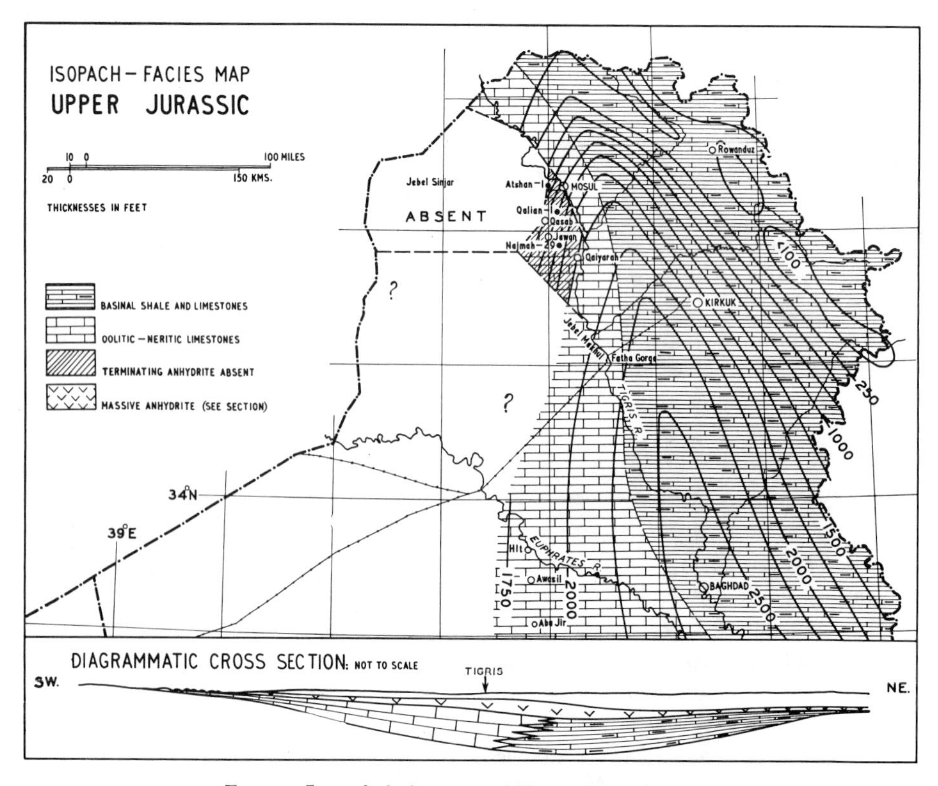

FIG. 3.—Isopach-facies map of Upper Jurassic rocks.

Little is known of conditions in the zone between the mountain exposures and the wells of the Tigris area. By analogy with conditions found in the Tithonian-Berriasian stages it is argued that the thick, oölitic limestones of Najmah and Qalian pass eastward into thick, basinal, euxinic sediments, which become progressively thinner as the mountain zone is approached. Source potentialities are seen in these postulated basinal sediments, thickness and nature of which remain to be established by drilling.

JURASSIC-CRETACEOUS TRANSITION: TITHONIAN-BERRIASIAN

The basinal configuration of the Tithonian-Berriasian stages (Fig. 4) reproduces that of the Upper Jurassic, and the facies are similarly distributed. The disparity in thickness between the Kurdistan sections and the Tigris area well sections is smaller. In Kurdistan the sediments comprise about 300–600 feet of radiolarian shales and thin-bedded limestones with abundant ammonites (Spath, 1950). The lower beds are heavily impregnated with bitumen.

The rocks of the broad, rapidly subsiding, shallow shelf, are chemical limestones

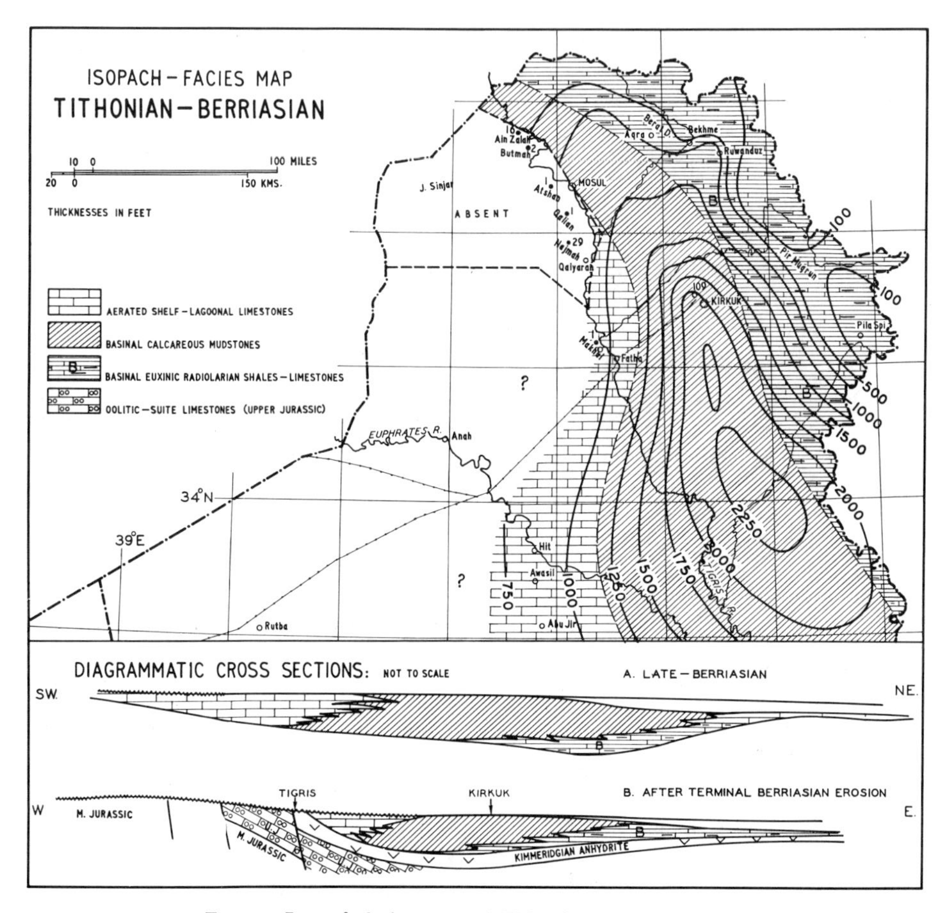

FIG. 4.—Isopach-facies map of Tithonian-Berriasian rocks.

and calcareous mudstones, with rare oölitic and pseudo-oölitic limestone intercalations, generally tight, impermeable and unfossiliferous. Oölitic intercalations increase in frequency and importance upward. Observed fossils are gastropods, rare algæ, and lituolids. The total thickness of the shelf facies is about 950 feet. Between these two extremes, a recent deep test on the Kirkuk structure has revealed a very thick mudstone-shale-limestone sequence, with an upper, unfossiliferous mudstone unit more than 2,000 feet thick, overlying about 450 feet of black, calcareous, Tithonian shales and limestones which contain a superabundant radiolarian fauna. The radiolarian rocks met in this well were partially impregnated with thick glutinous bitumen, but yielded small amounts of very light, asphaltene-free oil from fractures.

The Tithonian sediments are regarded as very important potential source rocks, without any apparent outlet to adjacent potential reservoirs. The thickness and plasticity of the overlying mudstone cover have probably been adequate to prohibit any upward migration of oil at Kirkuk, and the lateral equivalents of the radiolarian sediments toward the southwestern margin of the basin are impervious chemical limestones and mudstones.

Important tectonic movements occurred at or about the end of Berriasian time. These were manifested in—

1. Tabular elevation of the area lying northwest and west of Mosul, on bounding faults of unknown position (one such fault is deduced along the course of the Tigris, north and east of Ain Zalah; others may be conjectured on gravimetric and other evidence).
2. Tilting and uplift of a block embracing the Najmah, Qalian, and Atshan areas (Fig. 3). (The existence and course of the north-south fault limiting the basinward side of this block are conjectured).
3. Pronounced eastward shift in the facies and isopach pattern between Tithonian-Berriasian and Valanginian-Aptian time.

Copious local deposits of sedimented bitumen in the basal beds of the ensuing Valanginian testify to breaching of important early oil accumulations at about this time.

CRETACEOUS

Valanginian-Aptian.—Stratigraphic relationships were rather complex during this period. The map (Fig. 5) is a schematic construction, rather than a true facies-isopach construction. Seven different formations are recognized within the interval; all have diachronous contacts with adjacent formations, and interdigitation between formations is common.

The thin deposition in the northwest is due in part to absence of sediments of the Valanginian and Hauterivian over the high tabular feature which was raised during the Jurassic-Cretaceous transitional period. This large uplift was not overlapped until late Barremian-Aptian time. The sediments then deposited were *Orbitolina* marls and marly neritic limestones, with rare local sand and silt concentrations.

In the southwest at Awasil the succession comprises a thin Valanginian oölitic-

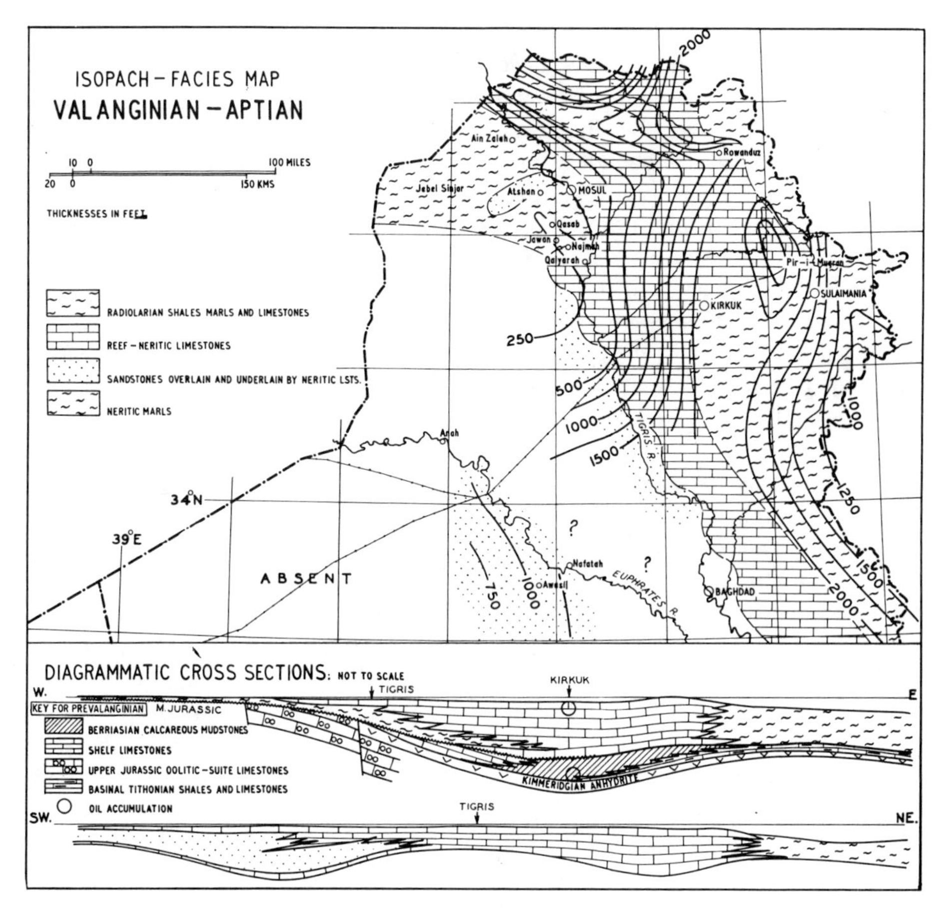

FIG. 5.—Isopach-facies map of Valanginian-Aptian rocks, schematic.

neritic limestone, thick Barremian-Hauterivian sandstones (correlative with the producing sands of the Basra fields), and a thin, dolomitized limestone of Aptian age. The southwestern limit, not shown on the map, must lie between Awasil and Muhaiwir, and is due to post-Aptian and pre-Cenomanian erosion. Rocks of Valanginian-Aptian age do not crop out in the southwestern part of the region, the pre-Albian outcrop being overlapped by transgressive sandstones of the Albian-Cenomanian interval.

Basinal marls, shales, and limestones, with abundant radiolaria, ammonites, and other planktonic fauna, occupy the entire eastern border of the region, and are also indicated in a narrow trough lying north of Mosul. The lower part of this sequence grades westward into neritic marls (and eventually into sands in the southwest). The upper part passes westward, rather abruptly, into neritic, reef-and-shoal limestones of great thickness.

The northwestern area represents an emergent uplift, later submerged, the southwestern sand-girded area indicates shelving approach to a continental mass, and the basin proper lay close to or beyond the northeastern borders of Iraq. The neritic limestones appear to have developed on an initially shallow, broad shelf, which subsided, in step with sedimentation, much more rapidly than did the basin area proper, or the high shelving areas in the west. The sharp, westward swing in the neritic-basinal facies boundary suggests control of facies by east-west faulting, and similar faulting is suggested by the approximately east-west trend of the basinal trough north of Mosul, and by the east-west course of isopachs west of Rowanduz.

The basinal sediments of the eastern parts of the region are notably bituminous. Depositional conditions were largely euxinic, and these rocks are considered to have been rich and effective source beds. The thick neritic limestones, with which these source beds are laterally and vertically juxtaposed, are locally of high porosity and permeability; they are classed as excellent potential reservoir and carrier formations.

Continuity of the neritic limestone zone southward from Kirkuk is inferred from reappearance of the same facies developments in Southwestern Persia (Kent, Slinger, and Thomas, 1951). But it may be that the sandstone units of the west pass eastward, gradationally and directly, into radiolarian sediments of a subordinate basin which may underlie the Mesopotamian plains.

Albian-Cenomanian.—Thickness and facies distribution for the Albian-Cenomanian units (Fig. 6) follow approximately the pattern which is found in the preceding Aptian stage. In the southwest, following an early uplift and peneplanation, an extensive marine sandstone blanket was deposited (correlative with the producing sands of the Burgan field). Basinal sedimentation continued in the east, and neritic limestones spread over the high northwestern area. In the center of the region, west of Kirkuk and south of Qalian, the Albian is represented by a thick sequence of marls, chemical limestones, and anhydrites.

In Cenomanian time the semi-lagoonal environment of the central area was modified, and tabular neritic limestones spread over the southwestern and western margins of the basin. Such limestones are thickly represented in the Awasil wells, and they occur as far to the southwest as the vicinity of Rutbah.

The isopachs represent actual thicknesses. After Cenomanian deposition was complete there were vertical adjustments of structural units of large dimensions, followed or accompanied by regression, and most of the region except the southeastern basinal area was subjected to erosion. Thinning of represented sediments on to the elongate northwest-southeast feature which runs through Mosul and Kirkuk is a result of erosional modification of an uplifted spur rather than of thin original deposition. The Makhul-Qalian area, in which lagoonal sedimentation occurred during Albian times, lost its veneer of Cenomanian neritic limestone during this erosion.

The broad pre-erosional uplift on northwest-southeast trend passing through Kirkuk merits comment, because all the "commercial" oil fields discovered in Northern Iraq lie on its flanks.

As in the Valanginian-Aptian, the basinal sediments are excellent candidate source rocks in some areas. They contain rich radiolarian and globigerinal faunas

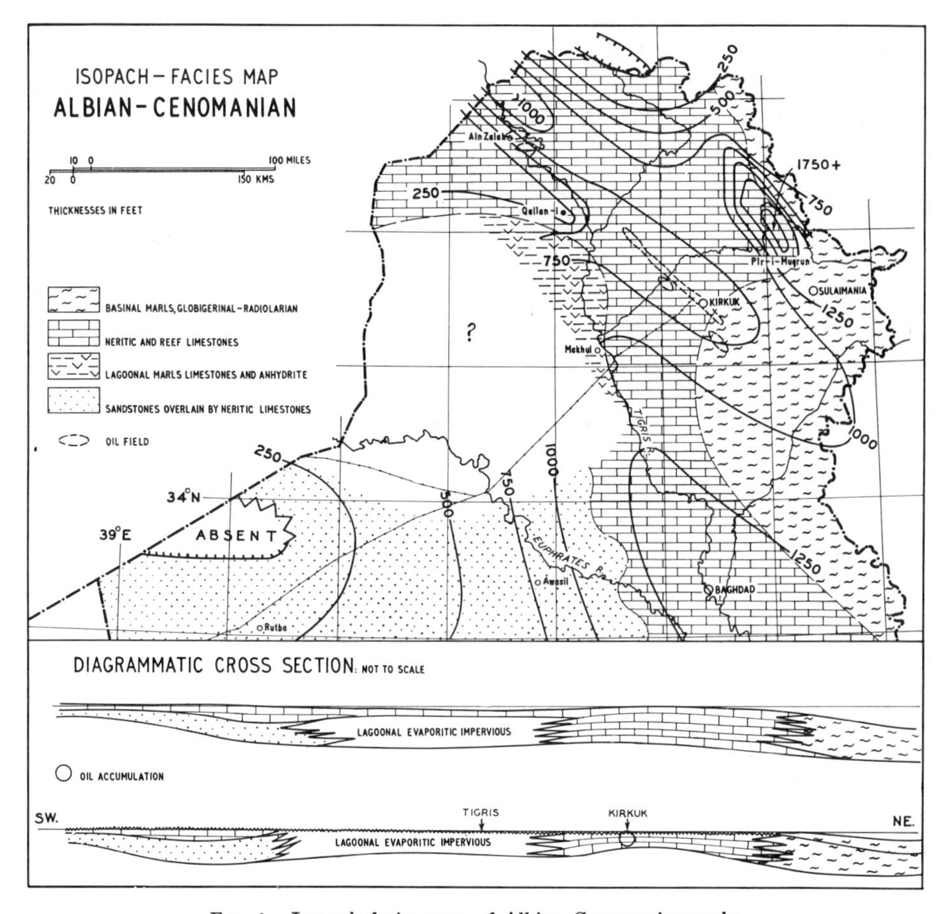

FIG. 6.—Isopach-facies map of Albian-Cenomanian rocks.

and are locally characterized by abundance of the problematical planktonic organism *Oligostegina* (especially near the boundary between the basinal and neritic facies). The neritic limestones provide excellent potential reservoir-carrier formations, especially where diagenetically modified during the terminating emergence. The impermeable Albian lagoonal-anhydritic sediments of the central area provided a limit to the possible migration of fluids toward the southwest within the neritic limestone zone.

Oil is found in porous, neritic, Albian limestone reservoirs in Kirkuk and Ain Zalah.

Turonian.—Turonian rocks are probably restricted to the southern and eastern parts of the region (Fig. 7). Sedimentation was continuous from Cenomanian to Senonian times in the southeast, where basinal globigerinal limestones and marls prevail. Neritic limestones are found in wells in the southwest, thinning westward; these units are not seen at outcrop. In Kurdistan, mid-Turonian *Oligostegina* limestones lie on eroded Albian neritic limestones and dolomites. Westward from Sulaimania through Kirkuk to Qaiyarah and Makhul similar *Oligostegina* sedi-

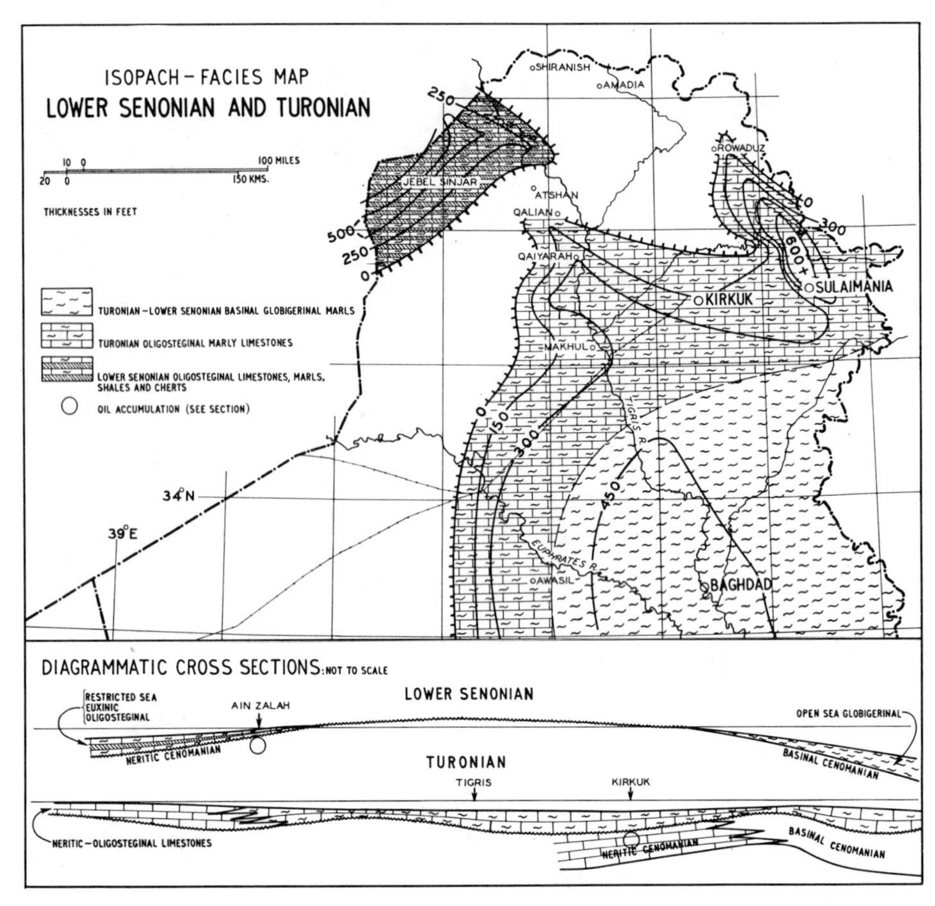

FIG. 7.—Isopach-facies map of lower Senonian and Turonian rocks.

ments, becoming progressively more marly and plastic, transgress over thin eroded Cenomanian on to eroded Albian lagoonal-anhydritic deposits of the central Tigris area. The Turonian is erosionally terminated, except in the basinal province, so that the original area of deposition was certainly larger than shown on the map. The absence of neritic rudist-bearing Turonian limestones is of interest, as such limestones are recorded from southwestern Persia (Kent, *et al.*, 1951), where their presence is paleontologically established, and from southeastern Turkey (Tasman, 1949), where published evidence for rocks of this age in reef facies is dubious.

The Oligosteginal rocks are locally bituminous and were deposited under locally euxinic conditions. They could be invoked as potential source beds. In the west they are sufficiently marly and plastic to have functioned as cap-rock seals to the underlying porous limestones.

Lower Senonian.—Deposits attributed to the lower Senonian interval are found in two widely separated areas (Fig. 7). Basinal globigerinal marls occur, in conformity with similar Turonian sediments, in the southeastern parts of the region, but these are cut out in the pre-upper Senonian unconformity over most of Kurdistan and the central and western areas. In the northwest, *Oligostegina* limestone with cherts and shales, carrying a restricted fauna of post-Turonian and pre-upper Campanian age, overlie eroded, Albian, neritic limestones. These sediments are restricted to a very small area, which probably represents a small graben-type basin, let down below sea level during the general emergence to which most of the region was subjected from late Turonian to the beginning of upper Campanian time. Precise age of these deposits is unknown. They were laid down under partially anaerobic conditions, and could be regarded as potential source rocks.

The top of the lower Senonian in the northwestern "basin" is taken at a minor break which marks the onset of upper Senonian transgression, and which introduces normal marine globigerinal and benthonic faunas. The limits of the area of existing sediments shown on the map probably correspond closely with the area of original deposition, as erosion at the mid-Campanian break was small.

Oil is produced from fractured cherts and *Oligostegina* limestones of this age in the Ain Zalah field, but the main producing reservoir lies in the Middle Cretaceous neritic limestones below.

Upper Cretaceous—upper Campanian-Maestrichtian.—The onset of full-scale Upper Cretaceous transgression, at some time during the upper Campanian, inaugurated an episode of thick and varied sedimentation (Fig. 8). Simultaneously with the commencement of transgression, the western part of the region, was segmented into three east-west aligned troughs, which presumably originated by faulting at depth. Globigerinal "basinal" sediments accumulated thickly in these steadily subsiding troughs, while neritic limestones were developed around their margins, and over the sinking horst-like residual shallows between them. A high feature of some nature was raised, to stand above the early strand line, in a zone coursing west-northwest/east-southeast through Shiranish, Aqra, and Rania. Meanwhile, globigerinal, "basinal" sediments transgressed from the west over the large area lying south of this high.

Shortly after commencement of transgression, uplifts of great magnitude occurred in the region to the east of the frontier, and a great bulk of Flysch-type clastic detritus, derived from this uplifted mass, was poured into a more or less linear, northwest-southeast-directed trough or "fore-deep." As transgression proceeded, the high ridge feature running through Aqra was overwhelmed, and very thick neritic limestones, including reef components, were deposited over it, and around its margins. The Flysch clastics advanced westward over globigerinal limestones and marls to the vicinity of Kirkuk. Lenticular reef-like masses of organic, detrital limestone developed locally within the clastic trough, and tongued eastward into the Flysch from the massive limestone barriers in the Aqra-Rowanduz area. Indigenous oil is recognized within the Flysch sediments, in several areas in Kurdistan.

The Flysch-type clastics include detritus of green rocks, and radiolarian cherts and limestones of Middle Cretaceous-Jurassic age. At the close of the transgression large sheets of radiolarites slid or were thrust into the northeastern borderlands of Iraq.

The normal development toward compression orogeny in the northeast was

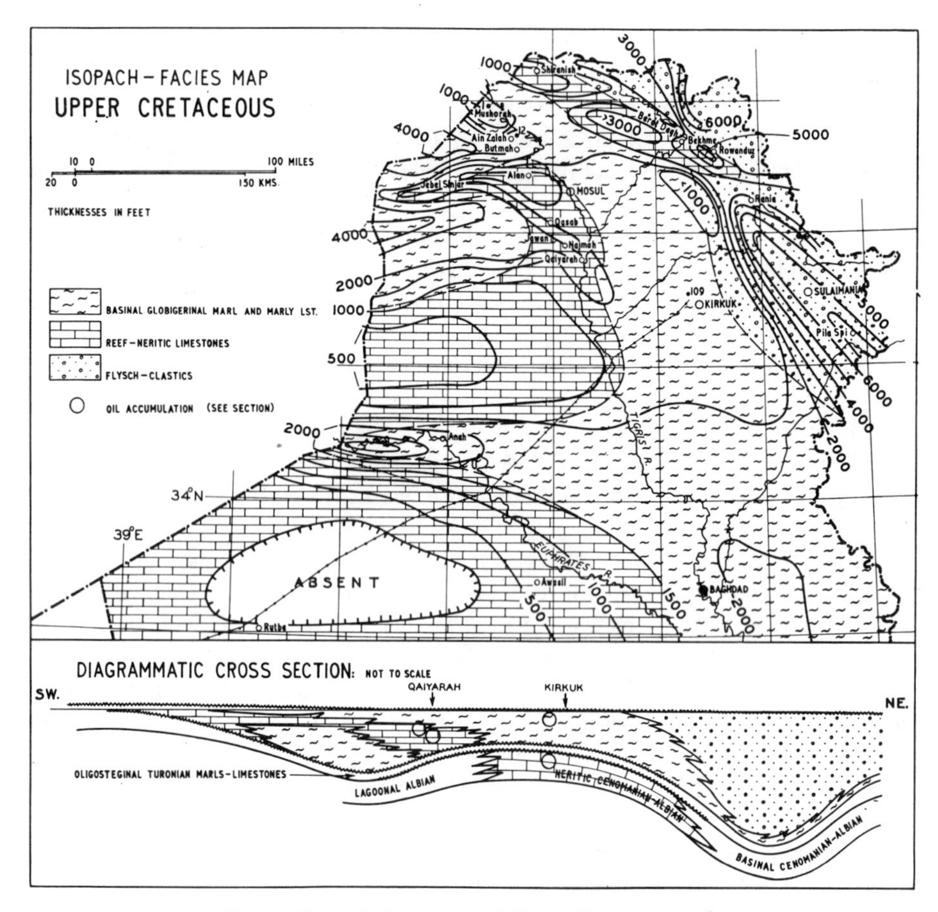

FIG. 8.—Isopach-facies map of Upper Cretaceous rocks.

cut short at the end of Upper Cretaceous times. Though slight folding adjustments of this date have been detected, and other evidence of compression is available, there is no well-defined system of frontal folds, ahead of the advancing orogen.

The portrayal in Figure 8 of the distributions of the various facies of this complicated interval is much schematized. The thicknesses shown are residual after an important episode of emergence and erosion at the end of Cretaceous time, which was preceded or accompanied by structural adjustments. The isopachs are not a true guide to original depositional thicknesses. The whole region was probably submerged by the end of Maestrichtian time. Absence of sediments of Upper Cretaceous age from the Ga'ara area in the southwest is probably due to early Tertiary erosion.

The globigerinal basinal sediments of the interval have been regarded as potential source rocks; they are richly fossiliferous, and locally bituminous. The neritic detrital limestones of the western margins, and of the Shiranish-Aqra-Rania high, and the Flysch trough, present excellent reservoir potentialities in many areas. Henson (1950a) has described some of the reef developments of Upper Cretaceous age in Northern Iraq, discussing their relationships to oil generation, migration, and accumulation. The neritic porous, Campanian-Maestrichtian limestones of the Qaiyarah area fields contain a large volume of heavy sulfurous oil. Light oil is produced from fractures in indurated, Maestrichtian, marly, globigerinal limestones in the Ain Zalah and Butmah fields, and small volumes of similar oil in similar limestones have been proved beneath the Tertiary producing reservoir at Kirkuk.

Bitumen pebbles in Maestrichtian Flysch conglomerates and reef-type limestones suggest escape to surface of large volumes of oil during the Upper Cretaceous. Although one source for this oil may be seen in the uplifted mass to the east, a nearer and more probable origin is at hand in the high Shiranish-Aqra-Rania ridge feature, which must have attracted migrating oil at this time, and which lacked any adequate seal until late in the Tertiary.

TERTIARY

Paleocene-Lower Eocene.—The Cretaceous-Paleocene transition period was one of widespread regression, which bared most if not all of the region, permitting very uneven erosion of different areas, but producing in most localities quite marked evidence of discontinuity in sedimentation. The northeastern area and its hinterland were elevated, and deposition was re-introduced by the downwarping of a broad linear northwest-southeast trough, crossing the region, to the southwest of the position of the Flysch trough of Upper Cretaceous time, and of the reef-crowned Shiranish-Aqra-Rania high of the Maestrichtian. Flysch-type clastics accumulated in great thickness in this subsiding trough, and Paleocene bounding reefs developed locally along its southwestern margins in some areas (Fig. 9).

In the northwest and central areas, emergent features inherited from the terminating uplifts of the Cretaceous Period, and probably further differentiated by faulting, provided a partial barrier, running from the south of Jebel Sinjar to the vicinity of Bai Hassan. Lagoonal limestone sediments, lenticular reef-type limestones and fore-reef shoals, and globigerinal marls, are intercalated with Flysch-type clastics and shales to the northeast of this barrier. Reef limestones are well developed over the Jebel Sinjar, and in the foothills running northwest-southeast between Koi Sanjak and the Persian frontier south of Halabja. Inter-relationships of different facies are intricate in the Flysch belt, and cannot be satisfactorily portrayed on a single map.

Southwest of the emergent barrier, and of the Koi Sanjak-Halabja reef belt, Paleocene and lower Eocene deposits are mostly in "basinal" globigerinal facies, and generally thin, with internal breaks which may reflect emergence of flat-lying island highs (or merely non-depositional submerged environments). The limits of the Makhul "island" are unknown on the west; it could extend over most of the "blind" area up to the Syrian frontier, where basinal sediments are shown on the map, or it could be much smaller in area than is indicated. Cherty, phosphatic marls with neritic-littoral limestone developments are found around the western margins of the Ga'ara uplift, in the southwest of the region.

In the Flysch zone the lower Eocene is represented by red beds, which overlap the underlying marine Paleocene and wedge out on to the slope of the northeastern land mass.

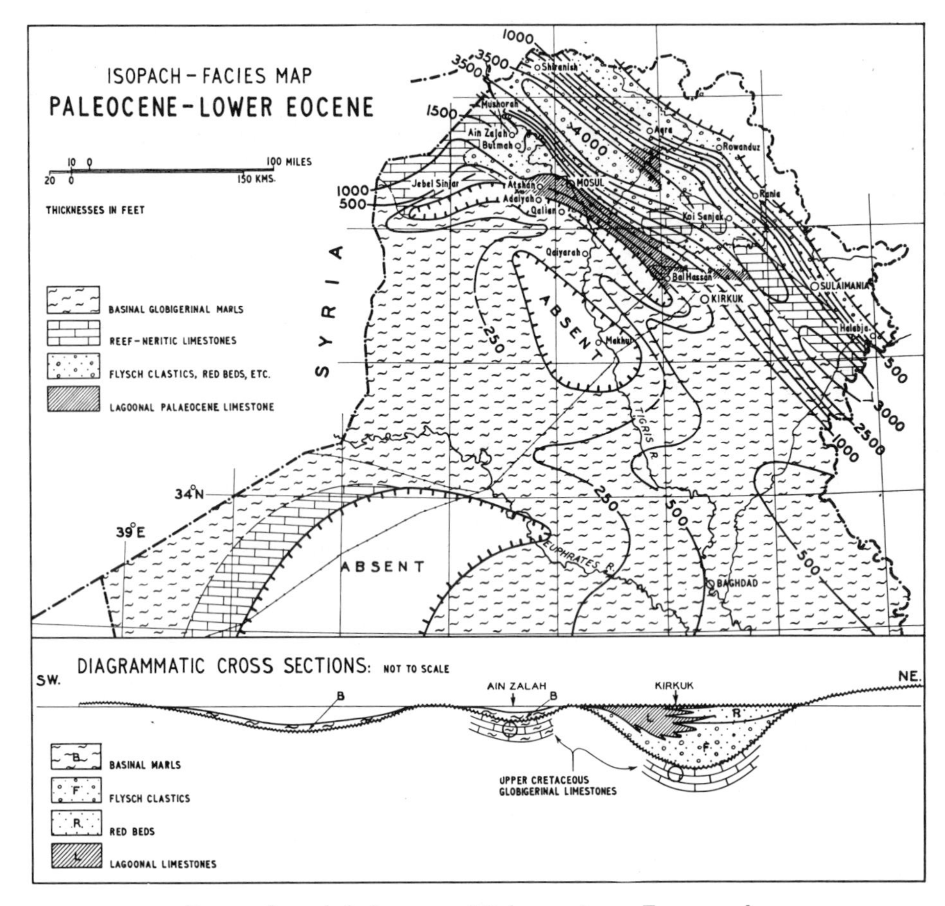

Fig. 9.—Isopach-facies map of Paleocene-lower Eocene rocks.

Indigenous oil indications are known in the Flysch zone, and the "basinal" globigerinal sediments can be favorably viewed as potential source rocks. But the principal function of rocks of this age interval in known fields is that of cap-rock marl seal to accumulations in upper Campanian-Maestrichtian fractured globigerinal limestones (Ain Zalah, Butmah, Kirkuk) and neritic limestones (Qaiyarah, Najmah, Jawan, Qasab).

Middle-upper Eocene.—The complications of the Paleocene-lower Eocene are lacking from the paleogeography of the middle and upper Eocene (Fig. 10). Introduction of Flysch clastics abated, after deposition of the lower Eocene red beds in the Kurdistan area. A broad basin, receiving globigerinal sediments and trending approximately northwest-southeast, occupied the central parts of the region. One margin of this basin is to be placed around the broad Ga'ara uplift in the southwest. Patchy neritic limestones, with *Nummulite* faunas, occur along this gently shelving coast, where thicknesses deposited were small. In the northeast the basin was limited against a steeper marginal slope. Nummulitic shoal limestone of considerable thickness developed outward from this slope, and protected a widespread

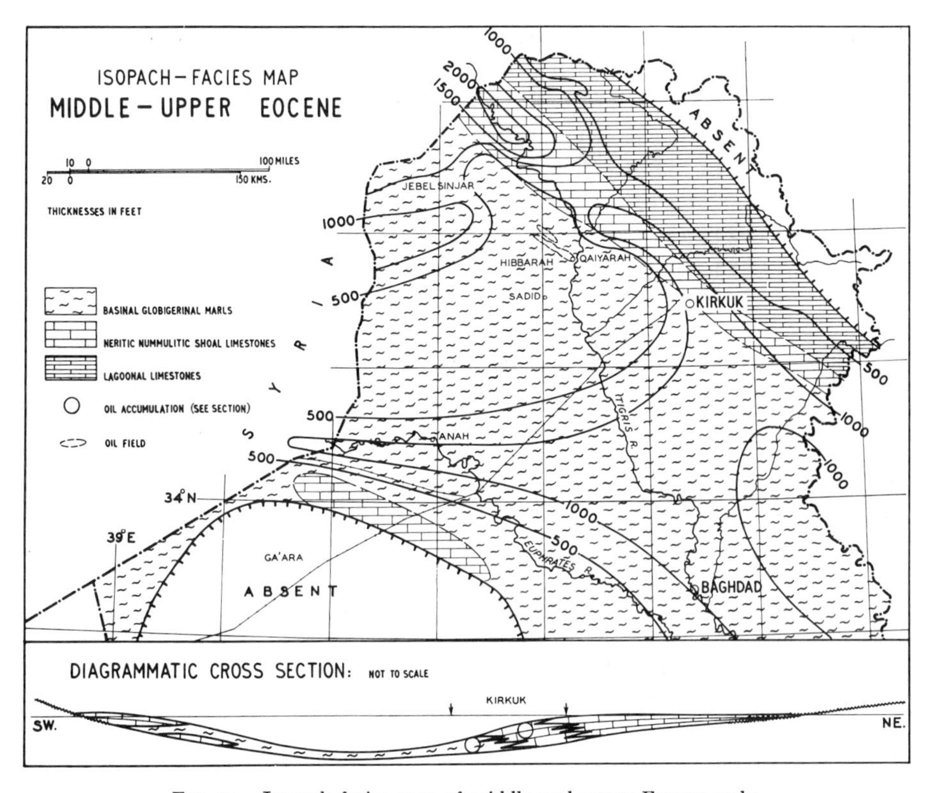

FIG. 10.—Isopach-facies map of middle and upper Eocene rocks.

shallow lagoonal area, in which chemically deposited limestones and dolomites were laid down. The lagoonal sediments extend northeastward into strand-line deposits, and interdigitate locally with red sands and silts close to the shore line.

Within the area of globigerinal marl deposition, internal troughs are inferred south of the Jebel Sinjar and in the Euphrates Valley through Anah. These prominent east-west features are regarded as persistent, in some form or other, from Upper Cretaceous time onward. A third east-west trough of thick basinal sedimentation, north of Sinjar, is complicated by the appearance of thick, shoal limestones at its eastern end. The area of thin sedimentation in the central area, west of Kirkuk, may be smaller than is shown, no control section being available between Sadid-Hibbarah and the Syrian frontier. Similarly the northeast-southwest trough connection lying south of Kirkuk and north of Baghdad is conjectural and may not exist as shown.

The lagoonal limestones of the northeastern zone are tight impermeable limestones, without great value as potential reservoir rocks, except perhaps where strongly fractured. The basinal globigerinal milestones are locally excellent source-bed candidates, as they are richly organic, with planktonic foraminiferal faunas throughout, and as they were deposited under anaerobic conditions in some areas. They are locally bituminous at outcrop in Persia (Lees, 1934), though not commonly conspicuously so where seen at surface in Iraq. The thick nummulitic shoal limestones of the linear zone which trends northwest-southeast through Kirkuk are exceptionally porous and permeable in parts. They are excellent potential reservoir rocks, and their potentialities are realized in the Kirkuk field, where they provide part of the accommodation for the 2,000 feet of oil column of this large accumulation. The underlying globigerinal limestones are also productive at Kirkuk.

In the fields of the Qaiyarah area, the thin middle-upper Eocene globigerinal marls provide part of the seal which separates the accumulations in the Upper Cretaceous and Miocene reservoirs.

Oligocene.—Only a very schematic and formalized account of the sediments of this time interval is attempted in Figure 11. The depositional history of these rocks has been studied in detail by R. C. van Bellen (1956), whose observations are drawn upon.

In general, the paleogeography and depositional history of the Oligocene follows quite closely the pattern set in middle-upper Eocene time. The northeastern margin of the depositional area lay much to the southwest of the position of the correspond-

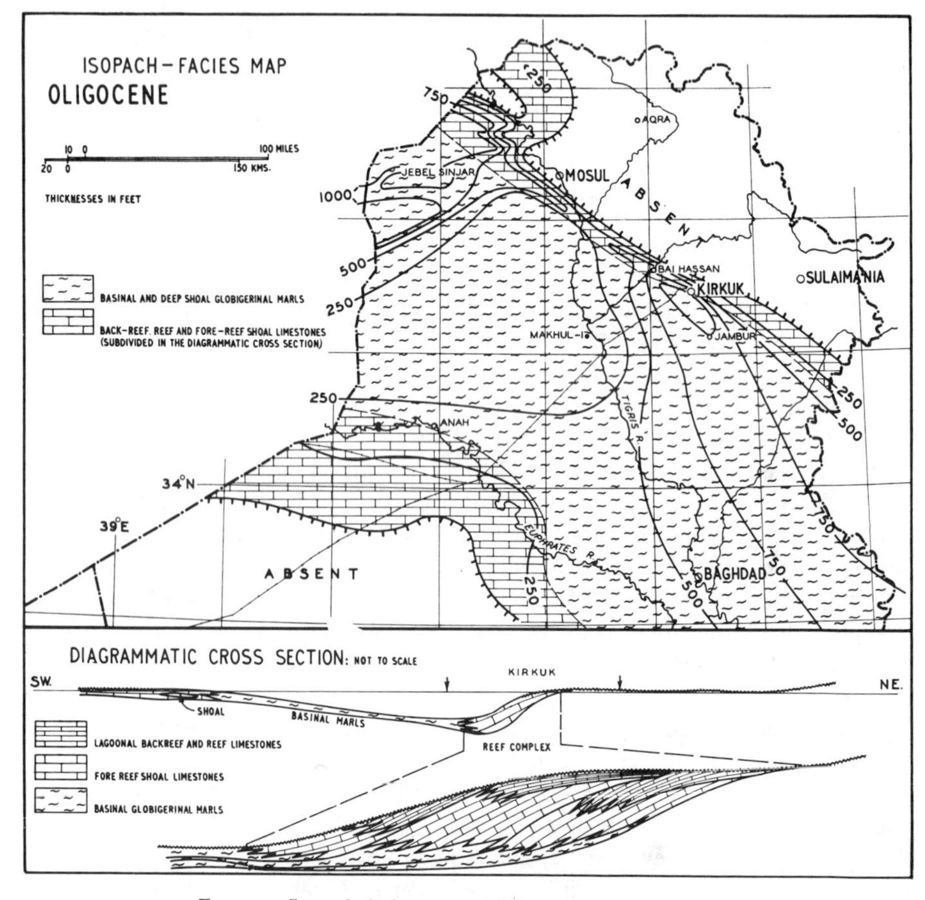

Fig. 11.—Isopach-facies map of Oligocene rocks, schematic.

ing boundary of the middle-upper Eocene stages. Reef limestones of Oligocene age, with back-reef limestones to the northeast, were deposited along a linear trend running across the entire region, approximately from northwest to southeast. The reef zone corresponds approximately with the basinal side of the belt of shoal limestones which were deposited in the middle-upper Eocene. The reef and back-reef zones of the Oligocene time are much narrower than the shoal limestone and lagoonal zones of the middle-upper Eocene. Upper Oligocene sediments were deposited after regression, and have much smaller areal distribution than those of the lower and middle Oligocene.

The reef-controlled sediments of the northeastern margin of the basin are repeated on the southwestern margins, where gentler shelving shores result in a much wider areal distribution of the facies belts than is found in the Kirkuk region. The broad, rather featureless basinal area between the reef-girded margins is occupied by globigerinal sediments. Anhydrites also occur in the central parts of the basin in the younger sediments of the interval.

The troughs south and north of Jebel Sinjar, noted from Upper Cretaceous through Eocene stages, are still evidenced in Oligocene to Miocene time by large thickness of basinal sediments. The persistence of the Anah trough, suggested by the 250-foot contour in Figure 11, may be illusory, as the position of this contour in the area north of Anah and west of Makhul is uncontrolled.

Oligocene sediments were subjected to erosion in some areas during the Aquitanian cycle, and later to extensive exposure following localized structural differentiation, during the time of deposition of the Euphrates limestone and basal Lower Fars sediments. The isopachs indicate found thicknesses, and offer an imperfect picture of depositional conditions. The disparities are greatest around the northeastern margins of the basin, where original deposits of back-reef Oligocene rocks were stripped off, before encroachment of the Lower Fars sea brought erosion to a close.

Most of the porous limestones housing the Kirkuk (Baba dome) and Bai Hassan fields are Oligocene reef or fore-reef limestones. The globigerinal basinal sediments have been considered the likely source for the Kirkuk oil, and they must certainly be included among the possible source beds of the region. In spite of low permeabilities, those parts of such rocks which lie above oil/water level in the Kirkuk field contribute greatly to the total oil-filled pore space in this reservoir. Drainage of oil from these tight sediments, within the present reservoir, is aided by the existence of an extensive fracture system (Daniel, 1954).

Euphrates limestone.—After the deposition of upper Oligocene rocks in the lower lying parts of the Oligocene basin, a further regression, dated approximately at the Oligocene-Aquitanian transition, preceded the commencement of lower Miocene sedimentation. The rocks of the lowest Miocene transgression, which are for the most part lagoonal limestones, comprise the Euphrates limestone formation.[5] This formation is thickly developed over most of the area in which globigerinal Oligocene sediments were laid down (Fig. 12). The Euphrates limestone overlaps the limits of Oligocene transgression in some areas, notably around the southwestern

[5] Formal definition published by R. C. van Bellen (1956).

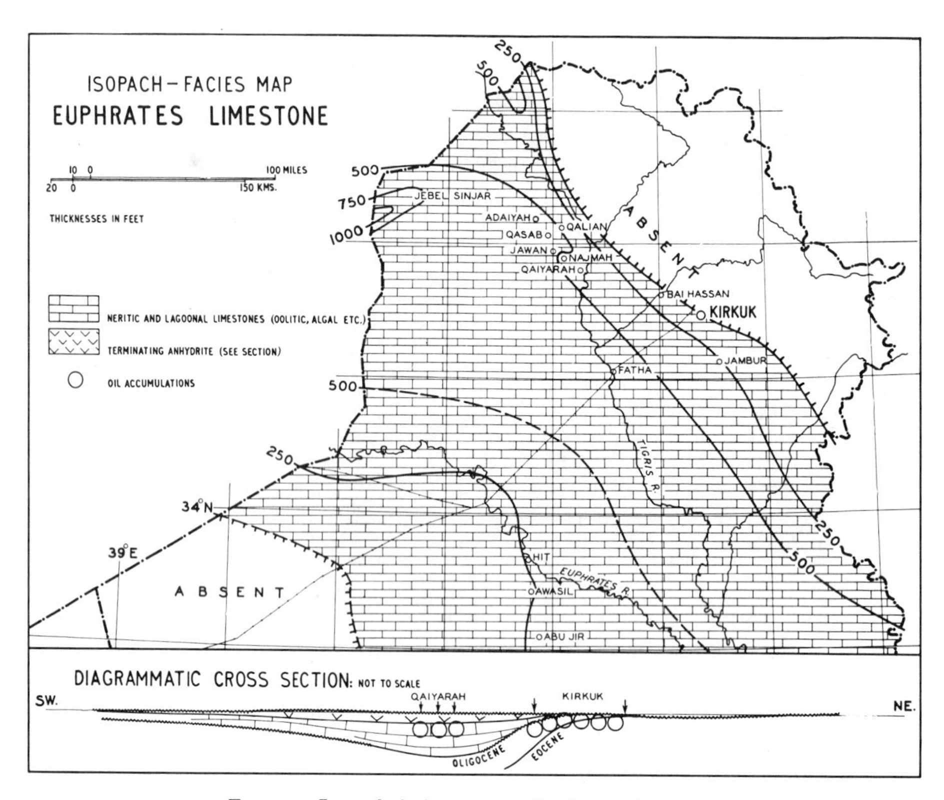

FIG. 12.—Isopach-facies map of Euphrates limestone.

margins of the basin. The main trough of sedimentation continues into northeastern Syria, where thick-bedded anhydrite and salt have been found in some wells. Anhydrites occur in Iraq also, thickening and proliferating towards the deeper parts of the basin.

Deposition was terminated by widespread withdrawal toward or at the end of the Aquitanian stage. In some localities the basal Lower Fars is transgressive over eroded Euphrates limestone.

The upper accumulations of the oil fields of the Qaiyarah area are contained in porous Euphrates limestone. The formation is very fossiliferous locally, corallinacid algae being especially abundant. But the characters of the sediments indicate deposition under aerated conditions, so that the unit is not considered to have contained important source rocks.

Lower Fars.—The precise age of the regression which followed the deposition of the Euphrates limestone remains moot. It may be dated within or at the end of the Aquitanian. During this regression much of Northern Iraq was exposed, and substantial erosion of earlier sediments preceded and accompanied the early stages of the Lower Fars transgression.

The Lower Fars sediments were deposited in an intermittently barred, basinal-lagoonal environment. They comprise rhythmic alternations of limestones, anhydrites, and silty marls and shales, with bedded rock salt appearing in the middle part of the series in the central part of the basin. The Lower Fars transgression extended much further to the northeast than did that of Euphrates limestone time, but the reverse situation arose in the southwest, where the Lower Fars converges to disappearance far to the northeast of the present limits of the Euphrates limestone.

Passing northeastward from the center of the Fars basin, the general characters of the series are modified progressively, first salt, then the bedded anhydrites, and eventually the limestones, falling out of the sequence. The entire series thins in the same direction, and also shows onlap convergence onto an eroded surface which is cut in Oligocene rocks in the southwest (as at Kirkuk) passing into Maestrichtian clastics in the northeast. The sediments of the northeastern zone, up to the "nappe front," are reddish marls, silts, and sands, with only rare subordinate limestones. But in the areas east, north, and northwest of Rowanduz, a prominent limestone unit, commonly conglomeratic at the base, underlies or lies within these clastics. This limestone, which thickens rapidly northeastward into a massive feature-forming unit along the frontier with Turkey, contains the same fauna, and is of the same age as part of the Lower Fars of the main basin. Thickness relations suggest that it marks the margin of a separate basin of "Lower Fars age" which may extend far into Turkey and Persia. The "nappe front" of the thrust sheets of the late-Tertiary orogeny over-rides the anomalous "undifferentiated Fars" rocks of this northeastern basin, preventing full inquiry into their distribution in Northern Iraq.

In the basinal area, the top of the Lower Fars is taken at the top of the highest bedded anhydrite, the overlying succession of shales, silts, and limestones being recognized as the "Middle Fars." The Middle Fars rocks, generally thin in Northern Iraq, are transitional between the dominantly chemical-lagoonal sediments of the Lower Fars and the entirely clastic Upper Fars. They are included with the Upper Fars and Bakhtiari rocks for the purposes of isopach portrayal in Figure 14. The entire Lower Fars (and Middle Fars) sequence becomes less clastic and more frankly lagoonal and marine from northeast to southwest.

The isopachs of Figure 13 relate only to the Lower Fars sediments of the basinal area, and to those parts of the "undifferentiated Fars" of the northeastern zone which are considered to be equivalent in age to the basinal Lower Fars. The thickness pattern is much simplified, and the actual basinal configuration is certainly far more complicated than is shown.

The flowage of salt under tectonic stresses in the central parts of the basin has produced marked disharmonies between the generally simple, anticlinal structures which have developed in the pre-Fars rocks, and the commonly complex, surficial structures produced in the Fars and overlying sediments. The nature and origin of such salt-facilitated disharmonies have been discussed lately by O'Brien (1950) and Lees (1952), and earlier by many other writers. Salt flowage and imbrication within the Lower Fars render difficult any evaluation of pre-flow and pre-imbrication thicknesses, especially where the only evidences available are those drawn from scattered wells drilled in the crestal areas of anticlines. Some of the thicknesses used

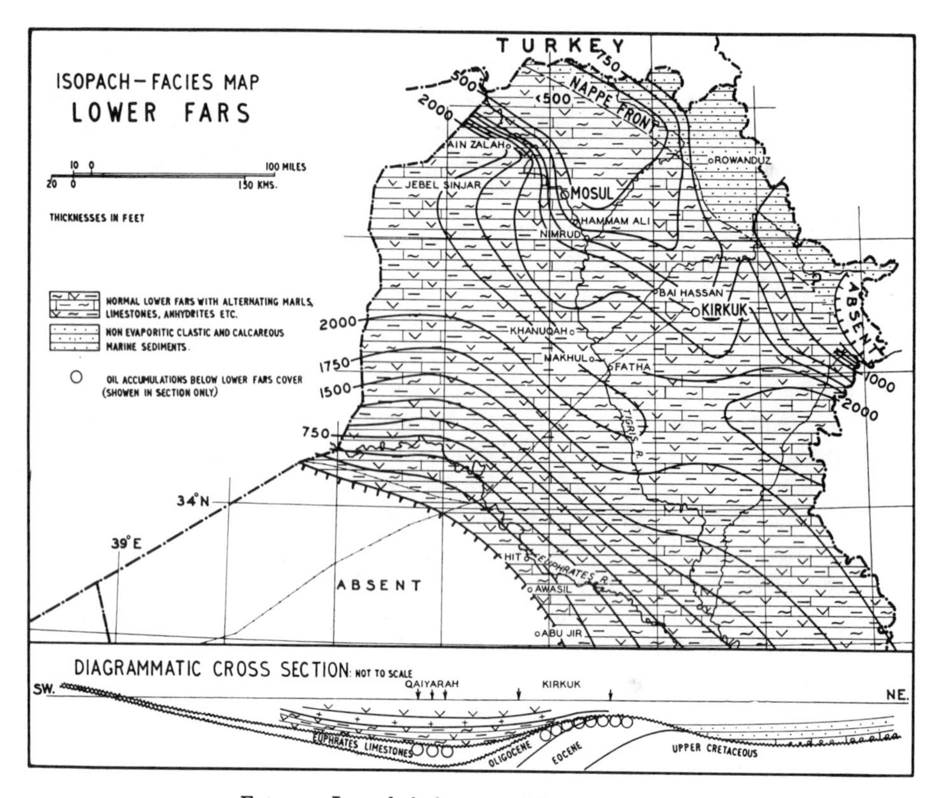

FIG. 13.—Isopach-facies map of Lower Fars rocks.

in drawing Figure 13 are estimated pre-folding thicknesses, compensated for presumable thinning or thickening in the salt section during folding, according to the positions of control wells on the anticlines. Individual thicknesses may be incorrect by several hundred feet in the central parts of the basin, but the general configuration of contours must be approximately as shown. The map may be in error in the area immediately to the east and southeast of Jebel Sinjar, where available measured thicknesses are inconsistent.

Some of the thin limestones and shales of the Lower Fars of the central parts of the basin show evidence of deposition under anaerobic conditions; they could be included among the potential source beds of the region. Some of the limestones are appreciably porous and permeable, and some contain oil where they overlie deeper accumulations, as at Kirkuk. A few contain small quantities of oils unlike those encountered in nearby reservoirs; such oils may be indigenous to the Lower Fars. In general the source capabilities and reservoir potentialities are so small as to be negligible. The principal role of the Lower Fars in thc oil accumulation process has lain in its ability to provide a plastic cap-rock seal of salt and anhydrite to retain oil in underlying reservoir formations. In this role it is of high significance. The Kirkuk, Bai Hassan, Jambur, and Qaiyarah area fields in Northern Iraq, and all the developed fields of Persia owe the preservation of their oil to Lower Fars cover.

It is the seal for about 15 per cent of the world's proven unproduced reserves.

Upper Fars and Bakhtiari.—After the brief episode of marine sedimentation which followed the deposition of the evaporitic Lower Fars, clastics entered the basin in large volumes from the rising orogen in the northeast. These deltaic-pedimentary clastics, which commence with red silts and marls (Upper Fars facies), pass upward gradationally into conglomerates and coarse sandstones (Bakhtiari facies). The major lithological divisions are not differentiable in terms of age. Subdivisions hinge on grade size of the clastics, and there is general increase in grade from southeast to northwest. The rare marine components of the central part of the basin dwindle and vanish in the same direction.

Isopachs of Figure 14 illustrate combined thicknesses of Middle and Upper Fars and Bakhtiari sediments, as measured into the principal synclines. But this is far from being a straight forward pre-folding thickness portrayal. It is known that the late-Tertiary folding, which reached its climax in late Pliocene time, was commencing already in the lower Pliocene, and that it developed intermittently throughout the duration of deposition of the Bakhtiari rocks. In the later stages of Bakhtiari deposition, the crests of the large mountain folds close to the advancing "nappe front" were undergoing erosion, while coarse sediments were being laid down thinly over the rising crests of the southwestern folds, and more thickly in the deepening synclines. A true isopach map should reveal closures of attenuation over every rising structure. Unfortunately, the information which would be required for such a map is not accessible. Figure 14 shows the areal thicknesses, after arbitrary elimination of crestal attenuation. As the entire Miocene-Pliocene sequence was prob-

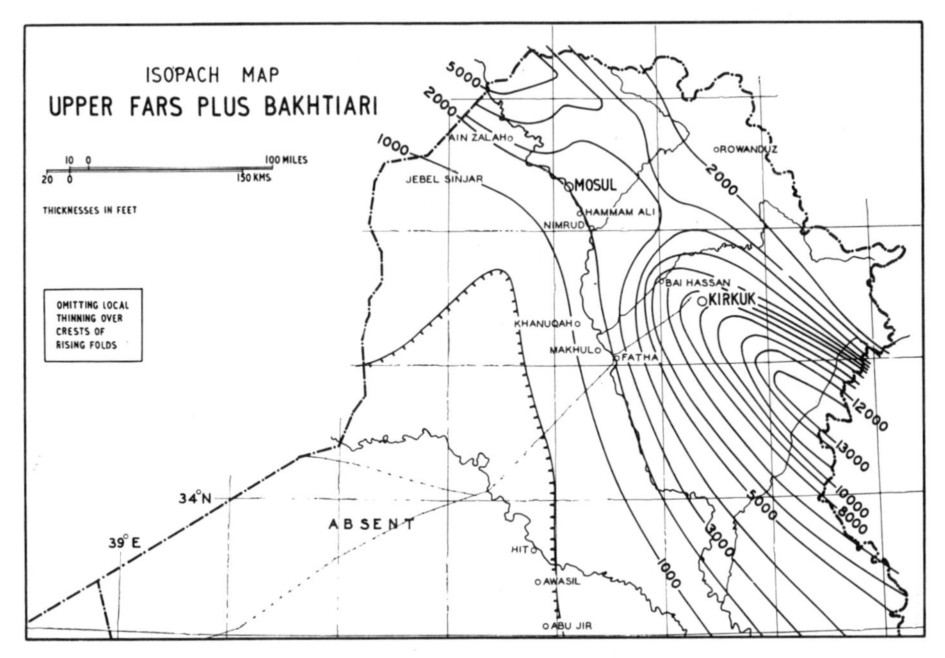

FIG. 14.—Isopach map of Middle and Upper Fars and Bakhtiari sediments.

ably limited by near-planar base levels of deposition, the contours also give rough measure of the subsidence of the folded zone which accompanied the late-Tertiary folding. The bold contour spacing and the smoothness of the areal gradients are false. Local gradients, resulting from contemporaneous uprising of the folds, cannot be assessed. They probably sufficed to control the oil-migration regime and to prevent long-distance regional movements, which the regional gradients would have encouraged but for the presence of growing structural traps.

THE OIL FIELDS

Lacking the results of a systematic chemical evaluation of possible source sediments, it has been accepted that any rock unit of appreciable thickness which was deposited in a more or less euxinic environment may be suspected of having generated and liberated oil. These potential source beds range from stinking, black, bitumen-saturated shales and limestones to rocks which are now finely recrystallized, off-white, lithographic limestones. Potential source beds are widespread and occur at many horizons in the sedimentary sequence of Northern Iraq.

The obvious candidates for the roles of primary reservoir formations are the thick coarse sandstones of Cretaceous age, which appear in the southwestern part of the region (Figs. 5 and 6), and the reef limestones, oölitic limestones, and neritic limestones of the shallow-water shelves, which occur intermittently through the dominantly calcareous sequence. The calcareous potential reservoir rocks are much more widely spread than are the sandstones, and they occur in the sediments of most ages considered, up to the middle Miocene Lower Fars.

Within the folded zone, at least, the intensity of fracturing encountered in some formations renders them capable of housing and yielding oil in large quantities, even though the rocks may be intrinsically impermeable (Daniel, 1954). Such fracturing must be considered, generally, to be of late-Tertiary origin, and hence not available to house primary oil accumulations of early date.

The cap rock requirements of impermeability, coupled with some degree of plasticity (a high degree of plasticity in the strongly folded zone), appear to be met in several formations, scattered through the stratigraphic sequence, which have wide areal distribution.

Amid the profusion of possibilities for accumulation which are perceptible in the known associations of presumable source, capable reservoir, and competent cap rocks, there are a few combinations that have resulted in oil accumulation, and many which have been barren. The profusion of source and reservoir possibilities has naturally tempted a multiplicity of speculations as to the source and place of origin of the known oils, the modes of their accumulation and the factors controlling their distribution. These problems require consideration now against the background of the developmental history of the region which has been summarized and simplified in the foregoing isopach and facies maps.

AIN ZALAH AREA

Ain Zalah field.—The Ain Zalah field, situated northwest of Mosul, is in a simple anticline, 12 miles long and 3 miles wide, which exposes Lower Fars limestones and anhydrites. Production is drawn from two "pays." The upper or "First

Pay" yields oil from fractures in unpermeated globigerinal limestones of Upper Cretaceous age. The "Second Pay" reservoir is of porous and fractured Middle Cretaceous limestones and dolomites, abetted by fractures in overlying lower Senonian-lower Campanian cherts, shales, and oligosteginal limestones. The two "pays" are separated by about 2,000 feet of barren, marly, globigerinal limestones similar to those in which the "First Pay" is developed. The geology of this field has been discussed by Daniel (1954).

The generalized stratigraphy is indicated on the cross section (Fig. 15). The seal retaining the "First Pay" oil is provided by basinal marls of Paleocene-lower Eocene age. This seal has not been entirely competent everywhere, for oil of type similar to that in the underlying fractured reservoir occurs sporadically in thin, lenticular, porous limestones, in the lower part of the Paleocene marl unit, in restricted areas over the crest of the structure.

The erosional unconformity which separates the Tertiary seal from the Upper Cretaceous reservoir has had no apparent function in aiding oil accumulation. No significant secondary porosity was developed in the Cretaceous limestones during exposure, and the fracture system, in which the oil is now found, was not developed until after the marly cover had been deposited.

Before the discovery of the "Second Pay" it was considered possible that the oil of the "First Pay" originated in the overlying basinal Paleocene-Eocene marls,

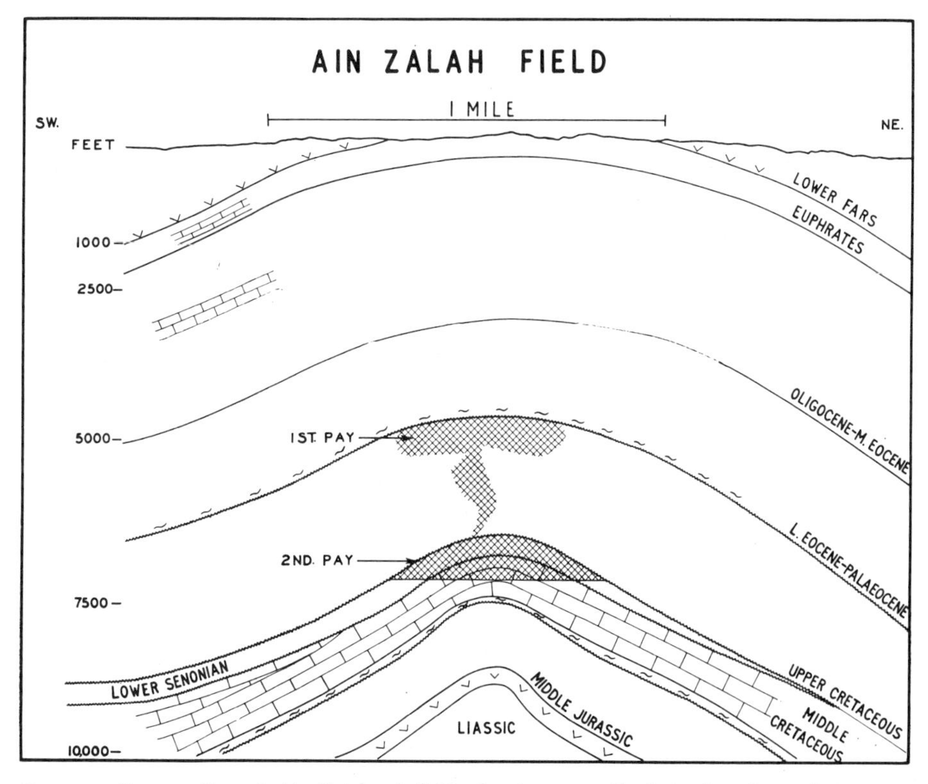

FIG. 15.—Cross section of Ain Zalah oil field, showing generalized stratigraphy and structure.

and that it had been expressed downward into available fractures, across the Paleocene/Cretaceous unconformity, during compaction of these Tertiary marls. This account was not very satisfying, as most of the open fractures which now house the oil were probably not formed until Pliocene folding commenced, by which time compaction rates must have been very small. The Paleocene-lower Eocene marls contain widespread porous limestone beds and thin silty beds which would have been more effective vehicles for reception of expressed oil than were the tight fractures of the Cretaceous limestone; whereas some of these contain oil, most do not, and the rare sporadic saturations can be accounted for by upward leakage from the "First Pay."

It has also been argued that the Upper Cretaceous limestones were the source for their own oil, which was retained within the rock mass until fractures opened to allow the present segregation. This possibility seems to be denied by the absence of any significant amount of residual oil in the pore spaces of the globigerinal limestones. No mechanism is apparent by which undersaturated oil, housed in very fine pores in an almost impermeable limestone, could be expelled into fractures developing late in the induration history of the rock.

Yet other origins have been suggested, involving the presence of primary accumulations of oil in Upper Cretaceous reef limestones down flank from the present field, or of similar reef accumulations in Paleocene sediments outside the explored limits of the structure. From the time of discovery most geologists who had experience of the field leaned to the view that the "First Pay" oil entered its present reservoir by migration from below.

With the proving of the "Second Pay" oil, the problem of origin of the "First Pay" accumulation was solved. The two oils are of the same gravity, and are chemically similar, such differences as exist being small and explicable. Moreover, the two "pays" are intimately connected, presumably by a fracture network, through the intervening 2,000 feet of seemingly barren marls and limestones. Production of either "pay" affects the other directly. It is now regarded as certain that the "First Pay" accumulation originated by upward migration of oil which escaped from the "Second Pay." Such migration has occurred extensively in the past, and migration of fluids is occurring rapidly at present from the "Second Pay" to the "First" as oil is produced from the latter. There is no likelihood that Paleocene-lower Eocene marls, or the Upper Cretaceous limestones themselves, have contributed importantly and directly to the "First Pay" accumulation, though either may have subscribed a very minor proportion of the oil there found (Daniel, 1954).

The origin of the accumulation in the "Second Pay" is more controversial. The principal reservoir formation is provided by the Middle Cretaceous limestones and dolomites. These are terminated by an erosional unconformity representing Turonian and early Senonian emergence. Though the early Ain Zalah wells did not reveal secondarily developed porosity below this break (Daniel, 1954, p. 785), later wells on this and other structures have shown marked dolomitization, leaching, and porosity enrichment below the contact, and extensive, if patchy, distribution of highly permeable rocks is presumable below the unconformity. The overlying sediments are the restricted-fauna, "basinal" shales, cherts, and limestones of the lower

Senonian-lower Campanian depositional cycle, which underlie basinal Upper Cretaceous marly limestones; the whole Upper Cretaceous sequence may be regarded as a possible source for the "Second Pay" oil. The Middle Cretaceous and Lower Cretaceous rocks are not euxinic at Ain Zalah; it is very unlikely that they have themselves produced any of the oil now found in the porous sections. The Lower Cretaceous marls and limestones rest on eroded Middle Jurassic limestones and dolomites which are classed among the potential source rocks. There is no evidence of porosity enrichment at this contact and the unconformity cannot have acted as a channel for migrating fluids. Arguable possibilities for the source of the "Second Pay" oil are therefore—

1. That it originated in overlying source beds and was expressed downwards into the porous Middle Cretaceous limestones and dolomites, across the unconformity, during compaction of the source beds.
2. That it entered the trap laterally, through the Middle Cretaceous reservoir-carrier formation, from some source area outside the present field limits.
3. That it originated in underlying source sediments and accumulated in the Middle Cretaceous reservoir following vertical migration.

The volume of overlying sediments which were deposited under more or less euxinic conditions is very large, and more than adequate to supply the small accumulation found in the "Second Pay." But it is very improbable that downward expression of oil can have occurred from the main mass of the Upper Cretaceous rocks, because the least porous beds in the section are found within the lower Senonian-lower Campanian. Within the lower Senonian-lower Campanian, expression would more probably have been lateral, toward the basin margin in the northeast (Fig. 7), than across the bedding planes into the Middle Cretaceous reservoir. But in the northeast, toward the convergence of Upper Cretaceous onto Middle Cretaceous, oil may have entered the latter from any part of the lower Senonian sequence, to reach the Ain Zalah trap after more or less extensive lateral migration in the main reservoir-carrier formation. Although downward entry of oil *in situ* is thought to be improbable, origin of the "Second Pay" oil in Upper Cretaceous and/or lower Senonian source beds within the general area must be regarded as a reasonable possibility.

The volume of euxinic Middle Jurassic and older sediments below Ain Zalah is also very great, and some of these rocks appear to have been effective source rocks. Disseminated residual bitumen is common in Middle Jurassic sediments of Ain Zalah. However, the Middle Jurassic rocks were subjected to prolonged exposure, during Neocomian times, following an earlier history of fairly deep burial. Any indigenous oil present in the upper part of the sequence should have been dissipated during this emergence, and any gas content must have been lost due to release of hydrostatic pressure. No oil should have remained in the Middle Jurassic sediments after subsequent deep burial, and if any under-saturated oil did survive, in housing of low permeability, no adequate mechanism is apparent for its expulsion into a later fracture system. The lower parts of the Middle Jurassic section appear to be sufficiently impermeable and plastic to have acted as fairly competent seals to prohibit upward passage of oil from deeper horizons. The lowest rocks encountered in a deep well on Ain Zalah were Liassic anhydrites. Live oil indications were observed in these, but the oil appears to have been of different type from that now found in the Ain Zalah "pays." Deeper oils found elsewhere are gas rich; they would be saturated with gas at the depth of the "Second Pay," yet the actual oil at this depth is very undersaturated. The Lower Cretaceous section includes marl units which should have offered at least a partial seal to oil migrating upward in the structure, but no oil accumulation appears below these potential trapping beds. From these considerations, none of which is conclusive, it is argued that the oil probably did not originate directly from underlying Jurassic or earlier source beds, as has been suggested by Daniel (1954).

The remaining alternative is that the "Second Pay" was fed laterally, *via* the Middle Cretaceous limestone carrier. The ultimate origin of the oil is not explained by adoption of this alternative, which smacks of the familiar device of thrusting awkward problems into areas where they cannot be investigated. The original source may still have lain within the overlying basinal sediments, or deep down in the Mesozoic section. But the paleogeographic constructions suggest a more probable source. Figure 5 indicates the presence of a localized Lower Cretaceous trough, trending northwest-southeast, which lies a little to the northeast of Ain Zalah. The sediments of this trough, where seen at its northeastern margin, were radiolaria-rich, euxinic shales and limestones. Neritic limestones spread over the trough in the Middle Cretaceous. Depositional gradients would have favored migration of oil generated in this trough, southwestward into the Middle Cretaceous high on which Ain Zalah was placed (Fig. 6). Later sedimentary loading in Upper Cretaceous time would have compressed the source sediments, expelling oil into the overlying neritic limestones, while Upper Cretaceous trough developments, west and southwest of Ain Zalah, would have imposed gradients prohibiting further migration in these directions (Fig. 8). Paleocene-lower Eocene events include further loading of the source basin by very thick deposits, and further steepening of gradients favoring migration, within the Middle Cretaceous carrier, into the Ain Zalah area (Fig. 9). The Paleocene-lower Eocene tilting would have served to concentrate accumulation at the western extremities of the high feature of Middle Cretaceous time (productive wells on Ain Zalah are restricted to the western end of the anticline).

Local structural uplift, of which there is inconclusive evidence, may have served to concentrate oil within an incipient Ain Zalah "high" during Upper Cretaceous time. Gradient changes imposed on Middle Cretaceous horizons by events during middle Eocene to lower Miocene time were relatively small. They abetted or were inadequate to modify significantly either the step gradients favoring upward migration of oil into the vicinity of Ain Zalah from the northeast, or those preventing migration out of the Ain Zalah area toward the southwest. With the onset of late-Tertiary folding, oil in the Middle Cretaceous carrier limestones must have been firmly locked within the confines of the rising anticlinal traps.

The rather complex accumulation history suggested above for the "Second Pay" oil is speculative and equivocable, but more satisfactory in many details than the alternatives of accumulation *in situ* from overlying or underlying source beds. The origin of the "First Pay" accumulation by upward migration from the "Second Pay" can be regarded as established.

Butmah field.—The Butmah field occupies one of two in-line domes of the Butmah anticline, adjacent to Ain Zalah, but lying to the southeast. Oil is produced from fractured Upper Cretaceous limestones in a setting comparable with that of the "First Pay" of Ain Zalah. In Butmah the equivalent of the "Second Pay" of the Ain Zalah field is oil stained but water logged. There is an active water drive in the "First Pay" of Butmah. The oil is similar to that from the Ain Zalah reservoirs. It is believed that primary accumulation occurred in the Middle Cretaceous limestone reservoir at Butmah, as at Ain Zalah, and that the whole of this initial accumulation has escaped upward into the higher fractured-limestone reservoir. The Upper Cretaceous marly limestones are somewhat thinner in Butmah than in Ain Zalah.

The mode of origin deduced for the oil accumulations of the Ain Zalah and Butmah fields is illustrated in Figure 16.

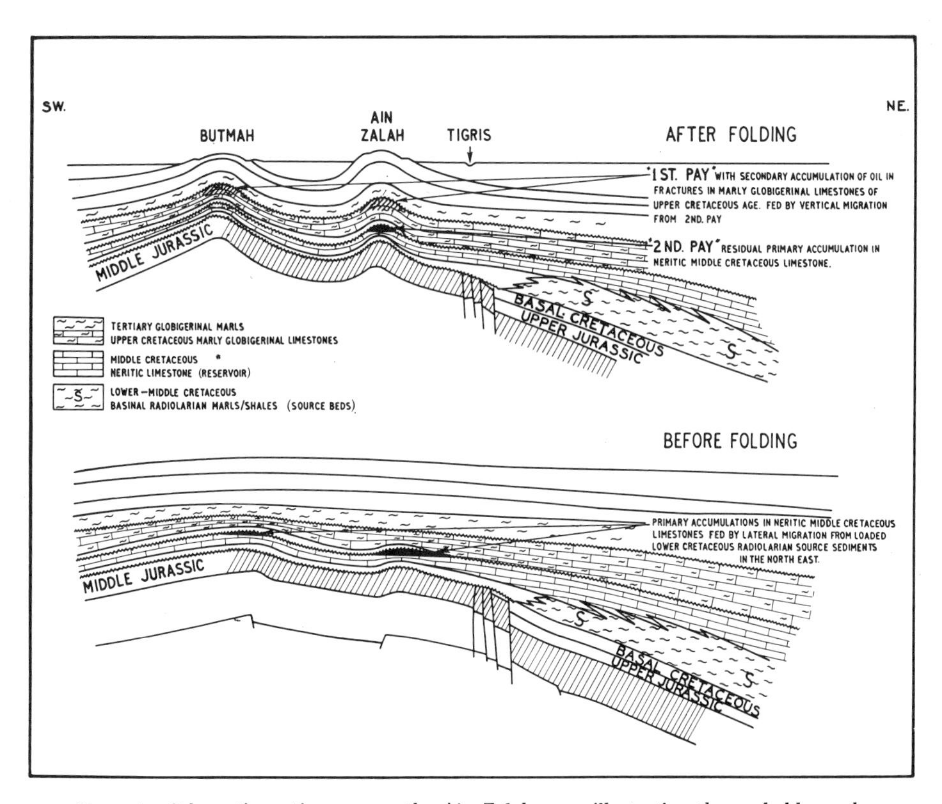

FIG. 16.—Schematic sections across the Ain Zalah area, illustrating the probable mode of accumulation of the oils of the Ain Zalah and Butmah fields.

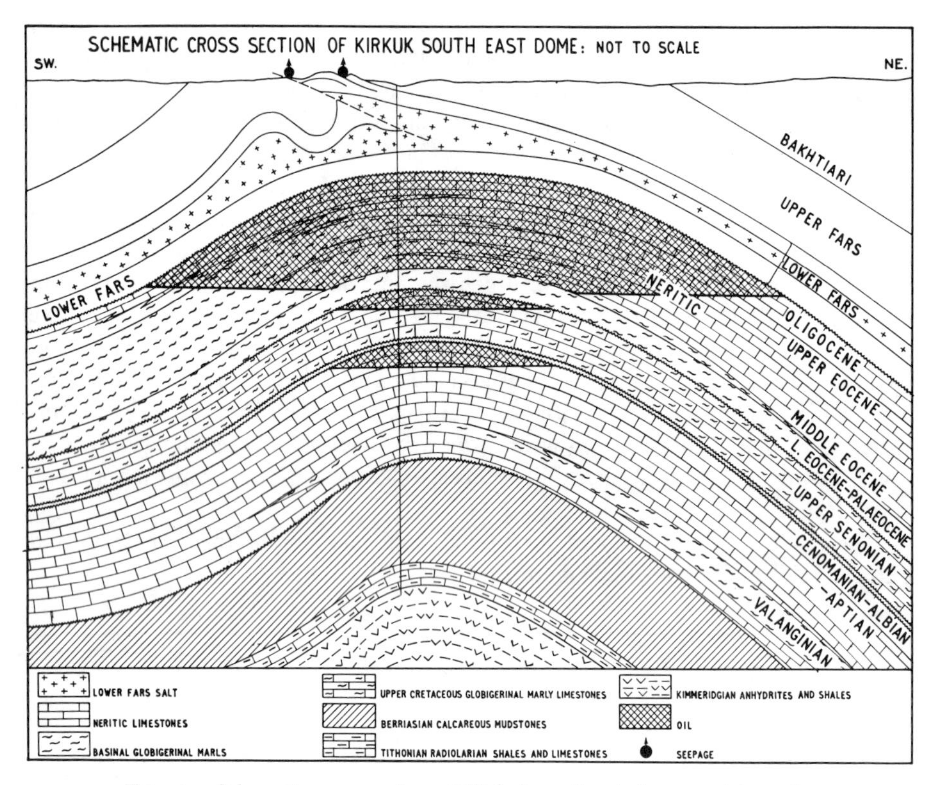

FIG. 17.—Schematic cross section of Kirkuk southeast dome; not to scale.

KIRKUK AREA

Kirkuk field.—The Kirkuk field is a sinuous anticline, some 60 miles long, divided by two prominent saddles into three major structural culminations. The structure is overthrust from the northeast in the exposed beds, due to sliding on the Lower Fars salt, but the fold at medium depth, below the salt zone of the Fars, is simple and almost symmetrical (Fig. 17). The geology of the field has been described recently by Daniel (1954). Production was 23.7 million tons in 1954, and producible reserves have been estimated at 1,000 million tons (McCollum, 1947).

The producing reservoir, named the "Main Limestone," is made up of back-reef, reef, fore-reef-shoal and globigerinal limestones of middle Eocene to lower Miocene age.[6] The "Main Limestone" is terminated by an erosional unconformity which cuts deep into the succession toward the northwestern end of the field. The overlying sediments of the Lower Fars converge by onlap in the same direction (Fig. 18). Thin limestones within the lower parts of the Lower Fars are porous and contain oil—they may be considered part of the reservoir but arc of negligible volume in comparison with the "Main Limestone." The field is capped by the salt and anhydrite beds of the Lower Fars "salt zone." The seal is imperfect, as there are

[6] A full account of the stratigraphy of the "Main Limestone" has been published by R. C. van Bellen (1956).

extensive gas and small oil seepages in the surficially thrust area, southwest of the crest of the southeastern "dome."

The axis of the anticline cuts obliquely across the shore line and facies trends of the different middle Eocene-Oligocene depositional cycles, each of which is characterized by passage, in a northeasterly direction, from globigerinal "basinal" sediments into nummulitic shoal limestones. The Oligocene cycles show further progression, toward the northeast, through reef limestones into back reef-lagoonal limestones (Fig. 18). A thin tongue of lower Miocene limestones enters the succession, below the basal unconformity of the Lower Fars, at the southeastern end of the field. The most porous and permeable constituents of the "Main Limestone" are the fore-reef-shoal sediments, and parts of the reef limestones. The back-reef limestones are porcelaneous and impermeable. The globigerinal sediments have variable porosity but generally low permeability. The reservoir is extensively fractured, and fluid and pressure connection is remarkably free through most of the field.

The cross sections shown in Figures 17 and 18, and the facies and isopach constructions for middle-upper Eocene and lower-upper Oligocene intervals (Figs. 10 and 11) illustrate that the porous, neritic limestones of the Kirkuk fold are ideally placed to have received any oils which were expelled from the contemporaneous basinal sediments. These potential source sediments lie to the southwest and south

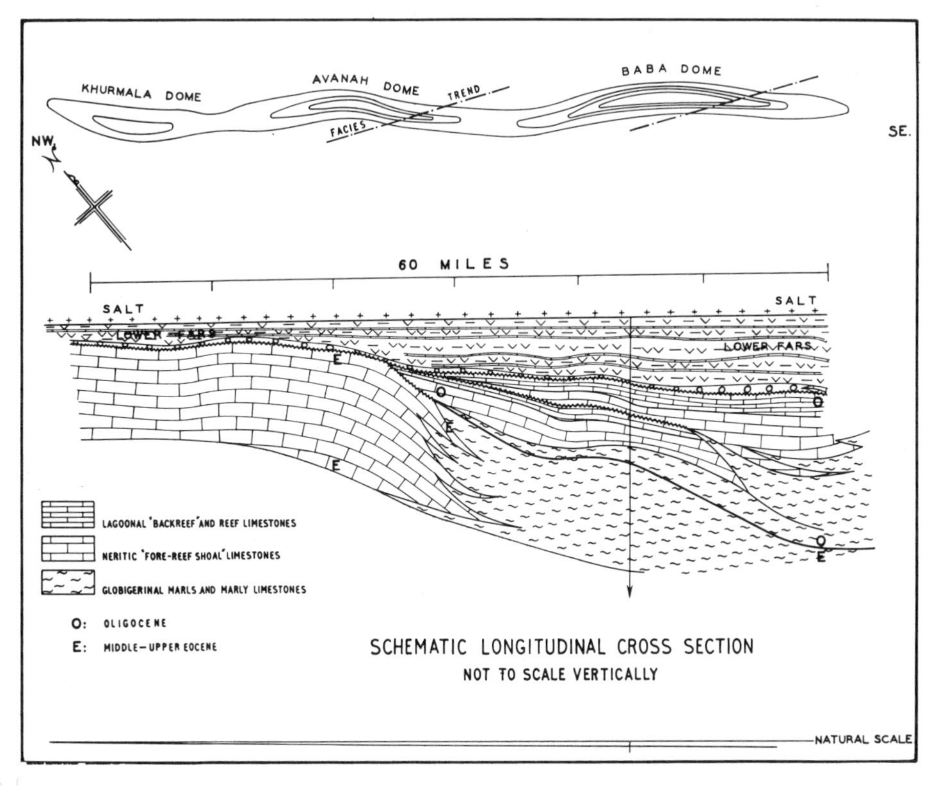

FIG. 18.—Schematic longitudinal cross section of Kirkuk fields.

of the present field, and also underlie and form a part of the southeastern dome. The straightforward and simple theory that the oil originated in nearby basinal sediments contemporaneous with the reservoir rocks has held sway for many years. It is apparent from the cross sections that possibilities for large-scale stratigraphic trap accumulations might have been realized, even without the added favor of folding, if the basinal source rocks did in fact yield any oil to the shallow-water limestones at any time after the deposition of the Lower Fars salt seal.

In questioning this obvious mechanism of origin, Henson (1950b) has pointed out the importance of the prolonged exposure of the northwestern end of the field, which preceded deposition of the Lower Fars cover. This emergence would have afforded ample opportunity for dissipation of any indigenous oil, or of any oil which was released from the basinal source beds before the deposition of the Lower Fars. Some sealed cavities in the "Main Limestone" contain oil which differs from that now found in the open reservoir, suggesting that an earlier accumulation did occur, that this was dissipated, and that the reservoir was again sufficiently water-permeated to permit cementation, before the accumulation of the oil which is now found (Henson, 1950a, p. 236). This and other evidence led Henson[7] to suggest that the Kirkuk "Main Limestone" reservoir has been filled by upward migration from a stratigraphic trap of large volume, which would be found to underlie part of the present field.

Weeks (1950) has contested Henson's arguments for deep origin of the oils, although agreeing that erosion and emergence prior to Fars deposition would have permitted loss of much indigenous and early-arriving migrant oil. According to Weeks, major oil migration from basin to reservoir may have awaited deposition of some thousands of feet of overburden upon the source sediments, so that major incursion of oil from the basin may have occurred after the deposition of the Lower Fars seal. Thus the Eocene and Oligocene basinal sediments could have been the source for all the Kirkuk oil. In addition, oil could have been furnished from source beds of which the shallow-water equivalents are lost in the pre-Fars unconformity, whereas the Lower Fars and Euphrates limestone could also have contributed appreciably to the large "Main Limestone" accumulation.

It has been suggested by Daniel (1954, p. 805) that large volumes of Oligocene and Eocene oils may have been held in porosity-wedge traps, in the vicinity of Kirkuk, during the pre-Fars emergence, and that these may have joined with oils newly liberated from the basinal source rocks, after the Tertiary folding created a fracture network which destroyed local seals. Similar destruction of seals at depth could have permitted oils from older reservoirs to add to the general accumulation.

In marked opposition to Weeks' contention that oil does not migrate across the bedding "as a rule," is the long-held and oft-reiterated view of Lees, that the oil of the Persian fields and of Kirkuk may have migrated upward through many thousands of feet of strata, until arrested beneath the almost perfect seal of the Lower Fars salt (Lees and Richardson, 1940, Lees, 1934, 1953a, etc.). According to this view the voluminous accumulations below the saliferous Fars may be mix-

[7] Unpublished Company reports, and Henson, 1950b, in discussion.

tures of oils derived from many underlying sources, each of which may have contributed only a small part of the volumes now found. There is no necessity to postulate exceptionally rich source rocks in order to account for the exceptionally large fields of the Iraq-Iran foothills zone.

The opinions summarized above cover a very wide field, and the divergences between them are so great that no general conclusions can be drawn without consideration of further evidence. Some contentions can be controverted in detail, or assessed in importance, from consideration of the relationships of the Tertiary sediments, their volumes, and their positions in the structure. Some of these arguments have been pursued by Daniel (1954). The major issue lies between those who seek to account for the oil by origin in nearby source rocks contemporaneous with or somewhat younger than the reservoir formations, and those who claim that oil entry has been by vertical migration from lower source sediments or from lower precedent accumulations. Clearly the Kirkuk field is voluminous enough to include oils of both lateral and vertical origins—a first necessity in the analysis of the Kirkuk accumulation is to determine the relative importance of lateral and of vertical migration.

A recent deep test well on the southeastern dome has provided evidence pertinent to this problem. The stratigraphy proved by this well is illustrated schematically in Figure 17. The principal interest of the test, in connection with the present theme, lay in the discovery of small oil accumulations—

1. In fractures in the upper part of a globigerinal, marly limestone sequence of Upper Cretaceous age.

2. In porous and fractured, dolomitized, neritic limestones of Albian age.

The Upper Cretaceous fractured reservoir is sealed by impervious Paleocene-lower Eocene marls which form the core of the "Main Limestone" reservoir. The seal for the Middle Cretaceous reservoir is provided by the lower part of the Upper Cretaceous marly limestones, and by a thin oligosteginal limestone of Turonian age. The accumulation in the Middle Cretaceous appears at the top of a very thick succession of Middle-Lower Cretaceous dolomites and neritic limestones, which yielded abundant evidence of one-time oil impregnation, many hundreds of feet below the base of the present accumulation.

These circumstances are closely comparable with those found in the Ain Zalah field. The parallel extends to the types of oil found, and even to the nature of the minor differences which do occur between the oils in the upper and lower accumulations of both fields.

By analogy with Ain Zalah it may be surmised that the oil in the Upper Cretaceous fractured reservoir at Kirkuk has originated by upward escape from the underlying, porous, Middle Cretaceous reservoir. It may be concluded, from the oil indications through the latter, that the Middle Cretaceous and Lower Cretaceous section has housed, at some time, an oil accumulation many times larger in volume than that which it contains at present. The suggestion is manifest that the vanished oil has ascended through fractures into the overlying "Main Limestone" reservoir, *via* an intermediate housing in the fractures of the Upper Cretaceous limestones.

This suggestion receives adequate confirmation from analytical data on the three Kirkuk oils.

Figure 19 illustrates for the three oils the specific gravities of distillation fractions of narrow-boiling-point ranges. The volumetric distillation curves themselves are closely comparable for the three oils. The striking agreement in specific gravities at all temperature ranges is sufficient to support the conclusion that the three oils are of common or closely comparable origin.

It is accepted that oils produced in similar environments at widely different times may be closely similar. It is argued that similarity of oils should increase with increasing similarity in the faunal and especially in the microfaunal constituents of the source sediments. Unless and until chemical inquiry reveals significant differences between the three Kirkuk oils, it is claimed that these originated from a common source, or else from source sediments which should evidence deposition in similar environments. The alternative view that the resemblances are purely fortuitous is unassailable except in terms of probability—the closeness of similarity

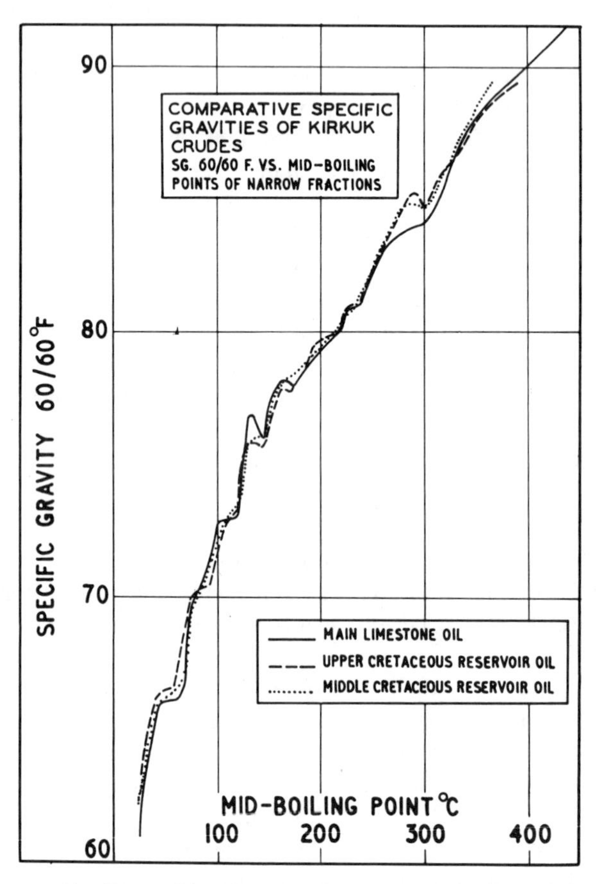

FIG. 19.—Specific gravities of narrow-boiling-range fractions of the oils from the three Kirkuk (Baba dome) reservoir formations.

illustrated by the specific gravities of narrow cuts must speak for itself in this connection.

If a common source for the three oils be admitted the problem resolves into that of locating this source. Several formations, or a single one, may be held responsible for the oil. But in any event the source cannot be of Tertiary age, as no opportunity has existed for feeding of Tertiary oil into the Middle Cretaceous reservoir. If there is a common source for the three oils, the "Main Limestone" accumulation must have originated by upward migration, either from the source directly, or through an intermediate accumulation in the Cretaceous.

The possible sources for the oil in the Middle Cretaceous reservoir are—

1. The basinal globigerinal limestones of the overlying Upper Cretaceous and/or the Turonian oligosteginal limestones.

2. The underlying rock sequence.

3. Distant sources including 1. and 2. above, entry of oil having been effected through the Middle Cretaceous carrier-reservoir.

The overlying basinal limestones of Upper Cretaceous and Turonian age are devoid of such conspicuous residual impregnation with oil as might be expected if they had functioned as very fruitful source rocks. It is concluded that they were not the directly responsible source sediments. Distant representatives of these units may have supplied an initial accumulation within the Middle Cretaceous, outside the limits of the structure.

The underlying section, proved in the deep test, includes massive neritic limestones and dolomites, underlain by neritic marls with porous limestone tongues, which follow a calcareous mudstone sequence some 2,000 feet thick. The neritic sequence cannot be considered a source-bed sequence; all the sediments were deposited under aerated conditions which must have permitted ready oxidation of any indigenous organic contents. Marl seals within the sequence do not appear to have retained any accumulations of oil, suggesting that no oil has passed upward through them.

The thick mudstone sediments suggest a euxinic depositional environment in which some oil may have been generated. But this sequence is almost unfossiliferous, and it is difficult to imagine voluminous oil generation in its rocks. Probably it retains most if not all of its original slender stock of indigenous oil, owing to its low permeability. The thick plastic mudstones provide now an impervious seal, to prevent or hinder upward migration of any oil which may be available in deeper sediments.

The competency of the mudstone unit as a cap rock is illustrated by the discovery below it, in the deep test, of a radiolaria-rich unit of shales, limestones, and cherts. This unit, which was deposited under euxinic conditions, was bitumen-soaked, yet productive of small quantities of a light, gas-rich oil, free from asphaltenes, which is completely unlike any of the oils in the overlying reservoirs. These productive radiolarian rocks are manifestly source rocks which still contain their original charge of oil, and which are themselves now sufficiently fractured to allow slow release of their oils when the overlying mudstones are penetrated. The lowest rocks encountered in the test were shales and interbedded anhydrites of upper Kimmeridgian age, which are also noteworthy for their sealing capabilities rather than for their source potentialities. The richly productive radiolarian sediments, which are of Tithonian age, are thus enclosed between two impressively competent sealing units. The equivalents of the radiolarian rocks, where seen in the Kurdistan mountain zone, are exceedingly bituminous.

The deep-test evidence indicates that there is little possibility of any appreciable volume of pre-Cretaceous oil having entered the structure by vertical migration, in the vicinity of the well, because of the barrier to such migration offered by the mudstone sequence. The mudstone section thins, and may disappear, toward the northwest end of the field, so that the prohibition is not an absolute one for the structure as a whole. Perhaps significantly, the northwestern dome shows anomalous conditions in the presence of a deep gas cap and of gas-saturated oil in the "Main Limestone." All the oils of the southeastern and central domes are undersaturated, though under producing conditions a small gas cap has developed in the southeastern one.

The contention of Lees that vertical migration may have extended from deep down in the sedimentary section appears to be incorrect for the main Kirkuk accumulations. The undersaturation of the three oils of the southeastern dome is further evidence that no extensive migration from great depth has occurred. Oil of deep origin would presumably have yielded gas to higher formation in which it accumulated, thus enriching any initially undersaturated oil body which might have gathered there before the arrival of the deeper-originating oils.

Though possible contributions of large volumes of oil from the directly overlying and underlying sections cannot be excluded, a more satisfactory account is that the original accumulation occurred in the Middle Cretaceous reservoir-carrier limestones, beyond the limits of the structure. The obvious candidates for the role of source-beds supplying these limestones are the thick, contemporaneous, basinal sediments, lying to the east of the field.

The Middle Cretaceous reservoir limestones of Kirkuk, and the underlying Lower Cretaceous neritic rocks, pass eastward at small distance into basinal radiolarian sediments, which are regarded on field evidence as rich potential source rocks (Figs. 5 and 6). The Middle Cretaceous porous limestones of Kirkuk pass southwestward into a dense, marly and evaporitic rock sequence, which must have prohibited migration in that direction. Late Cenomanian and early Turonian events raised a broad structural high feature coursing through Kirkuk on a northwest-southeast trend. At this time, also, emergence resulted in development of localized secondary porosity in the upper parts of the potential reservoir. In mid-Turonian time the region was resubmerged, and a thin plastic marl unit was laid down over the eroded surface of the Middle Cretaceous neritic limestones (Fig. 7). The earliest sediments of the Upper Cretaceous depositional cycle were also plastic marly limestones, which contributed to the original seal overlying the reservoir limestone.

In the Kirkuk area, marly limestone sedimentation continued through Upper Cretaceous time, but in the "basinal" area, to the east, very thick Flysch-type clastics were deposited, imposing heavy loads on the underlying source sediments (Fig. 8). The depositional load advanced from northeast to southwest across the basin, thus favoring continuously the expulsion of fluids into the reservoir-capable neritic limestone zone, and particularly into the broad, northwest-southeast "high" of Cenomanian time on which the Kirkuk structure lies (Fig. 6). In conjunction with this advancing load, a narrow localized trough sank or was impressed to accommodate the clastics. This trough also migrated from northeast to southwest, producing progressive, transient, steep gradients, which further favored the southwestward migration to which fluids within the basinal sediments were predisposed by the compressional factors. Oils which were expressed from the Middle and Lower Cretaceous source rocks into the permeable, neritic, carrier limestones, suffered lateral migration within these carriers, in response to changing regional gradients. Upward escape was forbidden by the overlying globigerinal limestone-and-marl succession, and escape to the southwest was ruled out by passage into impermeable anhydritic sediments.

By the end of the Cretaceous the entire basinal area of Middle and Lower Cretaceous time had been buried beneath at least 2,000 feet of rapidly deposited sediments. The main parts of the basin, lying directly east of Kirkuk, had been subjected to overburden loads corresponding to depositional thicknesses two or three times as great. By this time most of the oil may have been expressed into the neritic carrier formations, to migrate, where prompted and where possible, into structural or up-dip, porosity-wedge traps. Much of this postulated oil would have concentrated on the Middle Cretaceous "high" feature running below Kirkuk (Fig. 6).

Cretaceous-Paleocene emergence and subsequent events would have sufficed to promote minor, gravity-prompted adjustments or shifts, within the reservoir-carrier. Paleocene-lower Eocene depositional history repeats that of the Upper Cretaceous, and may have resulted in expression of additional oil from the source beds (Fig. 9). Gradients favored further persuasion of the oil southwestward. Middle Eocene to lower Miocene sedimentation (Figs. 10-12) probably accompanied only minor gradient changes within the Middle Cretaceous.

Lower Fars to late Pliocene conditions (Figs. 13 and 14) introduced a sharp northeasterly gradient, favoring migration away from the porosity-reduction facies boundary of Middle Cretaceous times, and also a marked northwesterly gradient upward from the center of the Bakhtiari "basin" toward Kirkuk. These latter day regional-gradient changes conspired to favor migration, within the Middle Cretaceous neritic limestones, into or through the general vicinity of Kirkuk. As the broad anticlinal structures were already forming at the commencement of Bakhtiari deposition, conditions were ripe for the concentration of very large volumes of Middle-Lower Cretaceous oils, in the Middle Cretaceous reservoir, within the rising Kirkuk structure.

The foregoing account is an outline of the most credible mode of accumulation of the original oil in the Middle Cretaceous reservoir of Kirkuk. It involves reference back to the original depositional basin, and investigation of the events to which this basin has been subjected. The account requires expulsion of oil from source sediments as they underwent compaction, due to sedimentary loading, and it demands more or less free lateral migration over a few tens of miles within neritic limestone carriers. It is not an unavoidable account, but it is the most convincing of the several alternatives which have been conceived and investigated.

If the account is accepted, and other alternatives are rejected, significant arguments can be brought to bear on the origin of the "Main Limestone" oil. The postulated source beds for the Middle Cretaceous accumulation are radiolaria-rich, ammonitiferous marls and shales, deposited in a markedly euxinic environment. This combination of faunal and environmental characteristics was not repeated, after the end of the Middle Cretaceous times, in the sediments of any part of the region. The close similarity in composition of the "Main Limestone" and earlier oils, already stressed, has been taken to indicate close similarity in the faunal and environmental factors controlling the nature of the source sediments. Subject to acceptance of this premise it may be concluded that the "Main Limestone" oil originated in radiolaria-rich, ammonitiferous sediments, deposited in a euxinic environment, and most probably in the same source sediments as fed the Middle Cretaceous accumulation. Further, it may be concluded that the Oligocene and Eocene basinal sediments, with their very different, rich, foraminiferal faunas, have contributed little or nothing to the existing "Main Limestone" accumulation.

The preferred account for the origin of the Kirkuk oils may be summarized as—

1. Primary accumulation in Middle Cretaceous neritic limestone carriers, in porosity traps, or incipient structural traps, at about the end of Cretaceous time.

2. Secondary readjustment into the rising Kirkuk structure during early phases of Mio-Pliocene folding.

3. Rupture of Turonian and Upper Cretaceous cap rocks, and escape of some oil into the fracture reservoir in Upper Cretaceous limestone.

4. Rupture of Paleocene-lower Eocene seal and vertical escape of most of the Middle Cretaceous accumulations into the Tertiary reservoir, leaving residual accumulations in the original reservoir, and in the fractures of the Upper Cretaceous limestone.

This account is illustrated schematically in Figure 20.

Other alternatives are apparent which do not absolutely deny the absence of oils of Oligocene-Eocene origin from the "Main Limestone" accumulation. Thus the Middle Cretaceous accumulation could originate from overlying Upper Cretaceous globigerinal limestones, and could have filled the Upper Cretaceous fractured reservoir, whereas somewhat similar oils, produced by somewhat similar globigerinal source beds of Eocene-Oligocene age could have supplied the "Main Limestone." The similarities between the several oils seem to the writer much greater than can be satisfactorily accounted for by this alternative, but the need for further chemical investigation is recognized.

If Eocene-Oligocene sources have supplied all or most of the "Main Limestone" accumulation, as advocated by Weeks (1950), then there has been no entry of appreciable volumes of oils of pre-Cretaceous origin into any accumulation. The Jurassic source-bed conditions and source organisms were completely different from

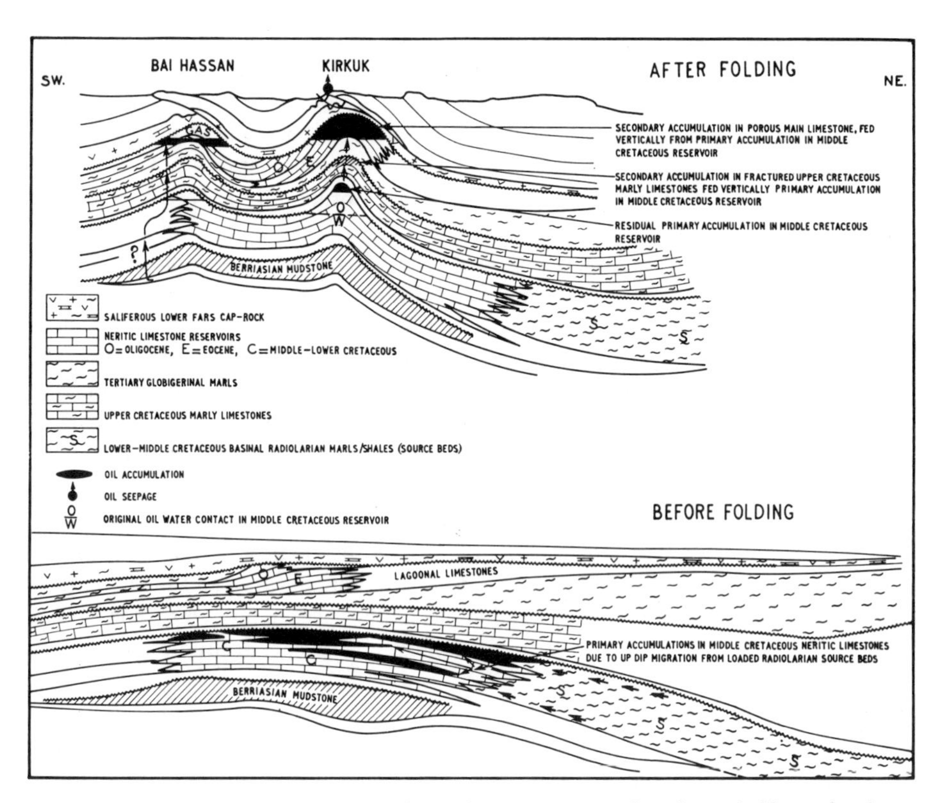

FIG. 20.—Schematic sections across the Kirkuk area, illustrating the probable mode of accumulation of the oils of the Kirkuk (Baba dome) and Bia Hassan fields.

those of Upper Cretaceous and later ages, and should have produced significantly different oils.

But if Jurassic earlier sources have contributed significantly to any accumulation they have contributed more or less equally to all, and the inference is strong that the three oils were originally one.

Bai Hassan field.—The newly proved Bai Hassan field lies parallel with the Kirkuk structure, some six miles to the southwest. This field is probably underlain by neritic limestone developments of Middle and Lower Cretaceous age (Figs. 5 and 6).

The reservoir unit now under development is equivalent to the "Main Limestone" of Kirkuk. The oil is saturated, and there is an extensive gas cap. The "Main Limestone" crest is much deeper than in the Kirkuk domes, but there are no surface seepages.

Apart from the presence of an exceptional gas content the "Main Limestone" oils of the two fields are closely comparable. Significantly, the resemblance between the Bai Hassan "Main Limestone" oil and the oil from the Kirkuk Middle Cretaceous reservoir is even closer than that between the two "Main Limestone" oils, except at the two extremes of distillation.

As in the case of Kirkuk it is postulated that the "Main Limestone" accumulation has migrated vertically upward from an earlier pool, which was housed in Middle Cretaceous neritic limestones, the probable origin being seen in basinal sediments of Middle-Lower Cretaceous age lying a few tens of miles to the east. The anomalously high gas content may be a result of later addition of gas, from a deeper source, from Eocene-Oligocene basinal marls, or perhaps within the "Main Limestone" from another anticline. The mode of origin of the accumulation is portrayed in Figure 20.

Jambur field.—The Jambur field, also under current investigation, lies almost in line with Bai Hassan, some 45 miles to the southeast. It is parallel with the Kirkuk structure but separated from the southeastern dome of Kirkuk by a deep syncline. The productive reservoir is equivalent to the "Main Limestone" of Kirkuk. Again there is a deep gas dome, and the oil is gas-saturated. Jambur lies farther into the Oligocene-Eocene basin than Kirkuk, and the pre-Fars break is very much smaller than on the Kirkuk structure. Jambur is thus better situated to have retained initial Oligocene-Eocene oils than is Kirkuk. The oil is very different from that of Kirkuk, being markedly less dense, and free from asphaltenes. The specific gravities of narrow boiling-point cuts, taken in fine-fractionation distillation, differ considerably from those of the corresponding fractions of the Kirkuk oils. At first sight it might appear that Jambur really does contain Oligocene-Eocene oil, and that the "Main Limestone" oil of Kirkuk, being chemically different, has a different origin.

However, the Jambur "Main Limestone" is permeated by copious bitumen residues, even high up in the gas cap of the present accumulation. Yet the produced oil is free from asphaltenes. It appears that there was originally an accumulation of heavier oil in the structure, and that the light oil with gas is a very late arrival. The ubiquitous bitumen was probably deposited from the original oil by solvent precipitation during invasion by the later, lighter oil. Again the suggestions may be made that the bitumen represents residues of an early-originating accumulation of Eocene-Oligocene oil which was inspissated during the pre-Fars emergence, and that the present oil arrived later, following genesis in the same basinal source beds, after deposition of Fars cover. This account does not bear close examination. The bitumen is too widespread, too deeply distributed, and too voluminous to be an inspissation residue, and similar "residues" are lacking in Kirkuk, where they might be expected to occur. It is improbable that the same source beds should yield first an asphaltene-rich oil, capable of leaving large volumes of heavy residuum in the formation, and, at a later time, a light oil, without asphaltenes, which would be incapable of dissolving any of the ubiquitous residuum of the first accumulation.

The distribution of bitumen in depth indicates that it was deposited from an oil column of considerable height, which could have come into place only after considerable development of late-Tertiary folding. The initial accumulation may be dated not earlier than Lower Bakhtiari "time." At this stage in development, the Eocene-Oligocene potential source beds had been buried under more than 4,000 feet of Miocene and perhaps early Pliocene sediments. Although the initial accumulation could have been of oils of Eocene-Oligocene source, it is also possible that it could have resulted from upward migration from an underlying Middle Cretaceous reservoir, as suggested for Kirkuk. Cenomanian neritic limestones probably underlie the northwestern end of the structure (Fig. 6). The later-arriving, light, gas-rich oil could be of Eocene-Oligocene source, but this again appears to be unlikely, because compaction due to sedimentary loading must have been very slight at the late stage in history at which the incursion must be placed. It is suggested, tentatively, that the light-oil invasion took place from below through fractures created by late-Tertiary folding, and originally from some such previously imprisoned source rocks as the radiolaria-rich Tithonian sediments which were encountered in the Kirkuk deep test.

Jambur is a new field, as yet imperfectly explored. The opinions advanced above are in high degree speculative, and their retraction may be required by new findings at any time. But Jambur does not fit comfortably into a simple history of Tertiary source feeding Tertiary reservoir. The peculiarities of this accumulation are more readily explicable by vertical than by lateral migration.

QAIYARAH AREA

The oil fields of the Qaiyarah area occur in the anticlinal traps of the Qaiyarah, Najmah, Jawan, and Qasab structures. There are two productive reservoirs, the upper one comprising the porous, neritic-lagoonal Euphrates limestone, of lower Miocene age, and the lower one being neritic limestones of Upper Cretaceous (upper Campanian-lower Maestrichtian age). The upper reservoir is more consistently porous and permeable than the lower, which shows marked variation in quality from dome to dome.

Qaiyarah, Najmah, and Jawan are successive individual culminations on a single sinuous fold axis, which trends approximately northwest from the western bank of the Tigris at Qaiyarah (Fig. 21). Qasab is a parallel anticline, with two domes, which is offset to the northeast, and separated by a broad, shallow, synclinal saddle, from Jawan.

The oils of the two superimposed accumulations are similar. Gravities range from 11.5° to 18° A.P.I. in the Upper Cretaceous reservoir, and from 11° to 19° A.P.I. in the Euphrates limestone reservoir. All oils are exceedingly sulfurous, the Euphrates limestone accumulation more so than the oil in the Cretaceous reservoir. The oils are density stratified, the heaviest lying in the deepest parts of the structure. The oil/water contact in both reservoirs is tilted, being highest in the northwest, in Qasab. Edgewater stringers appear in the Euphrates limestone accumulation, high above the bottom-water level, along the southwestern flanks of the Najmah and Jawan domes.

There are small gas domes in the Euphrates limestone pools in all structures, but the only gas dome found in the Cretaceous pools is in Qasab. The Cretaceous oils in the Qaiyarah and Najmah fields are undersaturated. The oil accumulations are continuous through the four domes in the Cretaceous reservoir, but the synclinal saddle between the Qasab and Jawan domes is lower than the oil/water level in the Euphrates limestone.

The cover for the Euphrates limestone accumulation is provided by anhydrites, limestones, and marls of the Lower Fars, which lacks salt in this area. The seal

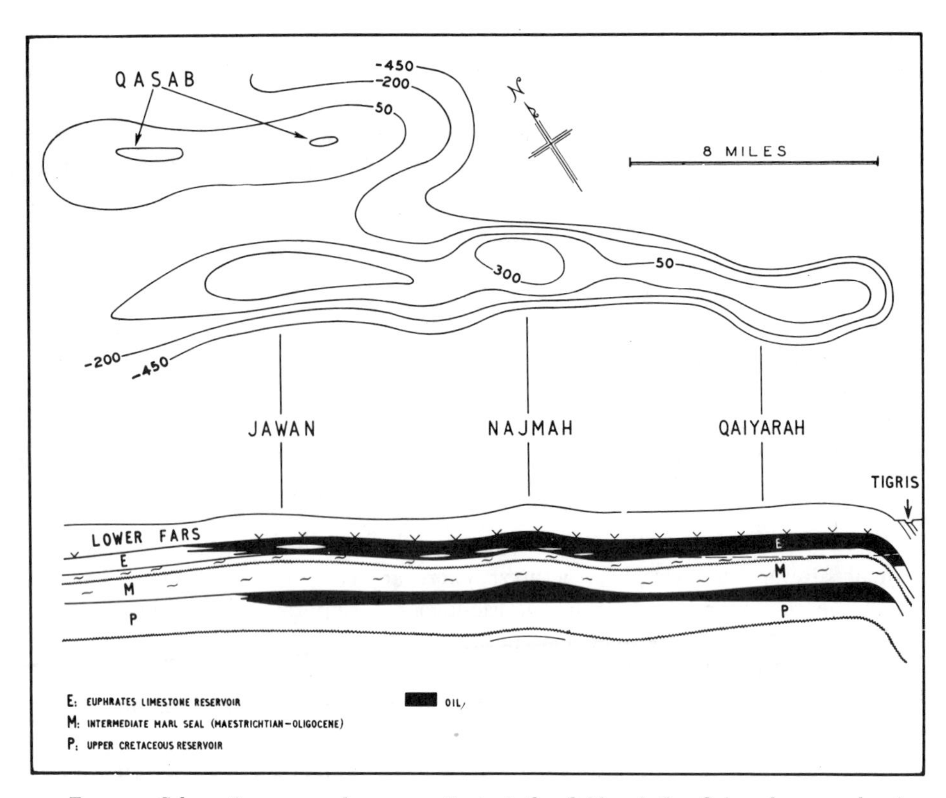

FIG. 21.—Schematic map and cross section of the fields of the Qaiyarah area, showing disposition of the two superposed accumulations. The oil/water contact surfaces in both reservoirs slope down to the southeastern pitch of the Qaiyarah field.

between the two accumulations is provided by thin Oligocene, Eocene, and Upper Cretaceous globigerinal marls. The Cretaceous neritic limestone reservoir is underlain by oligosteginal limestones of Turonian age (Fig. 7) which transgress over eroded Albian evaporitic limestones and shales. A thin Aptian-Barremian succession of marls and limestones intervenes between the base of the Albian and the top of the eroded Upper Jurassic in the Najmah dome. No wells have passed through the Albian on the other three domes.

The origin of the oils of the Qaiyarah area fields is controversial. Indigenous origin for the Euphrates limestone oils has been argued, as has a history of generation in the underlying globigerinal sediments. Indigenous origin is improbable, as the rocks were deposited in well aerated waters. The underlying globigerinal sediments are generally rather unusually rich in benthonic faunal components, suggesting that they also were not deposited under euxinic conditions. The Cretaceous limestone oils have also been reckoned indigenous, or derived from overlying basinal sediments or from underlying basinal Turonian or lagoonal Albian formations. None of the suggested source rocks is significantly bituminous. If the similarity of the oils in the two reservoirs is due to origin from a common adjacent source, the parent sediments must be the intervening globigerinal marls of Eocene-Oligocene and Upper Cretaceous age, which are not impressive for their source potentialities. They are also quite inadequate in volume to have supplied from the resources of a small area the major accumulations which are now found. Early-arising gradients, which might have aided drainage of large areas of source beds, were lacking from this area, as was any localization of heavy depositional loading. The most satisfactory account for the origin of the Euphrates limestone oils is that they have originated from precedent accumulations in the underlying Upper Cretaceous limestone reservoir, or from a yet deeper source which fed both reservoirs. The tilted oil/water contact and the intermediate waters in the Qaiyarah-area fields suggest that entry occurred in the southeastern plunge, and independently in the southeastern plunge of Qasab, and that the entering oils have never adjusted themselves entirely to the hydrostatic conditions, perhaps because of their high viscosity and ready emulsification.

For the Cretaceous limestone accumulations, origin in contemporaneous basinal source beds has been suggested. But isopach and facies constructions suggest that, in this event, entry into the system should have been from the northeast, north, northwest, or west, whereas gravity distribution, gas and oil distribution, and the tilted oil/water contacts infer entry from the southeast or east. The possibility of origin from deeper sources, directly underlying the whole area, appears to be ruled out by a deep test well on the Najmah fold, which penetrated a thick Albian-Liassic section without encountering any significant porous reservoir rocks or obvious source beds.

A clue to the origin of the Qaiyarah heavy oils is afforded by bitumen deposits and impregnations in the basal Lower Fars sediments of the Hit-Awasil area, on the Euphrates, and of the Fatha Gorge, on the Tigris, south of Qaiyarah. These impregnations and deposits resulted from extensive seepage in these areas, in the period immediately following deposition of the Euphrates limestone. The seepage oils were very heavy and very sulfurous, approximating to soft bitumens. Vast quantities of sedimentary bitumens and anomalous sediments were spread over large areas as leakage proceeded into the shallow Lower Fars sea.[8] Eventually the seepages were choked by these deposits themselves, and by the later sediments of the Lower Fars. In the Awasil area, Pleistocene erosion has cut through the thin seal, and the seepages are again operative, pouring forth several thousands of tons of bitumens and heavy inspissated oils each year, together with large volumes of sulfurous waters and low-pressure sulfur-laden gases.

The Awasil area and Fatha area seepages are seemingly identical in origin, and carry identical bitumens. Those of the Awasil area are linearly arranged, and correspond in position to a line or zone of pre-Maestrichtian faulting. The modern seepages at Fatha are linked by similarity of oils with those of Qaiyarah area, and with the reservoir oils of the Qaiyarah area fields.

In the Awasil area, a deep well has shown partial impregnations of Maestrichtian limestones, and of Middle and Lower Cretaceous sandstones, by heavy sulfurous

[8] The Miocene activity of the seepages of the Awasil area, and the nature of their rejuvenation in the present erosion cycle, were first remarked by W. T. Foran, in unpublished reports of the British Oil Development Company.

oils and bitumens, which are identical with those of the seepages. Structural closure is probably lacking in these reservoirs. The sporadic distribution of impregnations in continuously porous rocks suggests that they are in relation to a fault zone, through which oil has ascended, irregularly, with occasional more or less accidental lateral incursions into porous rocks. The well penetrated a thick Kimmeridgian anhydrite section, and yielded, from below this, some dense oil, with very high sulfur content, which was superficially indistinguishable from the seepage oils of Awasil or Qaiyarah.

In the Awasil area the heavy sulfurous oils are clearly of pre-Kimmeridgian origin, despite their wide distribution in sediments of several younger ages. Their appearance in overlying reservoir formations, and their copious seepage to surface in Lower Fars times, and at the present time, have been facilitated by faulting of early date, perhaps several times repeated, which has broken the Kimmeridgian anhydrite seal, along a line coursing more or less north-south from Hit to Abu Jir. On the evidence of regional correlation, the actual primary reservoir for the heavy oils is almost certainly the upper part of the underlying Jurassic, corresponding approximately in stratigraphic position with the fabulously productive "Arab Zone" of the Arabian fields.

In the Qaiyarah area, as in Awasil, the tectonic setting is one of block movements, with north-south bounding faults of late Jurassic or earliest Cretaceous age. The Kimmeridgian anhydrite cover is absent from the area west of the Tigris, due to Neocomian uplift and erosion. The underlying Upper Jurassic, including the presumed porous reservoir unit, has been eroded off the Qaiyarah-Qalian-Atshan tilted block but is believed to remain, intact below anhydrite cover, in the down-fauted block lying east of the Tigris (Figs. 3 and 4).

It is argued that the heavy oils of the Qaiyarah fields originated in basinal sediments which occupy a north-south trough passing below Kirkuk (Fig. 3). Oil was expressed westward from this trough, during terminal Jurassic and early Cretaceous time, under the impulse of heavy loading (Fig. 4) and encouraged by steepening westward gradients. This oil accumulated in contemporaneous neritic-oölitic limestones, beneath Kimmeridgian anhydrite cover, against the face of an uplifted block, along the Tigris River line. Much oil escaped to surface, and remaining oil may have been partially inspissated, in post-Kimmeridgian to Hauterivian time. Some of the retained oil, later buried beneath Cretaceous-Pliocene sediments, migrated upward, through and across the fault zone, into the reservoir-capable limestones of the Qaiyarah-Qasab area. Entry into these reservoirs was from the eastern ends of the Tertiary fold structures. Vertical movement of oil may have commenced during early Lower Fars times, or earlier, and may have continued intermittently through most of the late-Tertiary period. Much Upper Jurassic oil escaped directly to surface, up the fault zone itself, and much continues to escape, through present-day seepages lying along the Tigris River line, at Qaiyarah, Nimrud, Hammam Ali, Khanuqah and elsewhere.

At present the existence of porous Upper Jurassic reservoir rocks, beneath the Kimmeridgian anhydrite cover, is only conjectured. Exploratory drilling now in progress will shortly confirm or deny this conjecture, strengthening or weakening the proffered account of the genesis of the oils of the Qaiyarah area accumulations.

The deduced history of the oil of these accumulations is illustrated, somewhat diagrammatically, in Figure 22.

PAST AND PRESENT DISSIPATION OF OIL

In addition to the known oil fields, North Iraq offers prolific evidence of past and now dispersed accumulations on a grand scale. There are also many active and some relict seepages which should be considered in a full inquiry into the factors controlling distribution of accumulated oil within the region. Full discussion of the seepage occurrences is outside the scope of the present superficial survey.

Impressive bitumen impregnations occur in the Berat Dagh, at Aqra, Bekhme, and elsewhere, sporadically distributed through a bank- reef- and shoal-limestone complex of Maetrichtian and upper Campanian age (Henson, 1950a). Impregnations are found in the crestal region, and also low down on the flanks and in the core of what is now a very large, steep-sided anticline. The reef-complex passes northeastward, by intertonguing, into the thick sediments of the Upper Cretaceous Flysch basin, and southwestward into globigerinal basinal sediments (Fig. 8). Some, and perhaps all of the oil which was responsible for the heavy impregnations

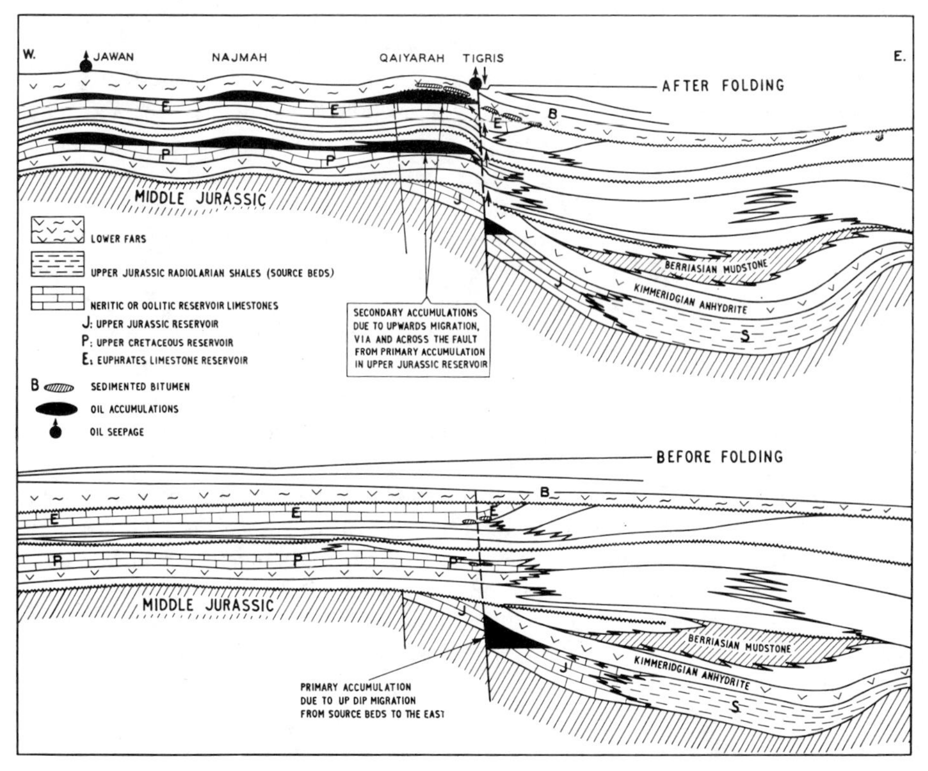

FIG. 22.—Schematic sections across the Qaiyarah area, illustrating the probable mode of accumulation of the oils in the Qaiyarah, Najmah, Jawan, and Qasab structures.

may have originated in either or both of the flanking basins, as suggested by Henson (1950a). The fore-reef tongues and porous parts of the massive limestones are locally bitumen saturated, but elsewhere free of hydrocarbons. It is supposed that large volumes of oil accumulated in localized porosity traps, prior to late-Tertiary folding. Doubtless there was great wastage by loss to surface during this primary accumulation period, for paleogeographic reconstructions indicate that the Berat Dagh area was never covered by any really competent seal. When late-Tertiary folding occurred, the local trap conditions were destroyed by fracturing, and the oil migrated into the main anticlinal trap, and to almost immediate dissipation at surface (Henson 1950a). Judging by volume and nature of the residual bitumen, the original porosity-trap accumulations were very large, and also the original oil was heavy and asphalt rich.

Berat Dagh impregnations.—The Berat Dagh impregnations appear to demonstrate that oil in quantity was generated in, and liberated from, sediments of Upper Cretaceous age. In the cases of the known oil fields the contributions from such sources are believed to have been negligibly small. The productivity of the Upper Cretaceous source sediments, even at Aqra, is not entirely proven, as the reef developments of this age rest directly upon porous and permeable Middle Cretaceous limestones, without any intervention of a seal formation, in the area where impregnation is heaviest. Possibly the copious bitumen is residual from secondary accumulations, within the Upper Cretaceous reef-limestone complex, of oils which entered the uplifted Aqra area through Middle Cretaceous carrier-limestones, and which were generated in basinal sediments of Middle-Lower Cretaceous age. The source area would have lain to the east and north of the Aqra-Bekhme range (Fig. 6). Heavy loading and steep gradients, imposed during Upper Cretaceous and Paleocene-lower Eocene time (Figs. 8 and 9), would have favored extensive migration into the Aqra area. Presence of residual oils in the Middle and Lower Cretaceous rocks shows that such migration did occur. Bitumen impregnation in the lower parts of the Upper Cretaceous also suggests filling from below, as oil entering laterally, before folding, would have ascended through the porous rock mass. Pebble-armored bitumen balls occur abundantly within the lower parts of the Upper Cretaceous section, demonstrating that oil in considerable quantity was seeping to surface early during Upper Cretaceous deposition, in the area where spectacular impregnation is now found in later beds.

Pila Spi area.—At Pila Spi, close to the Persian frontier southeast of Sulaimania, oil and bitumen permeate the Maestrichtian Flysch, the overlying Eocene-Paleocene clastics, and the lagoonal middle-upper Eocene limestone, in a setting which is now anticlinal. The accumulation may have predated much of the late-Tertiary folding, but if it did not, the total oil column in this now dissipated oil field must have been many thousands of feet thick. But the nature of the seal for such a large accumulation is obscure. Post-Eocene sedimentation was thin, and the Lower Fars is atypical in this area, in that salt and anhydrite are lacking. The significance of the Pila Spi impregnations is that the oil most probably originated within the Upper Cretaceous Flysch, or perhaps from an unknown Upper Cretaceous reef accumulation now buried beneath the very thick Flysch sediments. The Middle Cretaceous sediments are in basinal facies in this area, and it cannot be argued that the oil has migrated upward from a primary accumulation in a neritic Middle Cretaceous limestone reservoir. Origin from an even deeper primary reservoir or source rock can still be argued, but no candidate is apparent for the role of reservoir rock. It is preferable to admit that the Upper Cretaceous Flysch basin has generated oil in large quantity, and that accumulation occurred in this locality because of special trapping facilities which remain obscure.

Bitumen pebbles occur in some profusion in Paleocene-Lower Eocene conglomerates in the Pila Spi area, indicating that large oil accumulations were undergoing dissipation, somewhere in the region, at the time of deposition of the conglomerates. The source for these water-born bitumens may be seen in such cap-less accumulations as that already described from Berat Dagh. The basal Paleocene pebble bed of southwestern Persia is also characterized by derived bitumen (Kent, *et al.*, 1951, p. 149).

Pir-i-Mugrun occurrence.—A third, impressive, bitumen-impregnated structure, now dissected, is Pir-i-Mugrun, west of Sulaimania. The rocks of this large anticline show passage from basinal globigerinal-radiolarian sediments of Lower-Middle Cretaceous age into massive, much dolomitized, rudist bank-reef limestones of the same ages. The lateral passage has been illustrated and discussed by Henson (1950a). The reef limestones and associated massive limestones are very bituminous, as are the fore-reef-shoal tongues which pass out into the basinal sediments. The basinal sediments themselves are classed as fertile source rocks. Pir-i-Mugrun provides a cross section of the facies interdigitation between basinal source rocks, which are believed to have supplied the Kirkuk accumulation, and the reservoir-carrier limestones, in which primary accumulations of the Kirkuk oil is deemed to have occurred, some 80 miles to the southwest.

Pir-i-Mugrun, and other very large mountain anticlines which now expose the neritic Middle Cretaceous limestones, were probably traps for large accumulations of migrant oil, within this reservoir formation, before Pliocene or later erosion destroyed their seals. Dissipation of such accumulations, as well as of accumulations of the Berat Dagh type, may account for the origin of detrital bitumen in the middle and upper Bakhtiari sediments.

The actual volume of wastage from oil accumulations of the Berat Dagh, Pila Spi, and Pir-i-Mugrun types cannot be guessed at, but it must have been very large. The periods of dissipation of such accumulations can be surmised, and the surmises checked to some extent by the distribution of sedimented hydrocarbons in the rock sequence. The picture which emerges for the mountain zone is one of three principal phases of wastage. The first phase corresponds to the Middle-Upper Cretaceous transition, when oils of Middle Cretaceous origin escaped from neritic reservoirs, through windows in the sealing sediments such as are found at Berat Dagh. The second phase corresponds to the terminal Cretaceous regression, when poor-quality seals covering the Upper Cretaceous reef complexes were eroded, and when tectonic fracturing may have disturbed primary porosity traps in the reef sediments. The third phase corresponds to the main anticlinal folding, which destroyed primary porosity traps and encouraged migration into rising structures, but which fractured the low-quality seals capping these structures, and which also laid the crests of the higher structures open to Pliocene erosion.

These three phases of wastage are but little evidenced in the foothills belt, where the known oil fields lie. The Middle-Lower Cretaceous oils had perhaps not entered this area during the time of post-Cenomanian erosion, and they may have been already in place, below an adequate Upper Cretaceous cover, during the Cretaceous-Paleocene emergence. Abundant bitumen in Middle and Upper Cretaceous limestones of some tested structures, where Upper Cretaceous cap rocks are lacking, may represent inspissation of oil during this emergence. The main folding episode certainly promoted vertical migration, and in many cases such migration may have continued to surface, but losses of this origin cannot be assessed. Some tested structures, which have only mediocre cap rocks, show strong bitumen impregnation in Tertiary reservoirs. Such impregnation may indicate slow loss of original oils during the folding, or passage of large volumes of oils through the unsealed traps to the surface during the post-folding period.

Two additional episodes of widespread dissipation, affecting the Upper Jurassic oils of the foothills and foreland zone, have already been considered in connection with the origin of the oils of the Qaiyarah area fields. The first of these episodes is illustrated by the emergence and erosion of tilted blocks, embracing the Qaiyarah, Qalian, and Atshan areas, at about the Cretaceous/Jurassic transition period. Large volumes of Upper Jurassic oil were dissipated at this time, evidence in the form of abundant bitumen pebbles being preserved, in shallow-water deposits of the transgressive Lower Cretaceous, in some areas of continuous Jurassic-Cretaceous sedimentation. Early Cretaceous dissipation of Upper Jurassic oils, similar to that argued for the Qaiyarah area, may have occurred elsewhere in the region west of the Tigris.

The second episode of wholesale dispersion of the Upper Jurassic oils, corresponding to the onset of Lower Fars sedimentation, resulted in losses to surface, conservatively estimated at hundreds of millions of tons, in the areas around Awasil and Fatha, where sedimentary bitumens are prominent at the base of the Lower Fars. The known areas of Miocene seepage may be only a small part of the total area in which such seepages occurred, and the residual bitumen may be only a small part of the oil which escaped into the shallow Lower Fars seas.

Active seepages of bitumen, heavy oil, and gas, all rich in sulfur, characterize the Awasil area, and the vicinity of the Tigris. Most of these seepages are considered to be oils of Upper Jurassic origin, though the reservoirs now being drained may be of much younger age. Some seepages of very heavy, sulfurous bitumens, occurring far from the Tigris, may mark reappearances of heavy sedimented oils, which were deposited with the Lower Fars after escape into the Lower Fars seas.

Other important seepages are prominent in the crestal regions of large folds which contain or have contained oil accumulations in the Tertiary limestones, and which are capped by saliferous Lower Fars. These seepages are distributed through the foothills belt between Kirkuk and the Persian frontier, mostly in locations closely related to surficial thrust faults which are accommodated in the Lower Fars "Salt Zone." Gas seepages are most common, commonly associated with heavy inspissated oils, and locally with very light oils, which are perhaps of condensate type. Secondary mineralizations of gypsum into aragonite, and of calcareous marls into "gach-i-turush," have been described from analogous seepage areas in Persia (Lees and Richardson 1940; Lees, 1953b). Similar mineralization products are found in association with many of the gas seepages of Northern Iraq.

The current rates of gas loss from some structures, and the long history of dissipation which must be allowed, require either that the initial charges of oil must be seriously depleted of their original gas content, or that additional gas must be entering the traps to make good the losses. On the other hand the volume of fluid oil lost to the surface from these seepages must be very small, in comparison with the capacity of the traps, unless the rates of oil escape are much smaller now than in the past. The conclusion is reached that original accumulations of oil in the pre-Fars Tertiary limestones of the salt-covered structures have probably survived the Pleistocene erosion cycle with little loss. But some redistribution of oil from structure to structure, across the spillways, may have followed gas-volume changes consequent upon gas seepage, upon changes in hydrostatic pressure, or upon entry of additional gas or gas-rich oil from depth.

The presence of a gas seepage does not confirm the existence of an underlying oil accumulation, and the rate of escape of gas is not directly related to the volume of oil which is present in the underlying trap. Only one field (Kirkuk) of the five known fields which contain oil of Middle Cretaceous origin is marked by surface gas seepages. Some structures with very large gas seepages may have very small oil columns, or may contain no oil whatever.

Seepages in the strongly folded mountain zone are for the most part associated with residues of accumulations of the Berat Dagh, Pir-i-Mugrun, or Pila Spi types, though a few may originate by vertical migration from surviving, unexposed but probably inspissated accumulations in reservoir limestones of Middle Cretaceous or older ages. Some so-called seepages are in reality merely exceedingly bituminous "source rocks," stripped of their liquid oils, which exude semi-solid bitumens locally under the high prevailing summer temperatures. The Tithonian radiolarian shales and limestones, and some similar sediments of pre-Kimmeridgian Upper Jurassic age are the most markedly bituminous of such rocks; they have sufficient hydrocarbon content in some areas to render them usable as fuel.

The evidences of relict seepage cannot be considered in detail. Many drilled structures west of the Tigris have revealed thick sections, heavily impregnated by bituminous residues, which indicate one-time saturation by fluid oils which have escaped to surface, or which have been slowly inspissated in depth, presumably by seepage loss, under circumstances which remain to be explained, instance by instance. "Dikes" of hard bitumens, cutting subvertically through Fars sediments, and also recorded from basinal marls of Upper Cretaceous to Oligocene age, present further unsolved problems. The significance of these occurrences of hydrocarbons are not entirely clear. But both the "dikes" and the relict pools may be cited as direct evidence of extensive migration across the bedding.

SYNTHESIS AND DISCUSSION

The known oil fields of Northern Iraq are believed to have drawn at least the bulk of their supplies from only three principal source formations. Each of the three formations which are regarded as productive source rocks comprises basinal sediments, deposited under more or less euxinic conditions, and characterized by radiolarian faunas.

The youngest of the recognized fertile source beds, which are of Middle-Lower Cretaceous age, have been productive of accumulations of medium gravity oils (30°–38° A.P.I.) with about 1–3 per cent sulfur. The oldest recognized source beds, considered to be of pre-Kimmeridgian Upper Jurassic age, are held responsible for heavier oil (10°–20° A.P.I.) with much higher sulfur content (6–10 per cent). The intermediate Tithonian source beds appear to yield very light oils, up to 48° A.P.I., with less than 1 per cent of sulfur, and without any significant asphaltene content.

The occurrence of a light-oil source between two groups of sediments which produce comparatively heavy oils denies any glib generalization relating gravity of oil to depth of burial or to age. In the case of the oils of Middle-Lower Cretaceous origin, the accumulations in the Upper Cretaceous fractured limestone reservoirs, in Ain Zalah and Kirkuk, are lighter than those in the Middle Cretaceous reservoirs, whereas in Kirkuk the "Main Limestone" oil is lighter yet. This upward improvement may relate in part to adsorption of surface-active heavy components of the oil during migration through a tortuous fracture system. In addition the "Main Limestone" oil of Kirkuk may have been diluted by small volumes of light oil of Upper Jurassic (Tithonian) origin, which may have entered the reservoir from the plunging ends of the structure.

Probable major contributions of Upper Cretaceous Flysch-basinal and globigerinal-basinal sediments to contemporaneous reef-type reservoirs are recognized, and accumulations of indigenous oils within the Flysch sediments are acknowledged. But known accumulations fed from such sources lie in the mountain zone and are now largely dispersed. Original oils were probably heavy, asphalt rich, and sulfurous.

Productivity of the Upper Cretaceous source beds may have been restricted to the Maestrichtian-upper Campanian reef belt, and to the Flysch trough itself. No oil field containing oils of such sources has been located in Northern Iraq, but the Raman and Garzan fields of southeastern Turkey are in a setting somewhat comparable with that of the Berat Dagh impregnations.

Oligocene, Eocene, and Paleocene basinal sediments, which are dominantly globigerinal marls, have probably subscribed little or no oil to any present oil accumulation considered. Indigenous Miocene oils may occur in the Lower Fars, but accumulations attributable to such source are negligible in volume.

The foregoing tentative interpretations arise from consideration of the geological and paleogeographical factors, only lightly held in check by reference to the chemical and physical properties of the oils. Detailed chemical investigation of reservoir, seepage, and residual oils, of bitumens and of source sediments is now necessary, in order to verify or deny the deduced source-to-reservoir relationships. Until the data from such investigation are available, the stated interpretations may be regarded as preferred hypotheses, and alternative, if less probable accounts remain arguable, though they cannot be pursued in the discussion which follows.

The absence of any significant contribution of oil from the globigerinal sediments of the Tertiary basins may be ascribed to innate inability of such sediments to generate oil, to the absence of conditions permitting liberation of oil generated, or to early loss of oils which were generated and liberated before adequate cover was deposited over the reservoir area. The Kirkuk accumulation appears to lack Paleocene-Oligocene oils, in spite of ideal spatial relationships of basinal and reservoir facies, and in spite of favorable post-Oligocene loading, gradient, and tectonic factors. The hostile incident of Oligocene-Lower Miocene emergence seems inadequate to account entirely for the deduced absence or near-absence of Tertiary oils. Hence, for Kirkuk, the failure of the Tertiary globigerinal sediments to yield oil in quantity suggests that they were incapable of generating oil in quantity.

The Upper Cretaceous, lower Senonian, and Turonian basinal globigerinal deposits also seem to have contributed little or no oil to any known oil field (though the Berat Dagh impregnations may have drawn upon Upper Cretaceous globigerinal sediments, as remarked above). Assessment of these rocks as unproductive is less positive than in the case of their Tertiary facies equivalents. One alternative working hypothesis for the origin of the primary accumulations in the Middle Cretaceous reservoirs accepts a source in the overlying, younger, globigerinal sediments, rather than in the Middle Cretaceous and older radiolarian sediments. This alternative hypothesis is rejected, at present, because it does not account with any perfection for the distribution of oil fields and of dry structures in Northern Iraq, whereas the thesis that origin lay in the Middle-Lower Cretaceous basin explains such distribution.

Stigmatization of the Turonian to Oligocene globigerinal-basinal sediments as unproductive in the area considered should not be taken to infer that these sediments are everywhere negligible as source rocks. The evidence is quite inadequate to sustain any such prejudice. The interpretations put forward as to sources of known accumulations are tentative, and even if they should be confirmed for the accumulations concerned, other source sediments could well have supplied other as yet unknown accumulations in other areas, or even in the same area. The parts of Northern Iraq which remain untested are large, and unexplored areas in adjacent territories are larger. Major accumulations, drawing on globigerinal source rocks of any age from lower Senonian to Miocene, could exist within the region. Nevertheless, it may be significant that despite the presence of thick basinal sediments of lower Senonian-to-Oligocene age in Syria, commercial oil has not been found, though adequate cap-rock seals have been proved in some good structures, where fracture production, at least, might have been expected. In this large and at present unproductive region the basinal Upper Jurassic, Tithonian, and Middle-Lower Cretaceous radiolarian source sediments of Northern Iraq are lacking.

Because of differences in areal distribution of sources, carrier-reservoirs, and caprocks associated with the different basins in Northern Iraq, and because of the different times at which primary migration and accumulation occurred, the differently originating oils are now found to occupy different and to some extent mutually exclusive parts of the region.

The distribution of the Kirkuk-type oils relates primarily to the distribution of the reservoir-carrier facies of the Middle Cretaceous, and to the stratal gradients and loads imposed during stages following immediately upon the deposition of the Lower-Middle Cretaceous source sediments. Where favorable reservoir facies of the Middle Cretaceous is absent, and where favorable gradients did not develop to induce primary migration, oil of this origin is lacking.

The Qaiyarah-type oils are distributed in close relation to the shelf-like margin of the Upper Jurassic, and to the position of erosional and tectonic discontinuities in the Kimmeridgian anhydrite cap rocks, and hence in part to the position of block-bounding faults of late Jurassic or very early Cretaceous date, as well as to buried faults of later origin.

The distribution of the very light oils, which are tentatively attributed to origin in the Tithonian basinal deposits, relates directly to the distribution of the source beds, in which the oils are still largely contained. Distribution within the Tertiary limestone reservoirs relates to the presence or absence of a thick plastic "shield" of Berriasian mudstone and to the nature and thickness of the overlying Cretaceous sediments. No primary accumulations of the light Tithonian oils have yet been encountered, and perhaps none exists within the region.

All the discussed present-day accumulations share a similar early history. In each case expression of the oils from the basinal source rocks has been westward or southwestward, into reservoir-carrier formations of the same general age as the source formations. In each case this expression has accompanied heavy depositional loading of the source sediments, and the development of steepening marginward gradients, following shortly after sedimentation of the source beds. In each case primary accumulation has occurred in porous limestones, developed along the shoreward margins of the basin, and primary concentration has been into structural traps, or into stratigraphic traps of some nature.

The geographical distribution of the known oil fields has thus been controlled dominantly by the paleogeographies of the depositional basins in which the source beds were deposited, and by the gradients and loads which have been imposed upon source beds and more or less contemporaneous reservoir formations. The primary accumulations have come into place as a result of more or less important lateral migration. The one partial exception to these generalizations is provided by the very light oils, of probable Tithonian source, which were denied opportunity for westward expulsion from the source sediments by lack of communication with any porous reservoir formation. In general, oils of this origin must have awaited liberation by fracturing during late-Tertiary folding. If any accumulations of such oils exist, they must overlie the source sediments themselves, in circumstances permitting vertical migration into higher reservoir formations.

The pre-Turonian emergence and erosion influenced the distribution of primary accumulations in the Middle Cretaceous reservoir by limiting the area in which porous reservoir rocks survived, and by causing development of localized secondary porosity. The pre-Neocomian emergence and erosion dictated, to a large extent, the whereabouts of stratigraphic and fault-trap accumulations of pre-folding age, within the Upper Jurassic primary reservoir. Other unconformities have had little or no direct control upon the distribution of existing oil fields.

Vertical migration has played a dominant role in the subsequent development of all known oil field accumulations. Every reservoir either owes its oil content to migration from below, or has been greatly or entirely depleted by escape to higher reservoirs, or to the surface. The dissipation of the one-time oil fields at Berat Dagh, Pir-i-Mugrun, and Pila Spi is also attributable to vertical loss.

The Ain Zalah field illustrates partial depletion of the primary accumulation in the Middle Cretaceous reservoir, and consequent accumulation of oil in the fracture housing of the "First Pay" as a result of vertical migration, possibly at the time or late-Tertiary folding (Fig. 16).

The Butmah field shows further development than does Ain Zalah. The depletion of the Middle Cretaceous limestone reservoir is entire, and all the original oil content is here found in the upper fractured reservoir, below Paleocene-Lower Eocene marl cover (Fig. 16).

In the Kirkuk field (southeastern dome) the process of upward migration has followed the same course as at Ain Zalah, but the initial accumulation was much larger, and the Paleocene-Lower Eocene marl cover was much thinner. Small accumulations remain in the primary Middle Cretaceous reservoir and in the intermediate Upper Cretaceous fractured-limestone reservoir, but the bulk of the oil has ascended into the overlying "Main Limestone." During the Pleistocene erosional cycle, gas seepage to surface commenced, and the "Main Limestone" oil is now in process of losing the greater part of its original endowment of dissolved gases (Fig. 20).

The Bai Hassan field probably originated in the same manner as did Kirkuk, but crestal seepage has not yet commenced and the oil column is gas saturated. Additional gas appears to have entered this structure from below, at some time after the main incursion of the oil of Middle Cretaceous origin. It is suspected that the source of the excess gas lay (or lies) in the Tithonian radiolarian-rich source beds, which underlie the Kirkuk field, but which are there denied upward passage by the thickness and plasticity of the intervening Berriasian mudstone shield. Some gas and light oil from this source may be creeping upward into the northwestern dome, and into the southeastern end of the southeastern dome of the Kirkuk structure.

Jambur probably originated, like Bai Hassan, as a "Main Limestone" accumulation of upward-migrated Middle Cretaceous oil. It has been modified by addition of a different and lighter, gas-rich oil, which has entered either vertically from an underlying source (again, suspectedly, the Tithonian radiolarian sediments), or laterally, across a spillway, from a filled structure lying to the southeast or east, which was itself filled from such source.

The heavy noncommercial oils of the two reservoirs of the four Qaiyarah area fields have a common origin in a deeper accumulation of Upper Jurassic source, housed in Upper Jurassic porous limestone reservoirs. It is argued that oil entered vertically into the Qaiyarah trap, and thence into Najmah and Jawan. The Qasab accumulations may have been fed separately from another part of the same fault zone. In addition to the lateral feeding there may have been vertical leakage from the lower to the upper reservoir, within the limits of the individual domes. The Tertiary reservoir is currently undergoing gas and some oil depletion by vertical seepage on the Qaiyarah, Jawan, and Qasab domes (Fig. 22).

Although this synthesis lays great stress upon the importance of vertical migration in controlling the stratigraphic distribution of present oil accumulations, it is not suggested that oil has ascended from indefinite depths to fill available anticlines wherever these are capped by a sufficiently competent seal. Instead, it seems that the known accumulations do not include any significant contribution from sources older than Upper Jurassic. If oil was generated in pre-Upper Jurassic basins, it has not ascended into known reservoirs in Cretaceous and Tertiary rocks, and therefore it may be sought, in the first instance, in primary accumulations in reservoir formations of similar age to the source sediments, in areas which may be remote from the known oil fields. Middle Jurassic and older oils may be found in pre-Upper Jurassic reservoirs, beneath some of the structures which are productive from Middle Cretaceous or younger reservoirs.

If the interpretations offered are correct, over 95 per cent by volume of the known oil field reserves of the region have come into place by substantial migration across the bedding. This finding is in sharp contrast with conditions holding in the fields of the Basrah area, and of Arabia, where vertical migration has been unimportant and local, influencing the history of perhaps only a fractional percentage of the total oil in place.

REFERENCES

BAKER, N. E. (1953), "Iraq, Qatar, Cyprus, Lebanon, Syria, Israel, Jordan, Trucial Coast, Muscat, Oman, Dhofar and the Hadramaut," *Sci. of Petrol.* Vol. 6, Pt. I, pp. 83–87.

———, AND HENSON, F. R. S. (1952), "Geological Conditions of Oil Occurrence in Middle East Fields," *Bull. Amer. Assoc. Petrol. Geol.,* Vol. 36, No. 10, pp. 1885–1901.

BARBER, C. T. (1948), "Review of Middle East Oil," *Petrol. Times* (June).

BELLEN, R. C. VAN (1956), "The Stratigraphy of the 'Main Limestone' of the Kirkuk, Bai Hassan, and Qarah Chauq Structures in North Iraq," *Jour. Inst. Petrol.,* Vol. 42, No. 393, pp. 233–63.

CUVILLIER, J., AND SACAL, V. (1951), *Correlations Stratigraphique par Microfaciès en Aquitaine Occidentale,* E. J. Brill, Leiden, Netherlands.

DE BOECKH, H., LEES, G. M., AND RICHARDSON, F. D. S. (1929), "Contribution to the Stratigraphy and Tectonics of the Iranian Ranges," *The Structure of Asia,* Methuen and Co., London.

DANIEL, E. J. (1954), "Fractured Reservoirs of Middle East," *Bull. Amer. Assoc. Petrol. Geol.,* Vol. 38, No. 5, pp. 774–815.

HENSON, F. R. S. (1950a), "Cretaceous and Tertiary Reef Formations and Associated Sediments in Middle East," *ibid.,* Vol 34, No. 2, pp. 215–38.

——— (1950b), "The Stratigraphy of the Main Producing Limestone of the Kirkuk Oilfield," *Report 18th Int. Geol. Cong.,* Pt. 6, p. 34, Proceedings, Section E, p. 34 (abstract) (discussion pp. 68–73).

——— (1951a), "Oil Occurrences in Relation to Regional Geology of the Middle East," *Tulsa Geol. Soc. Digest,* Vol. 19, pp. 72–81.

——— (1951b), "Observations on the Geology and Petroleum Occurrences of the Middle East," *Proc. 3d World Petrol, Cong.,* Sect. 1, pp. 118–40.

KENT, P. E., SLINGER, F. C., AND THOMAS, A. N. (1951), "Stratigraphical Exploration Surveys in South-West Persia," *ibid.,* Sect. 1, pp. 141–61.

LEES, G. M. (1934), "The Source Rocks of Persian Oil," *Proc. World Petrol. Congress* 1, 1933, pp. 3–5.

——— (1950a), "The Middle East," *World Geography of Petroleum,* by Wallace E. Pratt and Dorothy Good, Amer. Geog. Soc. Spec. Pub. 31.

——— (1950b), "Some Structural and Stratigraphical Aspects of the Oilfields of the Middle East," *Report 18th Int. Geol. Cong.,* Part 6, pp. 26–33.

——— (1951), "The Oilfields of the Middle East," *Petroleum Times,* July 27, 1951, pp. 651–54, 671.

——— (1952), "Foreland Folding," *Quart. Jour. Geol. Soc. of London,* Vol. 108, Pt. 1, No. 429, pp. 1–34.

——— (1953a), "The Middle East," *Science of Petroleum,* Vol. 6, Pt. 1, pp. 67–72.

——— (1953b), "Persia," *Sci. of Petrol.,* Vol. 6, Pt. 1, pp. 73–82.

———, AND RICHARDSON, F. D. S. (1940), "The Geology of the Oil-field Belt of S. W. Iran and Iraq," *Geol. Mag.,* Vol. 77, No. 3, pp. 227–52.

McCollum, L. F. (1947), in "Mid-East Shows U. S. Way in Unitization Economy," *Oil Forum,* July 1947, p. 174.

O'Brien, C. A. E. (1950), "Tectonic Problems of the Oilfield Belt of South-West Iran," *Report 18th Int. Geol. Cong.,* 1948, Pt. 6, Proceedings, Section E, pp. 45–58.

Spath, L. F. (1950), "A New Tithonian Ammonoid Fauna from Kurdistan, North Iraq," British Mus. (Nat. Hist.) *Geology Bull.,* Vol. 1, No. 4, pp. 96–137.

Tasman, C. E. (1949), "Stratigraphy of Southeastern Turkey," *Bull. Amer. Assoc. Petrol. Geol.,* Vol. 33, No. 1, pp. 22–31.

Ternek, Z. (1953), "Geological Study of Southeastern Region of Lake Van," *Bull. Geol. Soc. Turkey,* Vol. 4, No. 2, pp. 1–32

Tolun, N. (1949), "Notes Géologiques Sur la Région de Silvan Hazru," *ibid.,* Vol. 2, No. 1, pp. 65–89.

Weeks, L. G. (1950), "Discussion of Papers by G. M. Lees, F. R. S. Henson, A. N. Thomas, C. A. E. O'Brien, E. R. Gee, and C. E. Tasman," *Report 18th Int. Geol. Cong.,* Great Britain, 1948, Pt. 6, Proceedings, Section E, pp. 70–73.

——— (1952), "Factors of Sedimentary Basin Development that Control Oil Occurrence," *Bull. Amer. Assoc. Petrol. Geol.,* Vol. 36, No. 11, pp. 2071–2124.

Wellings, F. E. (1954), "Middle East Oil Sources and Reserves" *The Post-War Expansion of the U.K. Petroleum Industry, Report of Inst. of Petrol.,* London, pp. 1–24.

BULLETIN OF THE AMERICAN ASSOCIATION OF PETROLEUM GEOLOGISTS
VOL. 49, NO. 12 (DECEMBER, 1965), PP. 2182-2245, 98 FIGS., 1 TABLE

STRATIGRAPHIC NOMENCLATURE OF IRANIAN OIL CONSORTIUM AGREEMENT AREA[1]

G. A. JAMES[2] AND J. G. WYND[3]
Tehran, Iran

ABSTRACT

The stratigraphy and correlation of Triassic to Plio-Pleistocene sediments within the Iranian Oil Consortium Agreement Area are discussed. Rock-stratigraphic units are named and defined. These are correlated with the Iraq, Kuwait, and Saudi Arabian stratigraphic successions.

The Agreement Area, situated northeast of the Arabian shelf and including part of the Zagros orogenic area, has been the site of more or less continuous sedimentation from Triassic to Plio-Pleistocene time. Regional disconformities occur at the top of the Aptian, the Cenomanian-Turonian, the Cretaceous, and the Eocene. A major angular unconformity produced by Mio-Pliocene folding occurs at the top of the Fars Group. Carbonate and shale deposition controlled by epeirogenic movements dominated until Late Cretaceous time when movements within the Zagros area began to influence sedimentation. Upper Cretaceous, Paleocene, Eocene, and Oligocene deposits are characterized by sharp facies and thickness changes as a result of orogenic movements in the Zagros area. Following deposition of the Oligocene-lower Miocene Asmari Formation the Agreement Area was part of a trough trending northwest-southeast. After initial evaporitic and marine phases, this trough was filled by clastics derived from the rising Zagros Mountains on the northeast. Conglomerates of the Bakhtyari, deposited unconformably upon the Fars Group, mark the end of this basinal filling.

INTRODUCTION

In October, 1954, oil-contract negotiations were initiated between the Iranian Government and National Iranian Oil Company on one hand and a consortium of eight international petroleum companies on the other. An agreement was reached which allocated to the latter the right to carry out oil exploration and production in a designated area of Iran, and to operate the refinery at Abadan. To perform these functions two operating companies were formed, the Iranian Oil Exploration and Producing Company and the Iranian Oil Refining Company.

The area defined for exploration and production (hereafter referred to as the Agreement Area) is a linear belt approximately 870 miles long and 120 miles wide which contains parts of Lurestan, Khuzestan, Fars, and Kerman Provinces. It is located along the southwestern part of Iran and is essentially the same area that formed the Anglo-Iranian Oil Company concession prior to 1951. For ease of discussion the Agreement Area is divided into Lurestan, Khuzestan, and Fars (Fig. 1).

The purpose of this paper is to propose formal names for Triassic to Plio-Pleistocene rock-stratigraphic units now recognized in the Agreement Area. The delimitation of these units has been based on recommendations and definitions proposed by the International Subcommission on Stratigraphic Terminology (1961) and the American Commission on Stratigraphic Nomenclature (1961).[4]

Fortunately, throughout the history of petroleum exploration in Iran rock-stratigraphic units

[1] Manuscript received, May 3, 1965.

The writers thank the directors of the National Iranian Oil Company and the Iranian Oil Exploration and Producing Company for permission to publish this paper.

They also acknowledge the many geologists and paleontologists on whose work this publication is largely based, including those early workers of the Anglo-Persian Oil Company and Anglo-Iranian Oil Company, who, under arduous and trying conditions, established a sound foundation of geological observations. Thanks are due to the staff of the Sedgwick Museum, University of Cambridge, England, who, since 1958, have supplied many of the megafossil determinations noted in this paper. Geologists of the National Iranian Oil Company, the Geological Survey of Iran, member companies of the Consortium, the Arabian American Oil Company, Kuwait Oil Company, and Iraq Petroleum Company are sincerely thanked for their useful criticism of this work.

Special appreciation goes to F. C. P. Slinger, previously head of the Geological and Exploration Division of the Iranian Oil Exploration and Producing Company and now with the British Petroleum Company. It was under his guidance that this paper was initiated.

[2] Senior geologist, review, Iranian Oil Exploration and Producing Company, Tehran, Iran.

[3] Senior paleontologist, Iranian Oil Exploration and Producing Company, Tehran, Iran.

[4] The terms *lower, middle,* and *upper* are used throughout the Middle East, where they have a definite paleontologic and time significance. This usage is not consistent with that of the A.C.S.N. (1961), whose practices are followed in this paper. Therefore, where the terms were employed in strictly a time sense, the words *early* and *late* have been substituted for *lower* and *upper*.

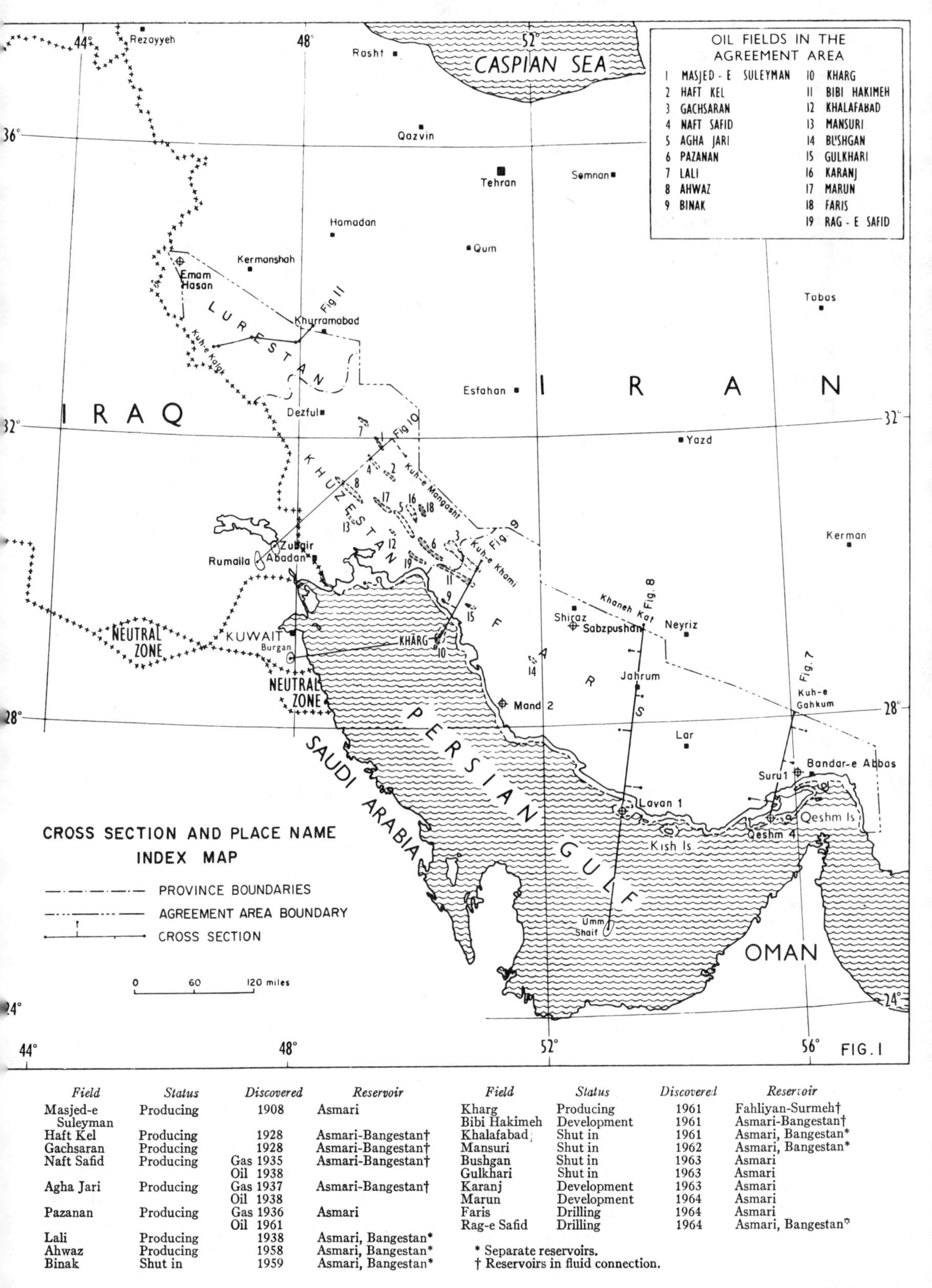

Field	*Status*	*Discovered*	*Reservoir*
Masjed-e Suleyman	Producing	1908	Asmari
Haft Kel	Producing	1928	Asmari-Bangestan†
Gachsaran	Producing	1928	Asmari-Bangestan†
Naft Safid	Producing	Gas 1935 Oil 1938	Asmari-Bangestan†
Agha Jari	Producing	Gas 1937 Oil 1938	Asmari-Bangestan†
Pazanan	Producing	Gas 1936 Oil 1961	Asmari
Lali	Producing	1938	Asmari, Bangestan*
Ahwaz	Producing	1958	Asmari, Bangestan*
Binak	Shut in	1959	Asmari, Bangestan*
Kharg	Producing	1961	Fahliyan-Surmeh†
Bibi Hakimeh	Development	1961	Asmari-Bangestan†
Khalafabad	Shut in	1961	Asmari, Bangestan*
Mansuri	Shut in	1962	Asmari, Bangestan*
Bushgan	Shut in	1963	Asmari
Gulkhari	Shut in	1963	Asmari
Karanj	Development	1963	Asmari
Marun	Development	1964	Asmari
Faris	Drilling	1964	Asmari
Rag-e Safid	Drilling	1964	Asmari, Bangestan*

* Separate reservoirs.
† Reservoirs in fluid connection.

have been employed which have stood the tests of usage for many years. These have been maintained with new names being applied where necessary to conform with recommendations of the International and American stratigraphic nomenclature commissions.

The correlations of formations in the Agreement Area with those of Iraq, Kuwait, and Saudi Arabia are based on publications by Owen and Nasr (1958), Steineke, Bramkamp, and Sander (1958), and the Lexique Stratigraphique International for Iraq (van Bellen *et al.*, 1959).

An index to the rock-stratigraphic units is found in Table I. Figure 3 provides a key to the lithologic symbols as used in the text figures.

For a detailed map, as well as stratigraphic and tectonic sections of the area under discussion, see: The British Petroleum Co. Ltd., 1956, Geological maps and sections: 20th Sess. Internatl. Geol. Cong., Mexico.

Oil Fields

Nineteen oil fields have been discovered in the Agreement Area. Five of these, Binak, Bushgan, Mansuri, Khalafabad, and Gulkhari, are of doubtful economic status.

The prolific Asmari Formation is the most important reservoir, accounting for the major share of production and reserves. Limestones within the Bangestan Group also contain considerable quantities of oil. In fields that have tested the Bangestan, about half have found it to be in fluid connection with the Asmari reservoir (Fig. 1). The only other reservoir from which a commercial amount of oil has been obtained is the Khami Group at the Kharg oil field.

Figure 1 lists the oil fields, their date of discovery, and producing formations.

Khaneh Kat Formation

Synonyms.—This rock-stratigraphic unit formerly has been called the Middle Triassic dolomites, Triassic dolomites, or included within undifferentiated Triassic.

References.—Kent, Slinger, and Thomas, 1951; The British Petroleum Company Ltd., 1956.

Type section.—The type section at Tang-e Daneh Qumbari, 1.5 miles southwest of Khaneh Kat on the Khaneh Kat structure, exposes 1,195 feet of dominantly dark gray, very fine-grained to porcellaneous, siliceous, thin to medium, evenly bedded dolomite (Figs. 12, 13, 14). Minor contemporaneous slump and brecciation features occur in the upper 400 feet. A conspicuous, massive, brown, porous, crystalline dolomite forms the top of the unit.

The base is the junction between the relatively prominent-weathering dolomite, and an unnamed unit of low-lying, yellowish-weathering shale with subordinate thin dolomite beds. The top is the junction between the uppermost massive dolomite of the Khaneh Kat and the low-weathering shale and thin-bedded dolomite of the Neyriz. Both lower and upper contacts apparently are conformable.

Regional aspects.—The formation is exposed in few places except in the more deeply eroded anticlines along the northeastern boundary of the Agreement Area. Toward the southwest a stratigraphically equivalent part of the section, exposed at Kuh-e Surmeh and drilled in the Mand 2 and Lavan 1 wells, is dominantly anhydrite and dolomite. Formal names have not yet been applied beneath the Neyriz in the latter localities.

In southwestern Lurestan-Khuzestan the Khaneh Kat apparently is not distinguishable.

Fossils and age.—At the type section the Khaneh Kat Formation contains a sparse, non-diagnostic microfauna which includes *Agathammina* sp., *Nodosaria* sp., and *Frondicularia* sp. *Aeolisaccus dunningtoni* Elliott occurs in the basal third of the section. Slightly above this level, in the middle part of the section, a few small *Trocholina* sp. have been noted.

Similar *Trocholina*-bearing limestones occur at Kuh-e Surmeh where they overlie gypsum and thin-bedded limestone which have yielded the following Lower Triassic megafossils: *Pseudomonotis aurita* Hauer, *Pseudomonotis ovata* (Schauroth), *Myophoria balatonis* Frech, and *Halobia parthanensis* Schafh.

At Kuh-e Mangasht, in the Bakhtyari Mountain front area of northeastern Khuzestan, the Middle Triassic ammonite *Paraceratites* has been recorded from limestones above the *Pseudomonotis* fauna. These limestones are here considered to be equivalent in age to the lower two-thirds of the type section at Khaneh Kat. The brecciated limestones of the upper third are, at present, considered to be of Late Triassic to Rhaetic in age.

Age of formation.—Early Triassic to Rhaetic.

TABLE I

INDEX OF ROCK-STRATIGRAPHIC UNITS
OUTSIDE THE AGREEMENT AREA

AGREEMENT AREA
ROCK-STRATIGRAPHIC UNIT INDEX

Neyriz Formation

Synonyms.—Previously the Neyriz Formation has been included within undifferentiated Triassic or called the Baluti Shale (the latter was borrowed from Iraq stratigraphic terminology).

References.—Kent, Slinger, and Thomas, 1951; The British Petroleum Company Ltd., 1956.

Type section.—Nine hundred fifty feet of the Neyriz Formation is present at the type section in Tang-e Daneh Qumbari which exposes the core of the Khaneh Kat structure (Figs. 12, 13, 15, 18). The basal third consists of thin-bedded, rubbly dolomite and greenish shale overlain at the top by a prominent, brown-weathering dolomite. Sandy siltstone dominates the middle part. The upper third is argillaceous, thin-bedded limestone to mudstone.

The formation is conspicuously low-weathering between the massive Surmeh above and the underlying Khaneh Kat. The lower contact is sharp but apparently conformable. The upper contact is gradational.

Regional aspects.—The Neyriz Formation is a widespread unit cropping out in a few deeply eroded anticlines in the Agreement Area. In the southwestern part of the Agreement Area a similar lithologic unit incorporating considerable anhydrite was encountered in the wells Mand 2 and Lavan 1.

Kent, Slinger, and Thomas (1951) have correlated the Neyriz with the Baluti Shale of Iraq. The Marrat Formation of Arabia also may be partly equivalent.

Fossils and age.—The dolomitic, shaly, and arenaceous lower two-thirds of the Neyriz Formation are almost devoid of fossils except for a few scattered occurrences of thin-walled ostracods and gastropods. The arenaceous foraminiferid *Orbitopsella praecursor* (Gümbel) occurs in the uppermost argillaceous limestones which are overlain conformably by the dolomitic limestones with *Lithiotis* and *Megalodon* of the lower Surmeh Formation.

Hudson and Chatton (1959, p. 78), in their description of the lower Musandam Limestone of Oman, show Liassic *Orbitopsella*-bearing limestone (group *a*) at the base of the formation. At Wadi Milaha the uppermost limestone of the group contains thick-shelled pelecypods of close affinity with *Megalodon* and *Lithiotis.*

Age of formation.—From regional correlations, the Neyriz Formation is considered to be of Early Jurassic (Liassic) age.

Adaiyah, Mus, Alan, Sargelu, Najmah, and Gotnia Formations

During 1963 and 1964, drilling at Emam Hasan by the National Iranian Oil Company and at Masjed-e Suleyman by the Iranian Oil Exploration and Producing Company (Consortium) penetrated a sequence of Jurassic sediments which have no lithological equivalents in Agreement Area outcrops, or in oil test wells previously drilled in Iran. The sequence was correlated with the Adaiyah, Mus, Sargelu, Najmah, and Gotnia Formations of Iraq which have been described and defined in the Lexique Stratigraphique International for Iraq (van Bellen *et al.*, 1959). For the time being these names are applied to their homotaxial equivalents in the Agreement Area (Fig. 16).

The Adaiyah Formation in Iran consists of approximately 200 feet of anhydrite with subordinate thin dolomite and dark shale. It is underlain by unnamed dark gray shale and limestone, and overlain by the Mus Formation. The latter is composed of about 180 feet of limestone and overlain by about 300 feet of Alan Formation anhydrite. The Sargelu is 500-700 feet of dark gray limestone and dark shale. A disconformity separates the Sargelu from the overlying pellety, algal limestone, about 60 feet thick, which makes up the Najmah Formation. The Najmah is succeeded by 450 feet of Gotnia Formation consisting of anhydrite and subordinate, dark gray shale. A disconformity may be present at the top of the Gotnia.

The relation between the Hith and Gotnia Formations in Iran is not fully understood because

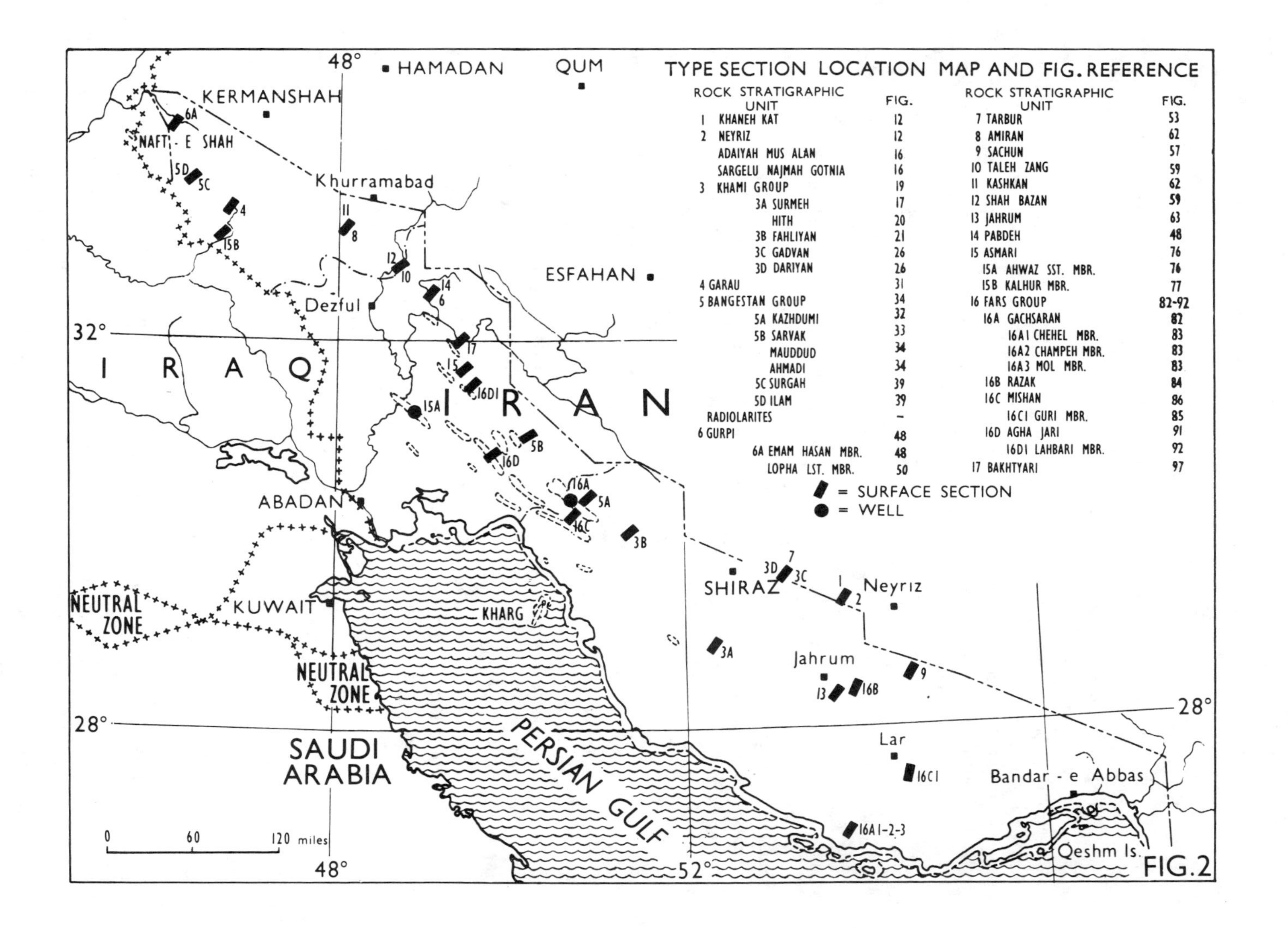
TYPE SECTION LOCATION MAP AND FIG. REFERENCE
ROCK STRATIGRAPHIC UNIT
FIG.
1 KHANEH KAT 12
2 NEYRIZ 12
ADAIYAH MUS ALAN 16
SARGELU NAJMAH GOTNIA 16
3 KHAMI GROUP 19
3A SURMEH 17
HITH 20
3B FAHLIYAN 21
3C GADVAN 26
3D DARIYAN 26
4 GARAU 31
5 BANGESTAN GROUP 34
5A KAZHDUMI 32
5B SARVAK 33
MAUDDUD 34
AHMADI 34
5C SURGAH 39
5D ILAM 39
RADIOLARITES –
6 GURPI 48
6A EMAM HASAN MBR. 48
LOPHA LST. MBR. 50
7 TARBUR 53
8 AMIRAN 62
9 SACHUN 57
10 TALEH ZANG 59
11 KASHKAN 62
12 SHAH BAZAN 59
13 JAHRUM 63
14 PABDEH 48
15 ASMARI 76
15A AHWAZ SST. MBR. 76
15B KALHUR MBR. 77
16 FARS GROUP 82-92
16A GACHSARAN 82
16A1 CHEHEL MBR. 83
16A2 CHAMPEH MBR. 83
16A3 MOL MBR. 83
16B RAZAK 84
16C MISHAN 86
16C1 GURI MBR. 85
16D AGHA JARI 91
16D1 LAHBARI MBR. 92
17 BAKHTYARI 97
= SURFACE SECTION
= WELL
HAMADAN
QUM
ESFAHAN
KERMANSHAH
Khurramabad
Dezful
NAFT - E SHAH
ABADAN
KHARG
KUWAIT
SHIRAZ
Neyriz
Jahrum
Lar
Bandar - e Abbas
Qeshm Is.
I R A Q
I R A N
PERSIAN GULF
SAUDI ARABIA
NEUTRAL ZONE
NEUTRAL ZONE
48°
52°
32°
28°
0 60 120 miles
FIG. 2

of lack of regional information. They are treated as separate formations in this paper because the Gotnia is considered to be a primary, relatively deep-water anhydrite, whereas the Hith appears to be largely an upper tidal flat, or supratidal, penecontemporaneous replacement phenomenon.

Fossils and age.—The Jurassic sequence Gotnia, Najmah, Sargelu, Alan, Mus, and Adaiyah, which was penetrated at Eman Hasan and Masjed-e Suleyman, provided little fossil evidence which could be used for dating these formations. A swarm of the pelecypod *Posidonia* occurs at the top of the Sargelu Formation. This *Posidonia* horizon has been recorded at outcrops in northeastern Lurestan (Sirvan area) and near Kermanshah. Associated ammonites indicate a Bajocian age for this horizon which, at Sirvan, is directly overlain by the Garau Formation containing the Hauterivian ammonites *Spitidiscus* and *Aegocrioceras*. At Sirvan, the Upper Jurassic sediments are either completely absent, or are a condensed sequence in the 200 feet of strata between the *Posidonia* and the *Spitidiscus* horizons. Lees recorded a similar section near Kermanshah (Arkell, 1956, p. 377).

Khami Group

The term Khami Limestone, originally designated by M. W. Strong and N. L. Falcon from outcrops at Kuh-e Khami, just north of Gachsaran, has for many years referred to a massive, feature-forming[5] limestone unit ranging in age from Early Jurassic to Albian. The Khami is here raised to group status and divided into five formations: the Surmeh, Hith, Fahliyan, Gadvan, and Dariyan.

The name Asbu Limestone which has been applied to the Khami Group in Fars Province is discarded (L. E. T. Parker, 1959, company report).

References.—Kent, Slinger, and Thomas, 1951; The British Petroleum Company Ltd., 1956; Slinger and Crichton, 1959.

Surmeh Formation

Synonym.—This unit has been included previously within the Khami Limestone.

Type section.—The type section was measured in the northwestern part of the northern flank of Kuh-e Surmeh. It consists of 2,205 feet of dominantly massive-weathering, feature-forming dolomite and dolomitic limestone (Fig. 17) which have been divided into three lithological units by P. E. Kent (1951, company report).

The basal 300 feet is a massive dolomitic limestone with abundant *Lithiotis* shells. The *Lithiotis* beds are a prominent and important marker. Overlying this is about 250 feet of relatively low-weathering, thin- to medium-bedded, marly to silty, gray to brown, dolomitic limestone, with abundant gastropods, pelecypods, ammonites, and worm casts. The upper 1,655 feet of the formation consists of massive, feature-forming, fine-grained to coarsely crystalline, cherty dolomite and dolomitic limestone.

The basal junction with the underlying shale and marly dolomitic limestone of the Neyriz Formation is gradational. A change from dolomite with chert nodules to oölitic and pellety limestone of the Fahliyan Formation defines the upper boundary.

Regional aspects (Figs. 13, 18).—The Surmeh is present as a feature-forming neritic limestone throughout Fars Province and the northeastern part of Khuzestan and Lurestan. It grades laterally into a dark shale, limestone, and evaporite facies in southwestern Lurestan and Khuzestan (Figs. 19, 20).

In many parts of coastal Fars the Surmeh Formation is overlain by the Hith Anhydrite. Where the Hith is not present, the top of the Surmeh is placed at the junction of the uppermost dark dolomite with the oölitic limestone of the overlying Fahliyan Formation.

Fossils and age.—The standard Jurassic stages of Arkell (1956, p. 10) are difficult to apply to the lower levels of the Surmeh Formation because of the lack of good ammonite markers. The 300 feet of basal dolomitic limestone contains the Liassic pelecypod *Lithiotis*. This useful marker can be traced in the basal Surmeh through most of the high Zagros Mountains area.

The succeeding 250 feet of limestone includes marly beds with rich molluscan faunas. J. A. Douglas identified the Bathonian echinoid *Acrosalenia* aff. *wylliei* Currie from this fauna. The upper part of this marly sequence coincides with the base of the zone of *Pfenderina trochoidea* Smout and Sugden (Fig. 22). The associated mi-

[5] **Feature-forming is used to signify resistant to erosion, or topographically prominent.**

FORMATION CORRELATION CHART

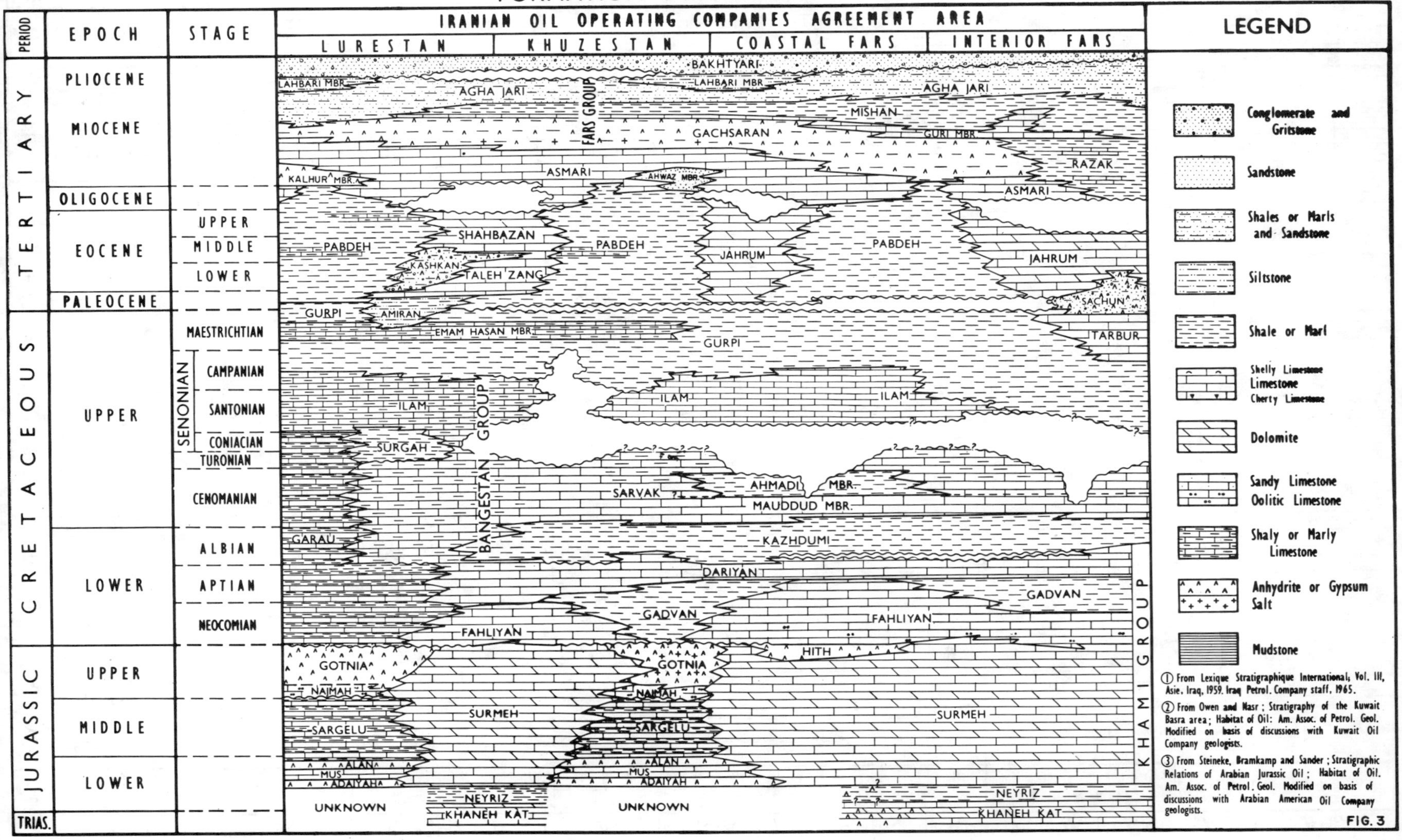

IRANIAN OIL OPERATING COMPANIES AGREEMENT AREA

LURESTAN | KHUZESTAN | COASTAL FARS | INTERIOR FARS

PERIOD | EPOCH | STAGE

TERTIARY: PLIOCENE; MIOCENE; OLIGOCENE; EOCENE (UPPER, MIDDLE, LOWER); PALEOCENE

CRETACEOUS: UPPER (MAESTRICHTIAN; SENONIAN: CAMPANIAN, SANTONIAN, CONIACIAN; TURONIAN; CENOMANIAN); LOWER (ALBIAN; APTIAN; NEOCOMIAN)

JURASSIC: UPPER; MIDDLE; LOWER

TRIAS.

BAKHTYARI
LAHBARI MBR
AGHA JARI
FARS GROUP
MISHAN
GACHSARAN
GURI MBR
RAZAK
ASMARI
KALHUR MBR.
AHWAZ MBR.
PABDEH
SHAHBAZAN
JAHRUM
KASHKAN
TALEH ZANG
SACHUN
GURPI
AMIRAN
EMAM HASAN MBR.
TARBUR
ILAM
BANGESTAN GROUP
SURGAH
SARVAK
AHMADI MBR.
MAUDDUD MBR.
GARAU
KAZHDUMI
DARIYAN
GADVAN
FAHLIYAN
HITH
GOTNIA
NAJMAH
SURMEH
SARGELU
ALAN
MUS
ADAIYAH
KHAMI GROUP
NEYRIZ
KHANEH KAT
UNKNOWN

LEGEND

Conglomerate and Gritstone
Sandstone
Shales or Marls and Sandstone
Siltstone
Shale or Marl
Shelly Limestone
Limestone
Cherty Limestone
Dolomite
Sandy Limestone
Oolitic Limestone
Shaly or Marly Limestone
Anhydrite or Gypsum
Salt
Mudstone

① From Lexique Stratigraphique International, Vol. III, Asie, Iraq, 1959. Iraq Petrol. Company staff, 1965.

② From Owen and Nasr; Stratigraphy of the Kuwait Basra area; Habitat of Oil: Am. Assoc. of Petrol. Geol. Modified on basis of discussions with Kuwait Oil Company geologists.

③ From Steineke, Bramkamp and Sander; Stratigraphic Relations of Arabian Jurassic Oil; Habitat of Oil. Am. Assoc. of Petrol. Geol. Modified on basis of discussions with Arabian American Oil Company geologists.

FIG. 3

FORMATION CORRELATION CHART OF IRAQ, IRANIAN AGREEMENT AREA, KUWAIT AND SAUDI ARABIA

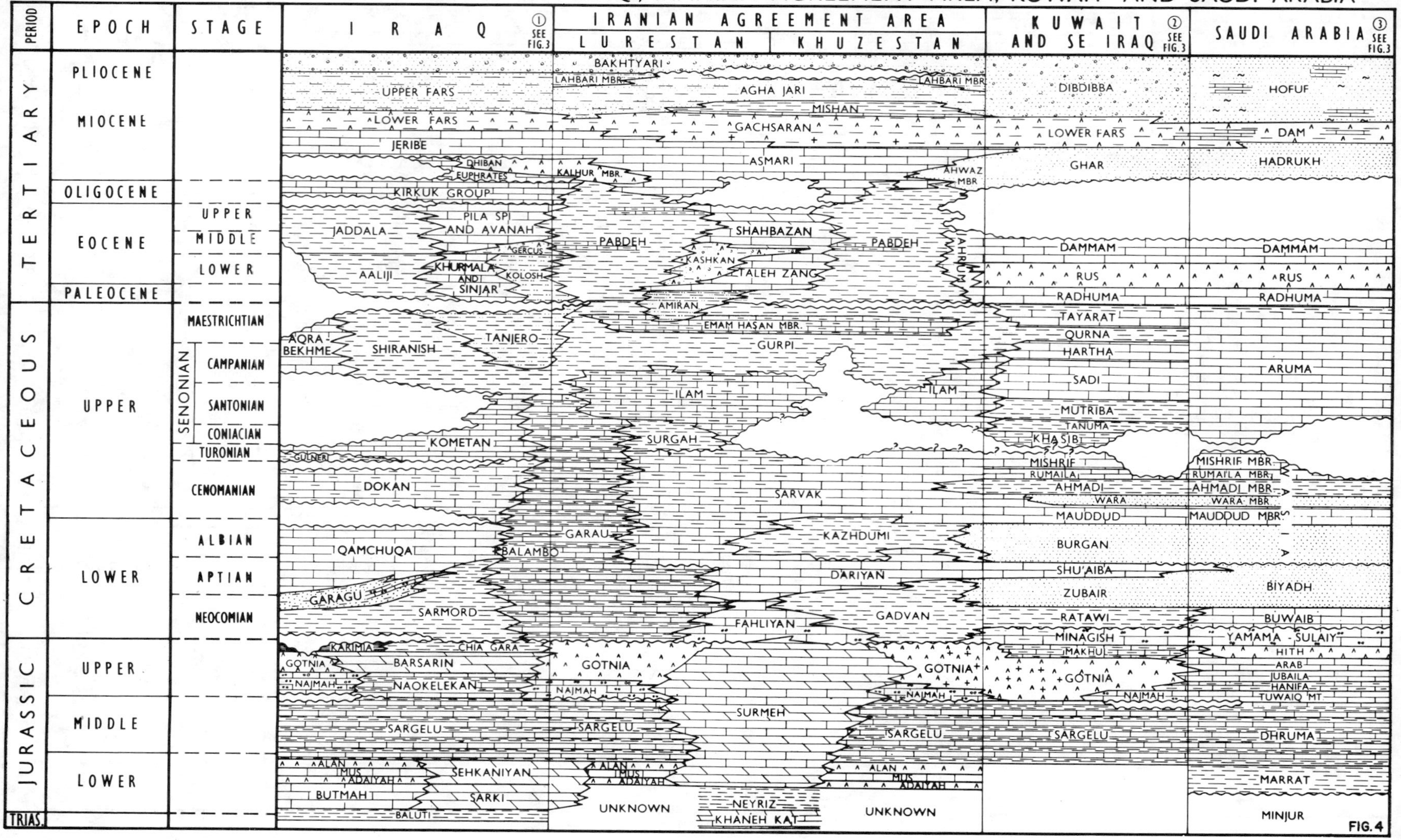

PERIOD
EPOCH
STAGE
IRAQ ① SEE FIG.3
IRANIAN AGREEMENT AREA
LURESTAN
KHUZESTAN
KUWAIT AND SE IRAQ ② SEE FIG.3
SAUDI ARABIA ③ SEE FIG.3

TERTIARY
PLIOCENE
MIOCENE
OLIGOCENE
EOCENE
UPPER
MIDDLE
LOWER
PALEOCENE

CRETACEOUS
UPPER
MAESTRICHTIAN
SENONIAN
CAMPANIAN
SANTONIAN
CONIACIAN
TURONIAN
CENOMANIAN
LOWER
ALBIAN
APTIAN
NEOCOMIAN

JURASSIC
UPPER
MIDDLE
LOWER
TRIAS.

UPPER FARS
LOWER FARS
JERIBE
DHIBAN
EUPHRATES
KIRKUK GROUP
JADDALA
PILA SPI AND AVANAH
GERCUS
AALIJI
KHURMALA AND SINJAR
KOLOSH
AQRA-BEKHME
SHIRANISH
TANJERO
KOMETAN
GULNERI
DOKAN
QAMCHUQA
BALAMBO
GARAGU
SARMORD
KARIMIA
CHIA GARA
GOTNIA
BARSARIN
NAJMAH
NAOKELEKAN
SARGELU
ALAN
MUS
ADAIYAH
SEHKANIYAN
BUTMAH
SARKI
BALUTI

BAKHTYARI
LAHBARI MBR.
AGHA JARI
MISHAN
GACHSARAN
ASMARI
KALHUR MBR.
AHWAZ MBR
PABDEH
SHAHBAZAN
KASHKAN
TALEH ZANG
JAHRUM
AMIRAN
EMAM HASAN MBR.
GURPI
ILAM
SURGAH
SARVAK
GARAU
KAZHDUMI
DARIYAN
FAHLIYAN
GADVAN
GOTNIA
NAJMAH
SURMEH
SARGELU
ALAN
MUS
ADAIYAH
UNKNOWN
NEYRIZ
KHANEH KAT

DIBDIBBA
LOWER FARS
GHAR
DAMMAM
RUS
RADHUMA
TAYARAT
QURNA
HARTHA
SADI
MUTRIBA
TANUMA
KHASIB
MISHRIF
RUMAILA
AHMADI
WARA
MAUDDUD
BURGAN
SHU'AIBA
ZUBAIR
RATAWI
MINAGISH
MAKHUL
GOTNIA
NAJMAH
SARGELU

HOFUF
DAM
HADRUKH
DAMMAM
RUS
RADHUMA
ARUMA
MISHRIF MBR.
RUMAILA MBR.
AHMADI MBR.
WARA MBR.
MAUDDUD MBR.
WASIA
BIYADH
BUWAIB
YAMAMA - SULAIY
HITH
ARAB
JUBAILA
HANIFA
TUWAIQ MT.
DHRUMA
MARRAT
MINJUR

FIG. 4

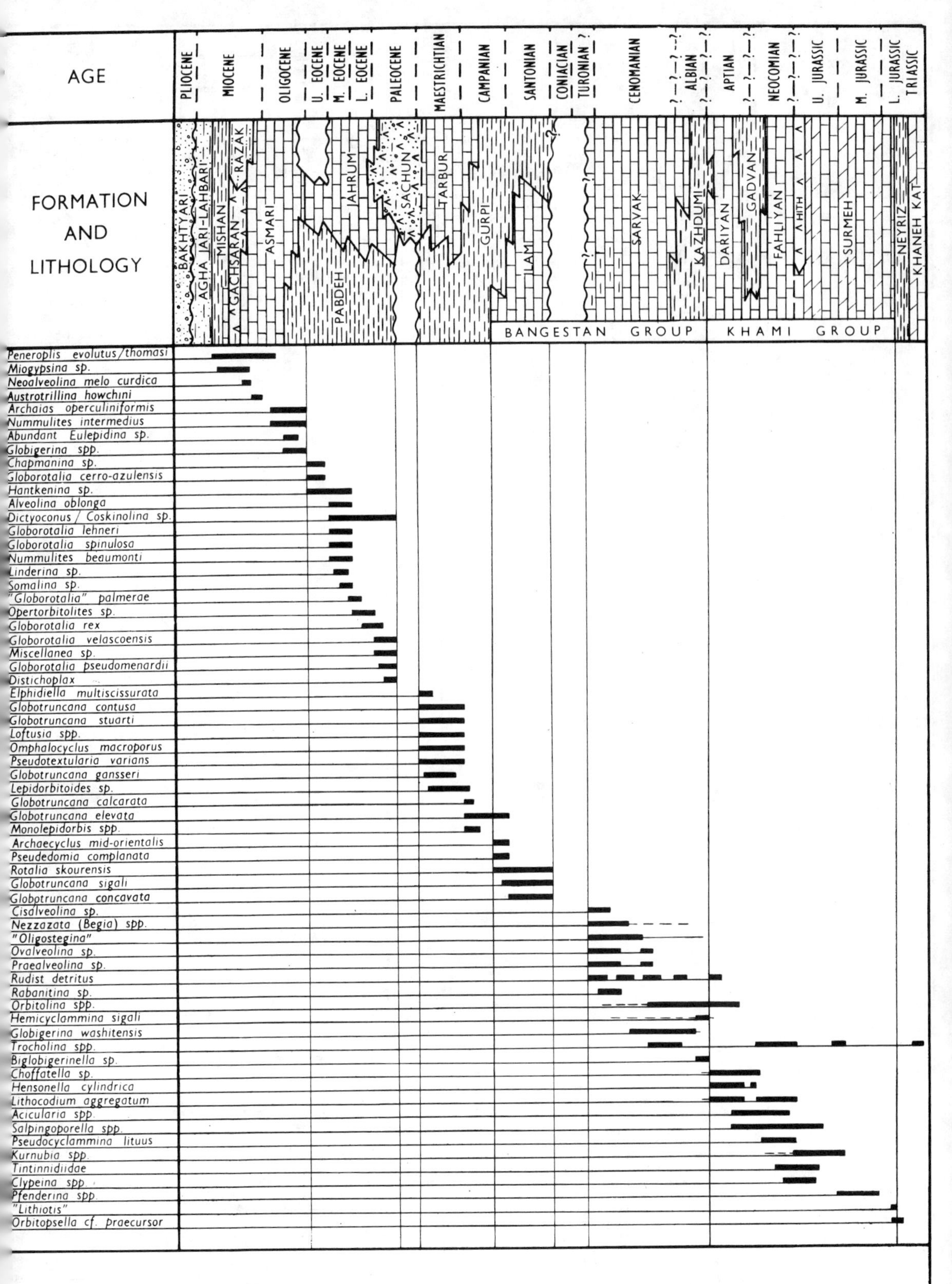

TYPICAL MICROFOSSILS OF KHUZESTAN AND FARS PROVINCES

FIG. 5

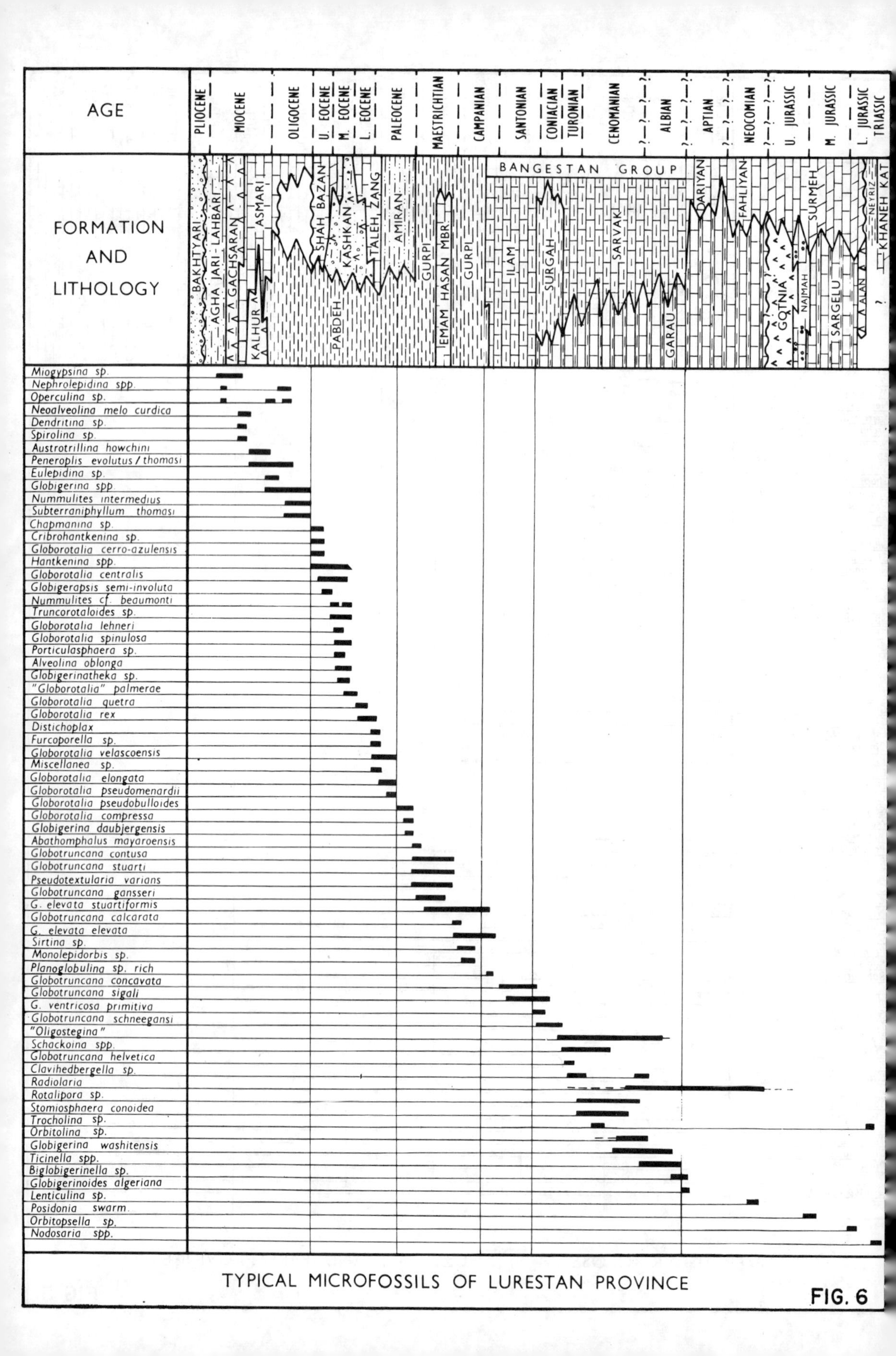

TYPICAL MICROFOSSILS OF LURESTAN PROVINCE

FIG. 6

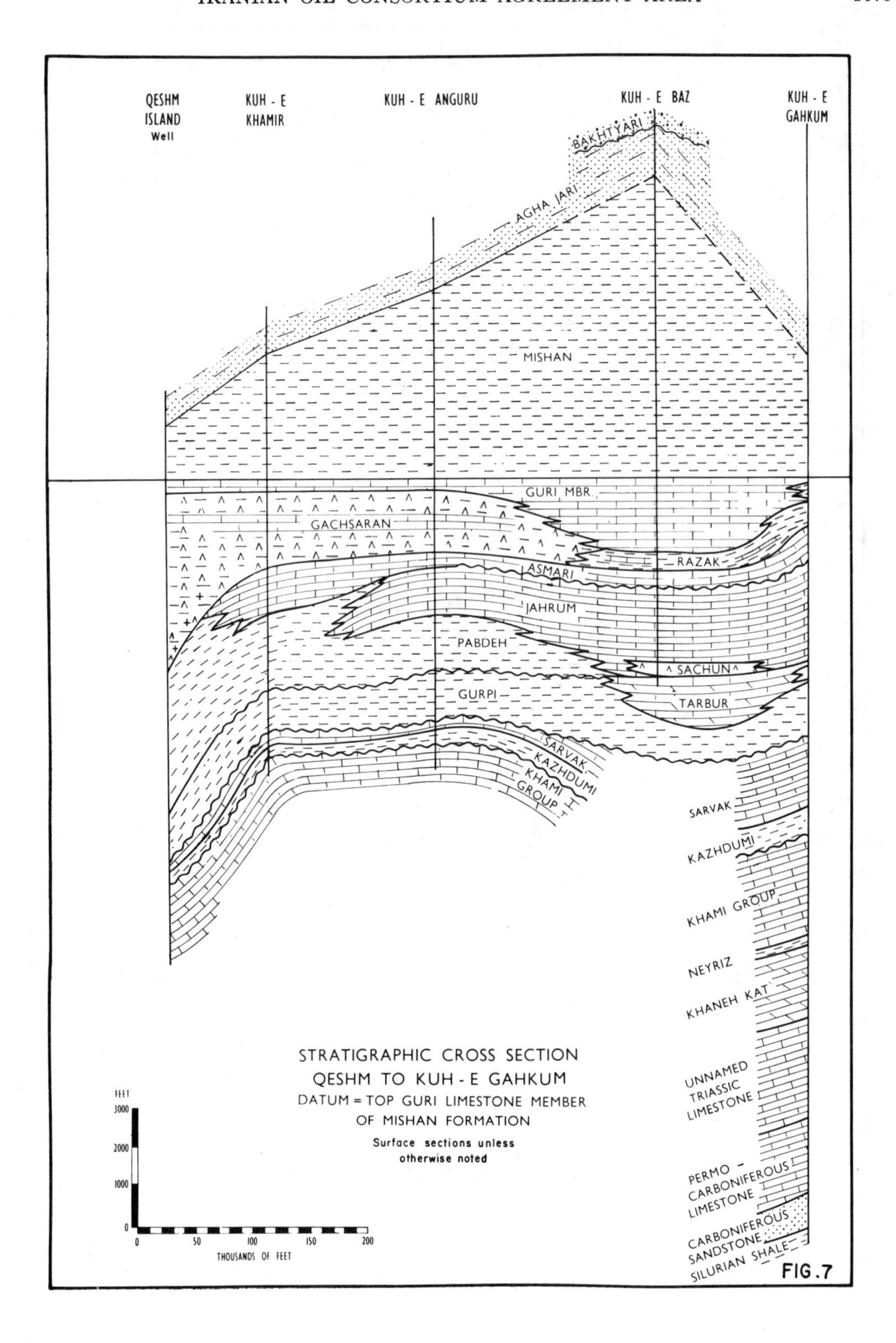

STRATIGRAPHIC CROSS SECTION
QESHM TO KUH - E GAHKUM
DATUM = TOP GURI LIMESTONE MEMBER OF MISHAN FORMATION
Surface sections unless otherwise noted

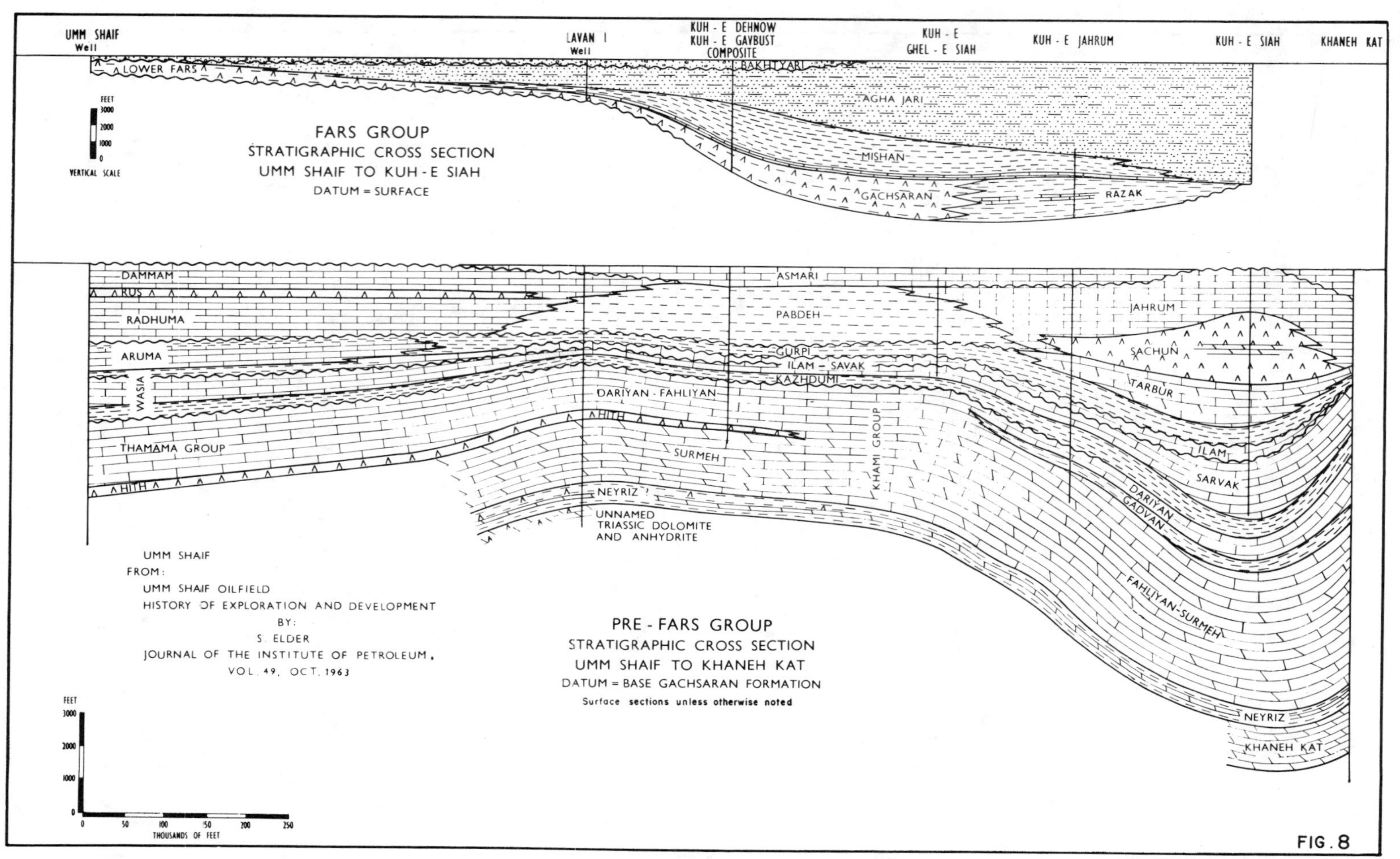
UMM SHAIF Well
LAVAN I Well
KUH - E DEHNOW
KUH - E GAVBUST
COMPOSITE
KUH - E GHEL - E SIAH
KUH - E JAHRUM
KUH - E SIAH
KHANEH KAT
LOWER FARS
BAKHTYARI
AGHA JARI
MISHAN
GACHSARAN
RAZAK
FEET
3000
2000
1000
0
VERTICAL SCALE
FARS GROUP
STRATIGRAPHIC CROSS SECTION
UMM SHAIF TO KUH - E SIAH
DATUM = SURFACE
DAMMAM
RUS
RADHUMA
ARUMA
WASIA
THAMAMA GROUP
HITH
ASMARI
PABDEH
GURPI
ILAM - SAVAK
KAZHDUMI
DARIYAN - FAHLIYAN
HITH
SURMEH
KHAMI GROUP
NEYRIZ ?
UNNAMED
TRIASSIC DOLOMITE
AND ANHYDRITE
JAHRUM
SACHUN
TARBUR
ILAM
SARVAK
DARIYAN
GADVAN
FAHLIYAN-SURMEH
NEYRIZ
KHANEH KAT
UMM SHAIF
FROM:
UMM SHAIF OILFIELD
HISTORY OF EXPLORATION AND DEVELOPMENT
BY:
S. ELDER
JOURNAL OF THE INSTITUTE OF PETROLEUM.
VOL. 49, OCT. 1963
PRE - FARS GROUP
STRATIGRAPHIC CROSS SECTION
UMM SHAIF TO KHANEH KAT
DATUM = BASE GACHSARAN FORMATION
Surface sections unless otherwise noted
FEET
3000
2000
1000
0
0
50
100
150
200
250
THOUSANDS OF FEET

FIG. 8

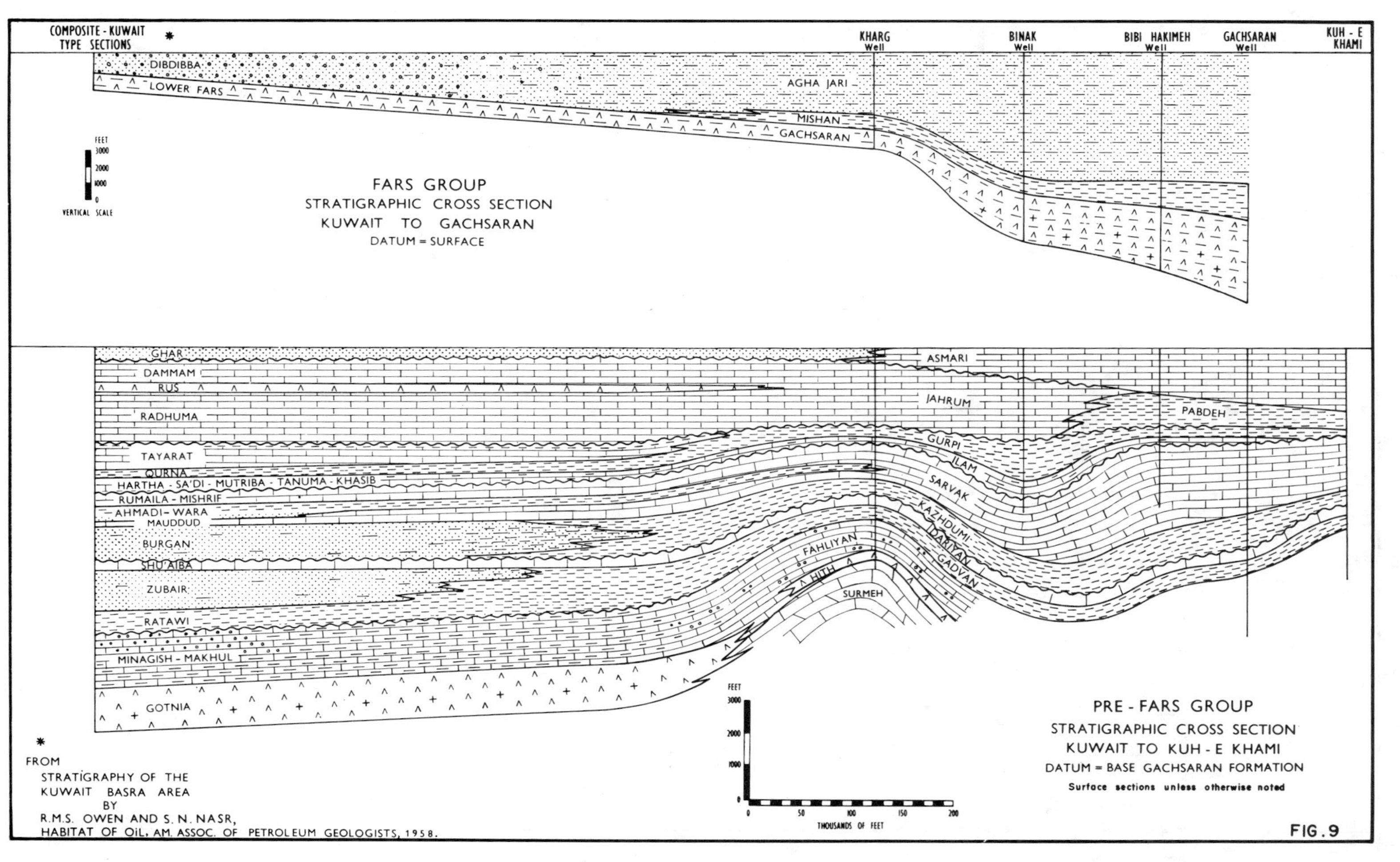
COMPOSITE - KUWAIT TYPE SECTIONS *
KHARG Well
BINAK Well
BIBI HAKIMEH Well
GACHSARAN Well
KUH - E KHAMI
DIBDIBBA
LOWER FARS
AGHA JARI
MISHAN
GACHSARAN
FEET
VERTICAL SCALE
FARS GROUP
STRATIGRAPHIC CROSS SECTION
KUWAIT TO GACHSARAN
DATUM = SURFACE
GHAR
DAMMAM
RUS
RADHUMA
TAYARAT
QURNA
HARTHA - SA'DI - MUTRIBA - TANUMA - KHASIB
RUMAILA - MISHRIF
AHMADI - WARA
MAUDDUD
BURGAN
SHU'AIBA
ZUBAIR
RATAWI
MINAGISH - MAKHUL
GOTNIA
ASMARI
JAHRUM
PABDEH
GURPI
ILAM
SARVAK
KAZHDUMI
DARIYAN
GADVAN
FAHLIYAN
HITH
SURMEH
THOUSANDS OF FEET
PRE - FARS GROUP
STRATIGRAPHIC CROSS SECTION
KUWAIT TO KUH - E KHAMI
DATUM = BASE GACHSARAN FORMATION
Surface sections unless otherwise noted
FIG. 9
* FROM
STRATIGRAPHY OF THE
KUWAIT BASRA AREA
BY
R.M.S. OWEN AND S. N. NASR,
HABITAT OF OIL, AM. ASSOC. OF PETROLEUM GEOLOGISTS, 1958.

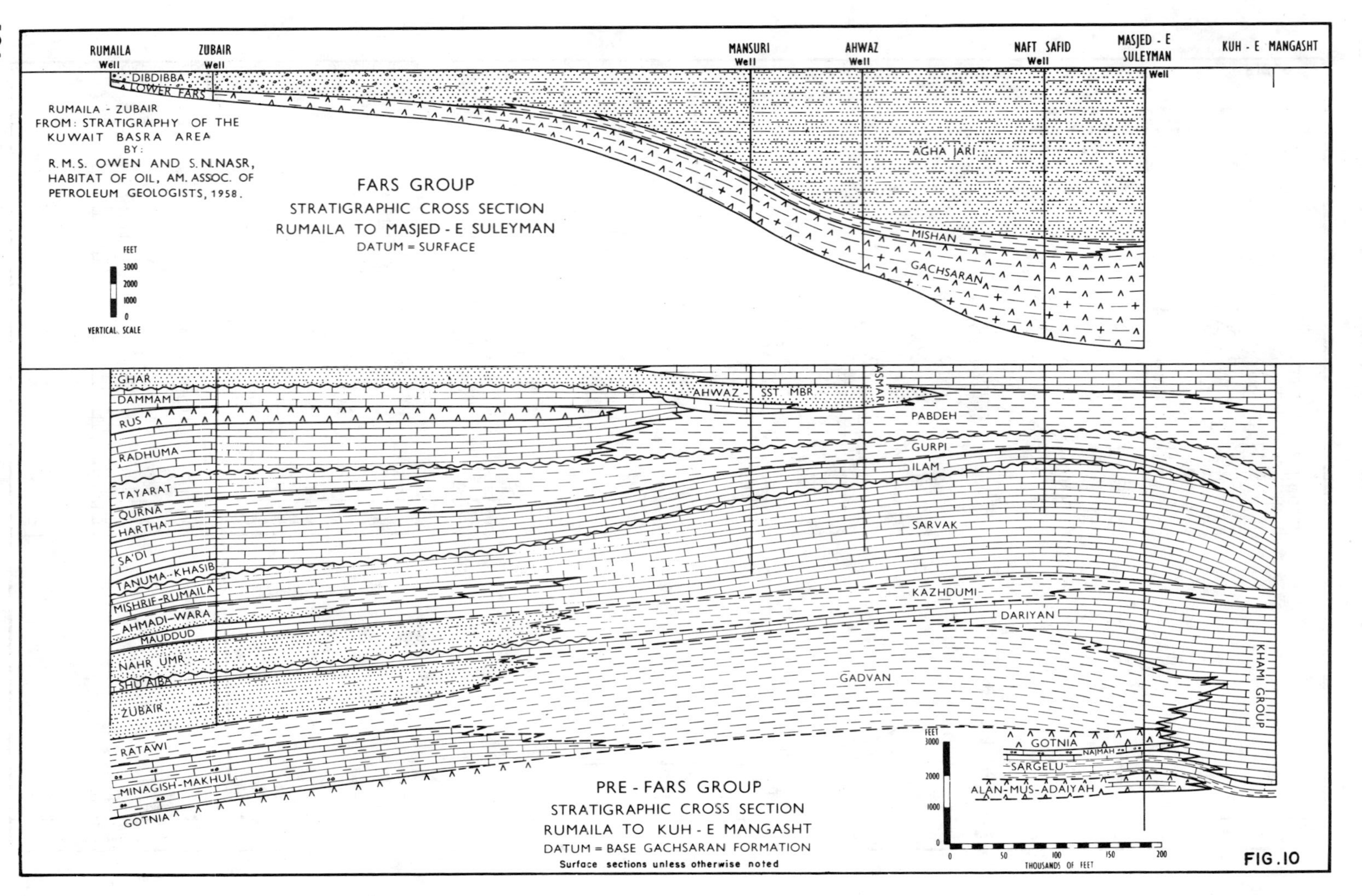

RUMAILA Well
ZUBAIR Well
MANSURI Well
AHWAZ Well
NAFT SAFID Well
MASJED - E SULEYMAN Well
KUH - E MANGASHT
DIBDIBBA
LOWER FARS
AGHA JARI
MISHAN
GACHSARAN
RUMAILA - ZUBAIR
FROM: STRATIGRAPHY OF THE KUWAIT BASRA AREA
BY:
R. M. S. OWEN AND S. N. NASR, HABITAT OF OIL, AM. ASSOC. OF PETROLEUM GEOLOGISTS, 1958.
FARS GROUP
STRATIGRAPHIC CROSS SECTION
RUMAILA TO MASJED - E SULEYMAN
DATUM = SURFACE
FEET
3000
2000
1000
0
VERTICAL SCALE
GHAR
DAMMAM
RUS
RADHUMA
TAYARAT
QURNA
HARTHA
SA'DI
TANUMA-KHASIB
MISHRIF-RUMAILA
AHMADI-WARA
MAUDDUD
NAHR UMR
SHU'AIBA
ZUBAIR
RATAWI
MINAGISH-MAKHUL
GOTNIA
AHWAZ - SST. MBR
ASMARI
PABDEH
GURPI
ILAM
SARVAK
KAZHDUMI
DARIYAN
GADVAN
KHAMI GROUP
GOTNIA
NAJMAH
SARGELU
ALAN-MUS-ADAIYAH
PRE - FARS GROUP
STRATIGRAPHIC CROSS SECTION
RUMAILA TO KUH - E MANGASHT
DATUM = BASE GACHSARAN FORMATION
Surface sections unless otherwise noted
FEET
3000
2000
1000
0
0
50
100
150
200
THOUSANDS OF FEET
FIG. 10

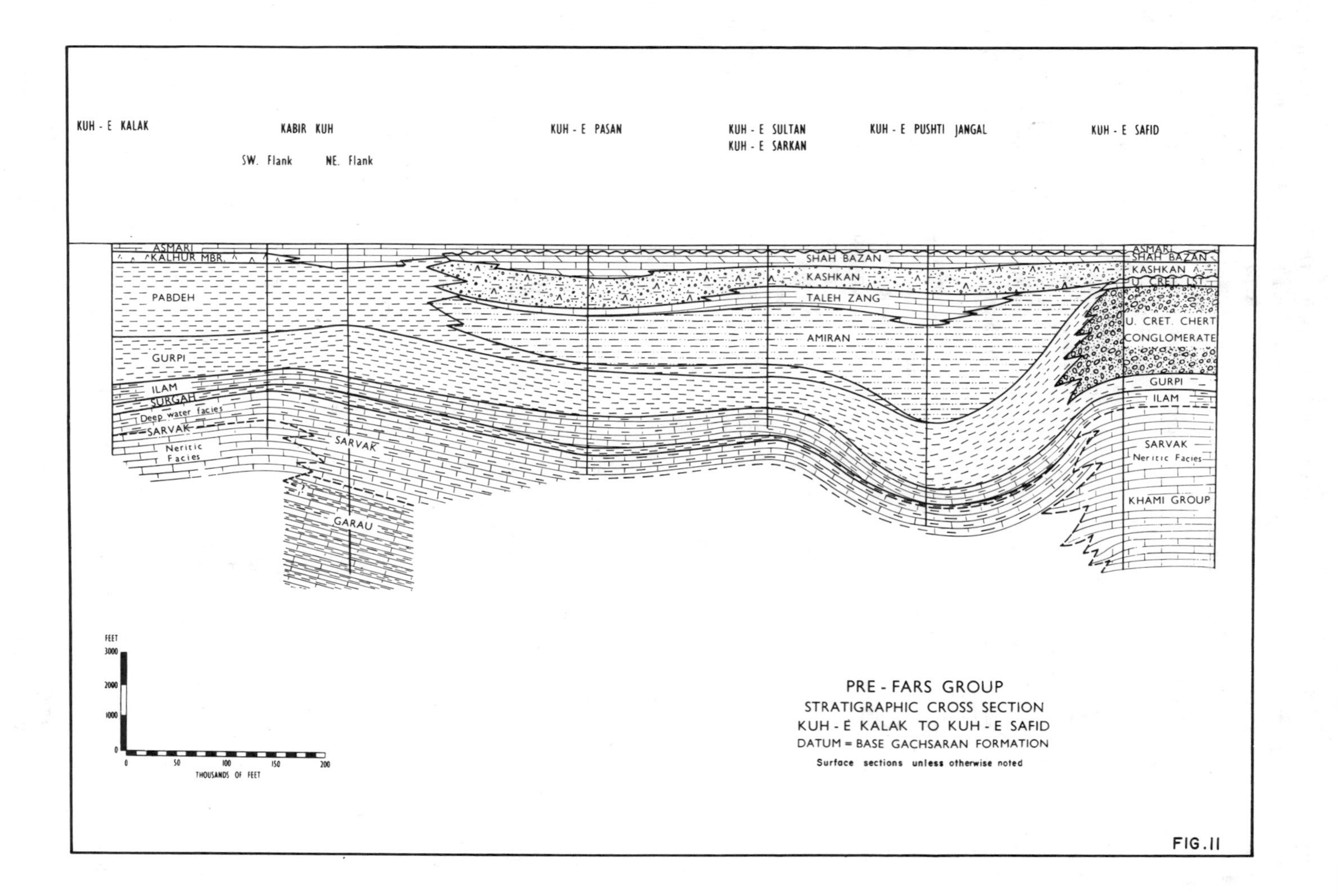
KUH - E KALAK
KABIR KUH
SW. Flank
NE. Flank
KUH - E PASAN
KUH - E SULTAN
KUH - E SARKAN
KUH - E PUSHTI JANGAL
KUH - E SAFID
ASMARI
KALHUR MBR.
PABDEH
GURPI
ILAM
SURGAH
Deep water facies
SARVAK
Neritic Facies
SARVAK
GARAU
SHAH BAZAN
KASHKAN
TALEH ZANG
AMIRAN
ASMARI
SHAH BAZAN
KASHKAN
U. CRET. LST.
U. CRET. CHERT CONGLOMERATE
GURPI
ILAM
SARVAK
Neritic Facies
KHAMI GROUP
FEET
3000
2000
1000
0
0
50
100
150
200
THOUSANDS OF FEET
PRE - FARS GROUP
STRATIGRAPHIC CROSS SECTION
KUH - E KALAK TO KUH - E SAFID
DATUM = BASE GACHSARAN FORMATION
Surface sections unless otherwise noted
FIG. 11

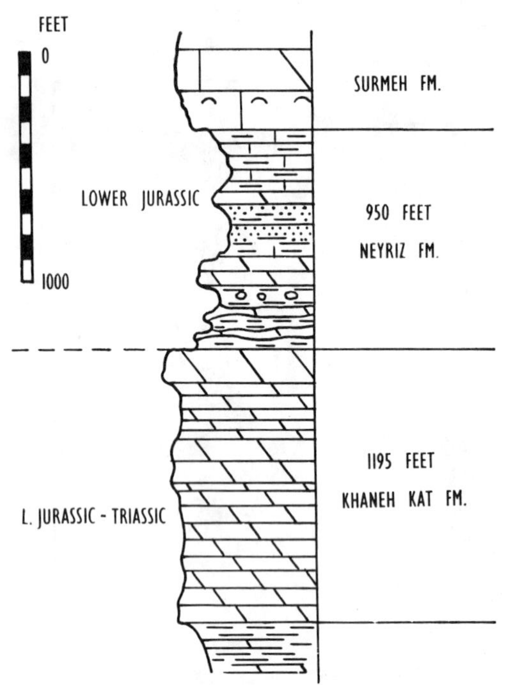

FIG. 12. Type section of the Khaneh Kat and Neyriz Formations. N 29°25′58″; E 53°37′33″

crofauna includes *Haurania* sp. and small *Pseudocyclammina* sp. This zone persists into the basal 100 feet of the massive, feature-forming limestone. A similar fauna occurs in the Fadhili Zone of the upper part of the Dhruma Formation of Arabia, where it is considered to be of Bathonian or Callovian age (Redmond, 1964, p. 257).

The succeeding *Trocholina palastiniensis* Henson zone at Surmeh is approximately 250 feet thick. "*Meyendorffina*" sp. occurs throughout the zone and *Kurnubia jurassica* s.l. (Henson) appears in the uppermost beds. This fauna is comparable with that of the Tuwaiq Mountain and Hanifa Formations of Arabia, although at Surmeh it is very much reduced in vertical extent. The lower part of the Tuwaiq Mountain Formation has been dated as middle Callovian, based on occurrences of the ammonite *Erymnoceras* (Steineke, Bramkamp, and Sander, 1958, p. 1306). Colonial corals, brachiopods, and mollusks are common in this part of the section in both Iran and Arabia. At Surmeh, a megafauna from the middle of the *Trocholina palastiniensis* zone includes *Lopha solitaria* (Sowerby), *Septirhynchia azaisi* (Cotteau), *Ovulastraea* sp., *Actinostroma* sp., *Isastrocoenia* sp., and *Milleporidium* sp. According to J. A. Douglas, this fauna is probably of Oxfordian age.

The 1,200 feet of dolomitic limestone above the *Trocholina palastiniensis* zone at Surmeh can be divided into two zones: a lower, with *Kurnubia jurassica* s.l. (Henson), and an upper, with *Clypeina jurassica* Favre. The ammonites *Torquatisphinctes* and *Virgataxioceras* indicate an early to middle Kimmeridgian age for the *Clypeina* zone. At the type section, the uppermost 300 feet of the Surmeh Formation is completely dolomitized.

In the interior Fars area, the Surmeh above the zone of *Trocholina palastiniensis* consists of cryptocrystalline limestone which may be divided, from the base upward, into a Radiolaria zone, a *Saccocoma* zone, and a *Calpionella* zone. The lower to middle Kimmeridgian ammonite *Torquatisphinctes* has been recorded from the base of the *Saccocoma* zone. The *Calpionella* zone limestones, of late Tithonian age, pass upward into similar limestones of the Neocomian Fahliyan Formation.

Age of formation.—Early to Late Jurassic.

Hith Anhydrite

The Hith, which is an important unit through much of the Persian Gulf area, has been named and defined in Saudi Arabia (Steineke and Bramkamp, 1952). In coastal Fars, the formation, which averages about 150 feet in thickness, overlies the Surmeh Formation and underlies the Fahliyan. Both contacts are considered to be conformable. From coastal Fars, the Hith wedges out toward the northeast, along the line shown in Figure 20.

The formation is believed to be an upper tidal flat, penecontemporaneous replacement anhydrite.

Lack of fossils has prevented any precise age designation of the Hith but it is assumed to be latest Jurassic and to mark the end of a cycle of deposition, although Banner and Wood (1964, p. 191) have suggested a middle Neocomian age for a similar anhydrite development in the Umm Shaif oil field.

Fahliyan Formation

Synonym.—The Fahliyan Formation was previously included within the Khami Limestone.

FIG. 13.—Type section of Khaneh Kat and Neyriz Formations. Kuh-e Khaneh Kat, interior Fars.
FIG. 14.—Even-bedded dolomites of Khaneh Kat Formation. Kuh-e Khaneh Kat, interior Fars.
FIG. 15.—Shales and rubbly, uneven-bedded dolomites of Neyriz Formation. Kuh-e Khaneh Kat, interior Fars.

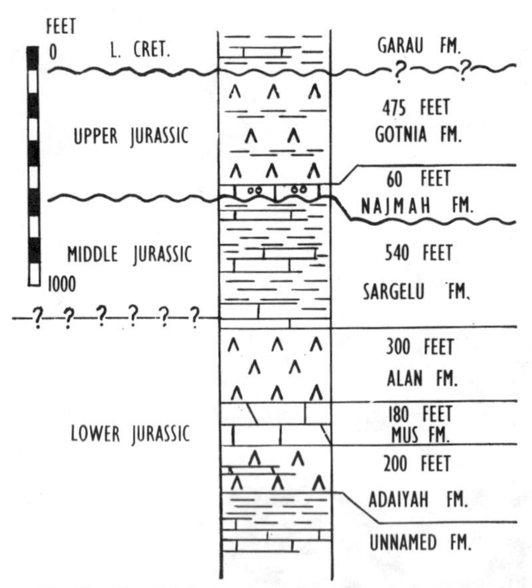

FIG.16. The Adaiyah, Mus, Alan, Sargelu, Najmah and Gotnia succession as encountered in the Emam Hasan and Masjed - e Suleyman Wells.

Type section.—The type section was measured near the village of Fahliyan on the southern flank of Kuh-e Dul. It is composed of 1,200 feet of gray to brown, generally massive, oölitic to pellety limestone with minor contemporaneous brecciation in the basal part (Figs. 21, 24).

Both the basal contact with the saccharoidal,

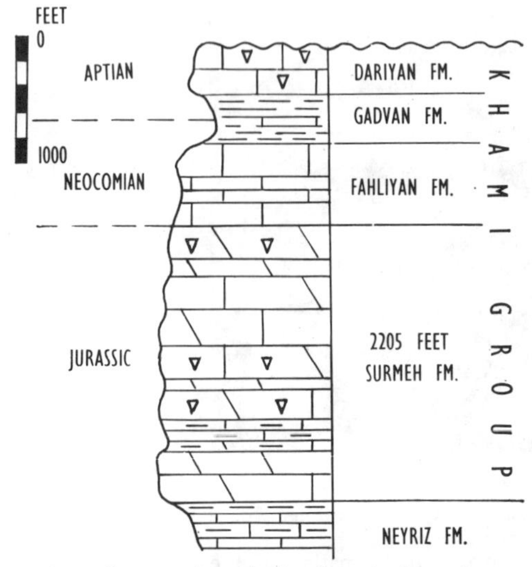

FIG.17. Type section of the Surmeh Formation.
N 28°34′08″ E 52°34′54″

FIG. 18.—Surmeh Formation overlying thin-bedded, lower-weathering Neyriz. Kuh-e Khaneh Kat, interior Fars.

dark brown dolomite of the Surmeh and the upper contact with the marls and thin-bedded limestone of the Gadvan are apparently conformable.

Regional aspects.—The Fahliyan is present throughout Fars Province and in northeastern Khuzestan and Lurestan. Toward southwestern Khuzestan and Lurestan the Fahliyan grades into the dark shale and limestone of the Garau Formation (Figs. 19, 20).

In coastal Fars the Fahliyan is separated from the Surmeh by the Hith Anhydrite. Where the Hith is not present the lower boundary is placed at the junction between the limestone of the Fahliyan and the dark-colored Surmeh dolomite.

Fossils and age.—The microfossils of the Fahliyan Formation include *Trocholina* sp., *Dukhania* sp., *Nautiloculina oölithica* Mohler, and the algae *Lithocodium aggregatum* Elliott, *Salpingoporella annulata* Carozzi, and *Acicularia* sp. *Pseudocyclammina lituus* (Yokoyama) occurs at or near the top of the formation (Fig. 23). In the type area the basal oölitic, pellety limestone contains

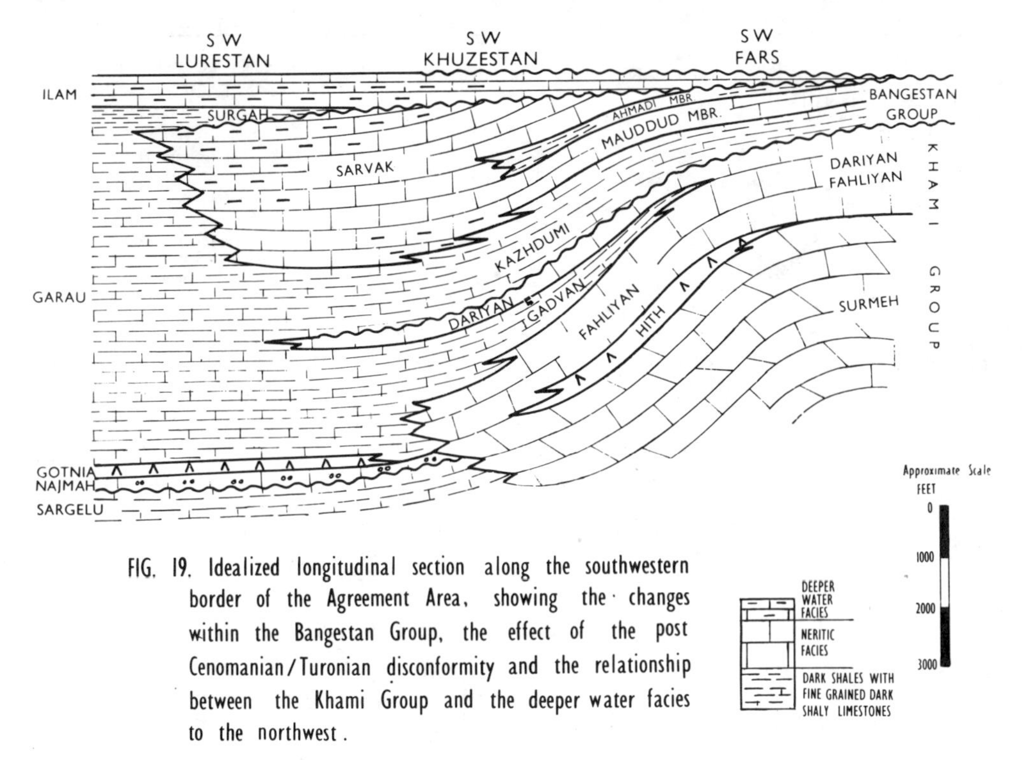

FIG. 19. Idealized longitudinal section along the southwestern border of the Agreement Area, showing the changes within the Bangestan Group, the effect of the post Cenomanian/Turonian disconformity and the relationship between the Khami Group and the deeper water facies to the northwest.

rolled arenaceous Foraminifera including valvulinids and "*Kurnubia*" sp. The ammonites, *Spiticeras indicum* (Uhlig), *Neocomites, Berriasella, Olcostephanus,* and the echinoid *Heteraster* cf. *couloni* Agassiz have been recorded from the Fahliyan of Fars and Khuzestan Provinces.

Age of formation.—Neocomian.

Gadvan Formation

Synonyms.—Previously the Gadvan Formation was included within the Ammonite shales, the Khami Limestone, or called the Aptian-Barremian marl (company reports). In the Gachsaran area it made up part of the unit called the Khami alternations.

References.—Kent, Slinger, and Thomas, 1951; The British Petroleum Company Ltd., 1956; Slinger and Crichton, 1959.

Type section.—The type section, located in the eastern end of Kuh-e Gadvan 26 miles east-northeast of Shiraz, exposes 350 feet of low-weathering, gray to green to brownish yellow marl or shale, and dark gray, argillaceous limestone (Fig. 26). Pelecypods, gastropods, and echinoids are common throughout.

Both the lower boundary with the Fahliyan and the upper contact with the Dariyan are gradational.

Regional aspects (Figs. 25, 27).—The Gadvan Formation is considered to be an extension of the Ratawi and a shaly, partial equivalent of the Zubair (Figs. 4, 9). A deep test at Masjed-e Suleyman indicated that the Minagish and Makhul Formations of Kuwait and southeastern Iraq may also have fingered out into shales of the Gadvan toward the northeast (Fig. 10).

The formation may be divided laterally into two facies. In Khuzestan and northwestern Fars Provinces it consists of dark shale and argillaceous limestone. Toward the southeast it grades into a relatively shallow-water facies, as shown by the abundant macrofauna and oxidized character of some of the sediments. In coastal Fars Province the formation is replaced by limestone (Figs. 8, 20).

Fossils and age.—Large *Choffatella* sp. and coarse pseudocyclamminids, with abundant mollusk and algal debris, are common throughout the formation. Some of the algae are similar to those forms found in the underlying Fahliyan Formation. Recorded megafossils include *Ancyloceras* sp., *Heteraster* cf. *couloni* Agassiz, *Heteraster*

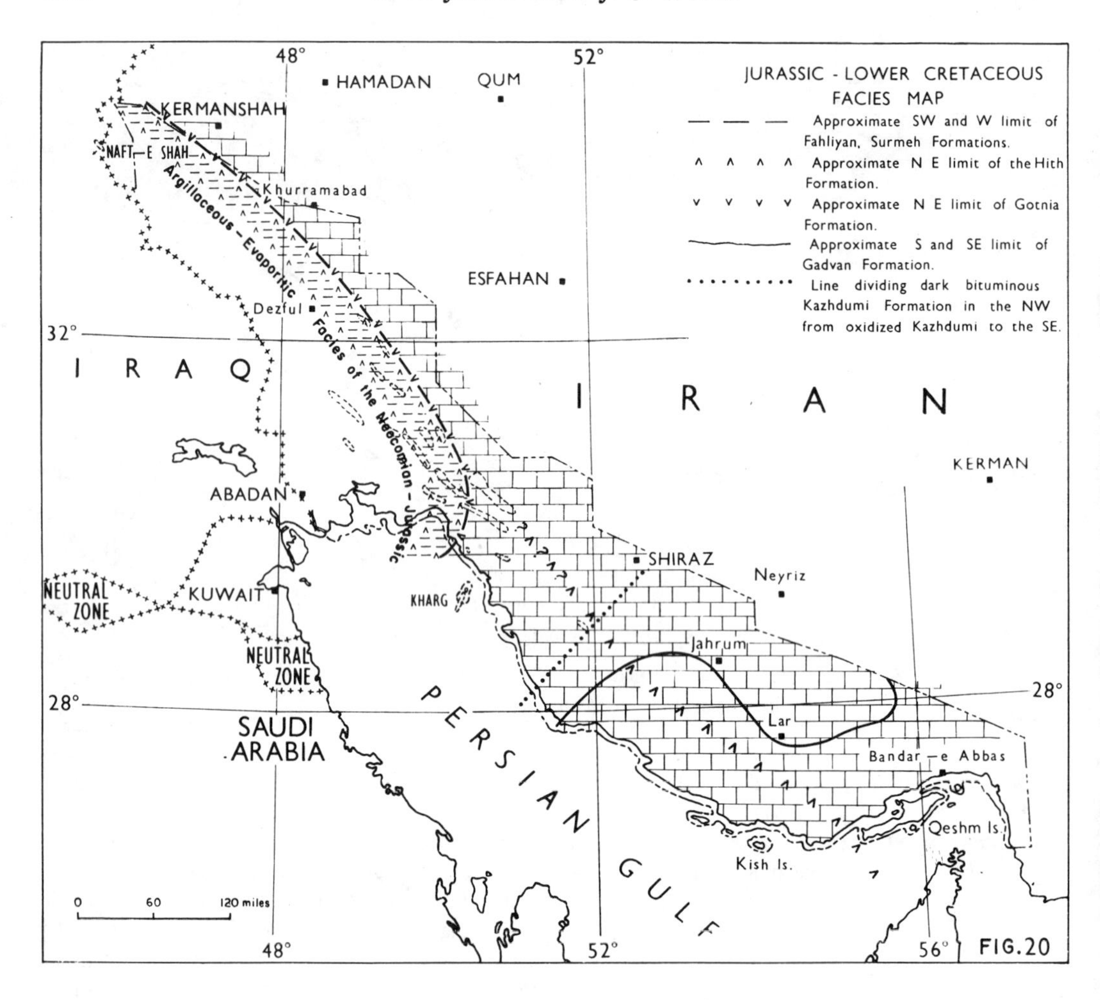

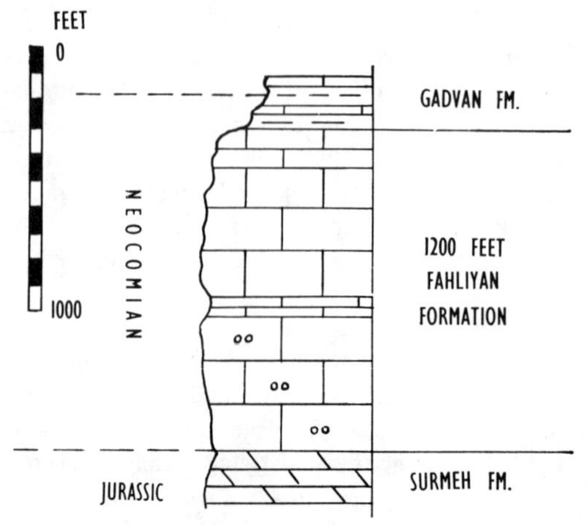

FIG. 21. Type section of the Fahliyan Formation.
N 30°11′19″ E 51°27′36″

FIG. 22.—Detrital limestone with *Pfenderina* sp. Surmeh Formation, Fars Province. Middle Jurassic. ×40.

FIG. 23.—Fine-grained limestone with *Pseudocyclammina lituus* (Yokoyama) (P), and small *Trocholina* sp. (T). Fahliyan Formation, Fars Province. Neocomian. ×50.

FIG. 24.—Type section of Fahliyan Formation. Kuh-e Dul, Fars Province. (Photo—A. J. Wells)

FIG. 25.—Section exposing Fahliyan, Gadvan, Dariyan, Kazhdumi, and Sarvak Formations. Near Fahliyan village, Fars Province. (Photo—A. J. Wells)

T
P
FIG. 23
SARVAK
KAZHDUMI
DARIYAN
GADVAN
FAHLIYAN
FIG. 25
FIG. 22
FAHLIYAN
SURMEH
FIG. 24

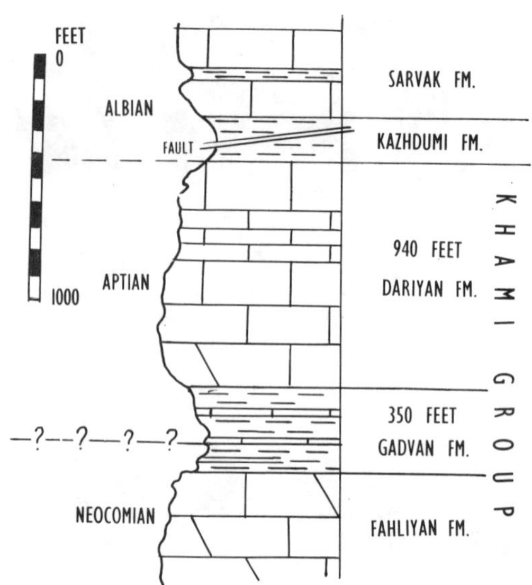

FIG.26. Type section of the Gadvan and Dariyan Formations. N 29°37'51"; E 52°57'12".

verruculatus Fourtau with common *Alectryonia*, and *Spondylus* sp.

Age of formation.—Late Neocomian to Aptian.

Dariyan Formation

Synonyms.—Previously the Dariyan has been called the *Orbitolina* limestone, the Albian-Aptian limestone, or included within the Khami Limestone.

References.—Kent, Slinger, and Thomas, 1951; The British Petroleum Company Ltd., 1956.

Type section.—The type section, measured at Kuh-e Gadvan just north of the village of Dariyan, consists of 940 feet of gray to brown, thick-bedded, feature-forming, *Orbitolina*-rich limestone (Fig. 26).

It gradationally overlies the low-weathering Gadvan Formation and conformably underlies the marl and thin-bedded limestone of the Kazhdumi Formation.

Regional aspects (Figs. 19, 25).—The Dariyan is considered to be an extension of the Shu'aiba of Saudi Arabia, Kuwait, and Iraq (Figs. 4, 9, 10). It is found through most of the Agreement Area, except possibly in southwestern Lurestan and adjoining parts of Khuzestan.

Where typically developed the formation is a prominent 400-1,000-foot-thick unit bounded by two low-weathering formations. Toward the Persian Gulf, however, the underlying Gadvan grades into limestones and the Dariyan merges with the Fahliyan (Figs. 8, 20).

In coastal Fars and Khuzestan there is generally evidence of a disconformity in the beds directly overlying the Dariyan; these beds are sandy, glauconitic, and slightly weathered. Toward interior Fars disconformity disappears, the top of the unit becoming younger in age.

In the northwestern part of the Agreement Area the Dariyan passes into the euxinic Garau Formation (Fig. 19).

Fossils and age.—The Dariyan contains a rich microfauna of conical and discoid *Orbitolina, Choffatella decipiens* Schlumberger, *Dictyoconus arabicus* Henson, *Pseudochrysalidina conica* (Henson), and the algae *Hensonella cylindrica* Elliott, (Fig. 29), *Lithocodium aggregatum* Elliott, and *Salpingoporella* sp. Rudist detritus occurs locally at the top of the formation.

Megafossils identified in the Dariyan include *Parahoplites, Douvilleiceras, Deshayesites, Hinnites, Exogyra,* and *Orthopsis.*

Age of formation.—Aptian.

Aptian-Albian disconformity.—The contact of the Dariyan with the overlying Kazhdumi is characterized by a strongly stained, oölitic, and glauconitic interval. Micropaleontological zonation of the Dariyan indicates that a considerable

---

→

Fig. 27.—Low-weathering limestone and shale of Gadvan Formation overlying massive Fahliyan limestone. Kuh-e Khami, north of Gachsaran.

Fig. 28.—Conical *Orbitolina* sp. with *Hensonella cylindrica* Elliott (H). Dariyan Formation, Fars Province. Aptian. ×35.

Fig. 29.—Fine-grained limestone with *Hensonella cylindrica* Elliott (H). Dariyan Formation, Fars Province. Aptian. ×50.

Fig. 30.—*Orbitolina* limestone from basal Kazhdumi Formation, Fars Province. Albian. ×40.

FAHLIYAN
GADVAN
FIG. 27
H
FIG. 28
H
H
H
FIG. 29
FIG. 30

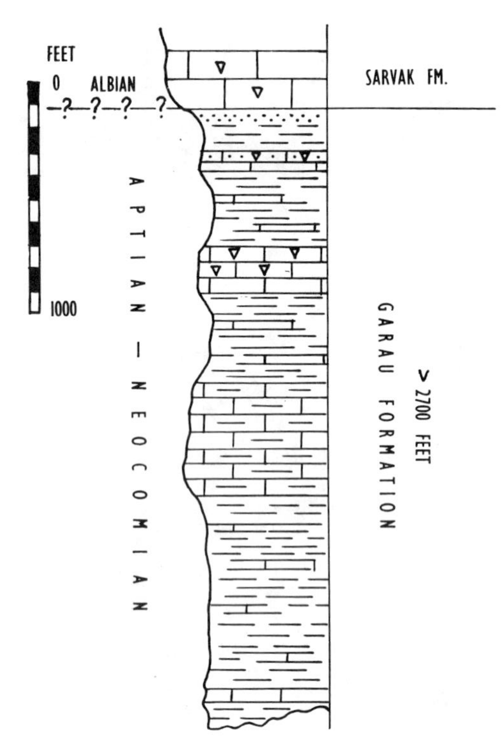

FIG.31. Type section of the Garau Formation. N 33°27'24"; E 46°44'21".

loss of limestone section occurs at the top of the formation from the type area southward toward the coastal Fars area. Several localities in coastal Fars show basal Kazhdumi marl containing middle to late Albian species of *Knemiceras* resting on strongly weathered, *Choffatella*-bearing Dariyan limestone of Aptian age.

Garau Formation

Synonyms.—This unit has not previously been named.

Type section.—The type section is on the northeastern flank of Kabir Kuh in southwestern Lurestan, about 6 miles northeast of the village of Arkwaz (Fig. 31). It may be divided into five units. The bottom 1,000 feet consists of dark gray to black, carbonaceous shale and dark, argillaceous, pyritic limestone, the base of which is not exposed. Overlying this is 500 feet of very fine-grained, dark gray limestone with carbonaceous shale partings. Four hundred feet of gray to brown, low-weathering shale with lesser amounts of thin, dark gray argillaceous limestone follows. Above this is 200 feet of very fine-grained limestone with black chert nodules. The upper 600 feet is made up of alternating gray shale and thin-bedded, fine-grained, shaly limestone. The top is sandy and glauconitic.

The base of the formation is not exposed. The top is the contact of the low-weathering dark shale and thin-bedded limestone of the Garau with the feature-forming limestone of the overlying Sarvak. A sandy, glauconitic zone at this contact suggests the presence of a disconformity.

Regional aspects.—The Garau is partly equivalent to the Balambo Formation of Iraq. In the Agreement Area it is known only from a few localities. The formation underlies the Sarvak in the type section but in the Emam Hasan well the Sarvak has graded laterally into dark shale and limestone of the Garau, extending the top of the formation into the Upper Cretaceous (Figs. 4, 19).

Fossils and age.— Radiolarian faunas are common throughout the Garau Formation, especially in the lower levels. The associated microfauna includes *Nannoconus* sp., *Lenticulina* sp., *Planomalina* sp., and rare textulariidae. *Globigerinelloides algeriana* Cushman and ten Dam occurs in the upper 600 feet of the formation at the type section.

Ammonite faunas have proved the most useful age indicators for the radiolarian-rich basal beds. *Berriasella* sp. has been identified from the base of the formation. From marl 500 feet above the base, the following late Valanginian fauna has been recorded; *Olcostephanus radiatus* Spath, *O. salinarius* Spath, *O. psilostomus* Neumayr and Uhlig, *Neocomites neocomiensis* (d'Orbigny), and *N. similis* Spath. At the type section the Garau is Neocomian to Aptian in age.

The well at Emam Hasan penetrated the Garau facies throughout the Coniacian to Neocomian interval. The microfauna from the Coniacian to Albian part of this interval at Emam Hasan is similar to that of the laterally equivalent Surgah and Sarvak Formations of Lurestan.

Age of formation.—Neocomian to Coniacian.

Bangestan Group

Slinger and Crichton (1959, p. 7) proposed the name Bangestan Limestone for the unit that had been called the mid-Cretaceous limestone, Rudist limestone, Hippuritic limestone, and Lashtagan Limestone. The term Bangestan is here elevated to group status and enlarged to embrace the Kazhdumi Formation as well as the Sarvak, Surgah, and Ilam Formations. The Kazhdumi is included within the group because it was deposited during the same cycle of sedimentation as the Sarvak. Furthermore, because of the minimal development of the Kazhdumi in some areas, it is useful to group it with the Sarvak and Ilam.

The inclusion of the important post-Cenomanian-Turonian disconformity within the Bangestan Group does not conform entirely with the recommendations of the Commissions on Stratigraphic Nomenclature. The Ilam and Sarvak, however, make up a practical rock-stratigraphic unit which has been profitably used for many years. This, in addition to the difficulties often encountered in defining the post-Cenomanian-Turonian disconformity, has prompted the present definition of the Bangestan Group.

References.—Pilgrim, 1904, 1908; Douvillé, 1902, 1910; Vredenberg, 1908; R. K. Richardson, 1924; de Böeckh, Lees, and F. D. S. Richardson, 1929; L. R. Cox, 1936; Lees and F. D. S. Richardson, 1940; G. M. Lees, 1938, 1953; Henson, 1950; Kent, Slinger, and Thomas, 1951; The British Petroleum Company Ltd., 1956; Slinger and Crichton, 1959.

Kazhdumi Formation

Synonyms.—The Ammonite shales and the Abbad Formation.

References.—Kent, Slinger, and Thomas, 1951; The British Petroleum Company Ltd., 1956; Slinger and Crichton, 1959.

Type section.—The type section was measured at Tang-e Gurguda in the mountain front just north of Gachsaran. It consists of 690 feet of dark bituminous shale with subordinate, dark, argillaceous limestone (Fig. 32). Glauconite is common, particularly in the lower 300 feet. The basal 100 feet contains numerous red, oxidized zones.

Contact with the underlying cherty limestone of the Dariyan is associated with red zones indicative of shallowing or a possible diastem. The upper contact grades into the marly, thin-bedded limestone of the basal part of the Sarvak Formation.

Regional aspects (Fig. 25).—The Kazhdumi is the shale equivalent of the Burgan-Nahr Umr of Kuwait and southeastern Iraq (Figs. 4, 9, 10). It is present throughout Khuzestan and Fars Provinces but grades into limestone toward Lurestan (Fig. 3).

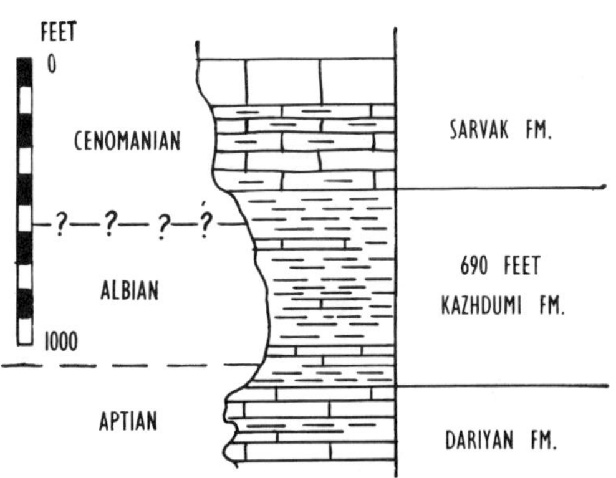

FIG.32. Type section of the Kazhdumi Formation. N 30°22′46″ E 50°54′16″

Like the older Gadvan it may be divided into two lateral facies (Fig. 20): a dark, bituminous shale and limestone facies; and a shale-limestone facies with oxidation and shallow-water features including red zones, laterites, and sandy, silty beds. The former facies is present in Khuzestan and northwesternmost Fars Province. The latter is present in the southeastern half of the Agreement Area where abundant physical evidence of a disconformity exists at the base of the formation.

Fossils and age.—At the type locality the Kazhdumi microfauna is predominantly planktonic. *Globigerina washitensis* Carsey and rich floods of "*Oligostegina*" occur throughout the upper 450 feet of the formation. The basal 240 feet contains a rich fauna of *Hedbergella* sp., *Ticinella* sp., *Biglobigerinella* sp., and *Planomalina* sp. with Radiolaria and spicules. *Hemicyclammina* cf. *sigali* Maync and orbitolines are also found in the basal beds (Fig. 30).

The rich ammonite fauna of the Kazhdumi includes *Knemiceras uhligi* (Choffat), *Knemiceras syriacum* (Buch), *Puzosia denisoni* (Stoliczka), *Spathiceras* sp., *Oxytropidoceras* sp., *Stoliczkaia* sp., and *Parahoplites* sp.

Echinoids identified in the Kazhdumi include *Douvillaster thomasi* Gauthier, *Salenidia boulei* (Lambert), *Codiopsis* sp., *Epiaster* sp., and *Pliotoxaster* sp.

Age of formation.—The Kazhdumi is generally Albian to early Cenomanian in age. A late Aptian age is indicated for the base of the formation by the presence of *Parahoplites* in a few localities

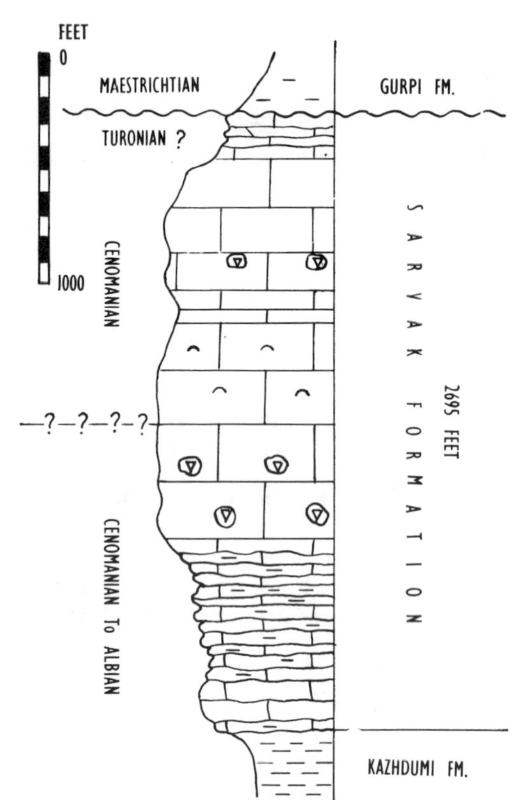

FIG.33. Type section of the Sarvak Formation. N 30°58′29″ ; E 50°07′11″

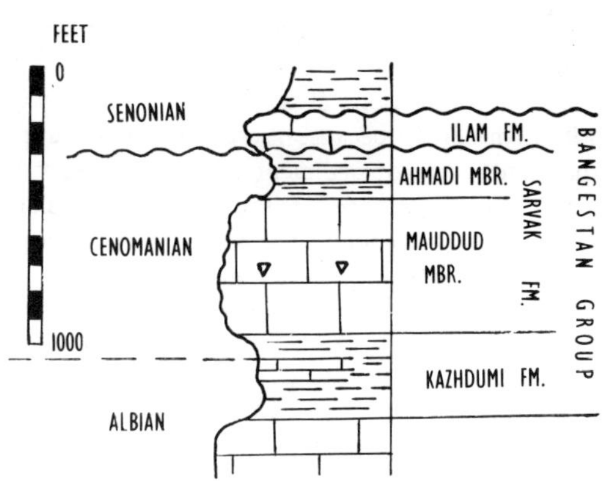

FIG.34. Typical Bangestan Group succession, coastal Fars Province.

where the Kazhdumi passes downward through a series of marl and limestone alternations into the underlying Dariyan.

Sarvak Formation

Synonyms.—Heretofore the Sarvak made up, or was part of, the mid-Cretaceous limestone, Rudist limestone, Hippuritic limestone, Lashtagan Limestone, or Bangestan Limestone.

References.—See Bangestan Group.

Type section.—The type section was measured at Tang-e Sarvak in the central part of the southern flank of Kuh-e Bangestan (Figs. 33, 35, 36). The lower 835 feet is composed of fine-grained, dark gray, nodular-bedded, argillaceous limestone with thin, dark gray marl partings (Fig. 37). Small ammonite impressions are common throughout. This grades upward into a massive, chalky white to buff limestone 360 feet thick with numerous brownish red, siliceous nodules.

Above the siliceous unit is 1,360 feet of very massive, tan to brown limestone containing abundant rudist debris. Large-scale cross-bedding is present in the basal part. The upper 140 feet of the formation is made up of 1-3-foot beds of rubbly, relatively low-weathering limestone with ferruginous red staining and leached zones of breccia. The top is very ferruginous, uneven, and rubbly.

The contact with the underlying Kazhdumi is gradational and conformable. The upper contact with the Gurpi Formation is uneven, and associated with a rubbly weathered zone.

Regional aspects.—The Sarvak Formation is developed in two major facies: one is a massive, feature-forming limestone deposited in the neritic environment, and containing rudists, gastropods, pelecypods, and a rich microfauna; the other is a deeper-water facies of thinner-bedded, fine-grained, dark-colored, argillaceous, "*Oligostegina*" limestones with a pelagic microfauna. The forma-

→

Fig. 35.—Southern flank of Kuh-e Bangestan exposing Sarvak, Ilam, Gurpi, Pabdeh, and Asmari Formations. Near Agha Jari, Khuzestan.

Fig. 36.—Sarvak Formation. Massive upper unit forming vertical cliff. Argillaceous limestone of basal Sarvak forming round shoulder overlying Kazhdumi. Kuh-e Bangestan, near Agha Jari.

Fig. 37.—Sarvak Formation; basal nodular, argillaceous limestone. Type section, Kuh-e Bangestan.

Fig. 38.—*Trocholina-Orbitolina* fauna of Mauddud Member of the Sarvak Formation. Cenomanian. ×35.

FIG. 35

FIG. 36

FIG. 37

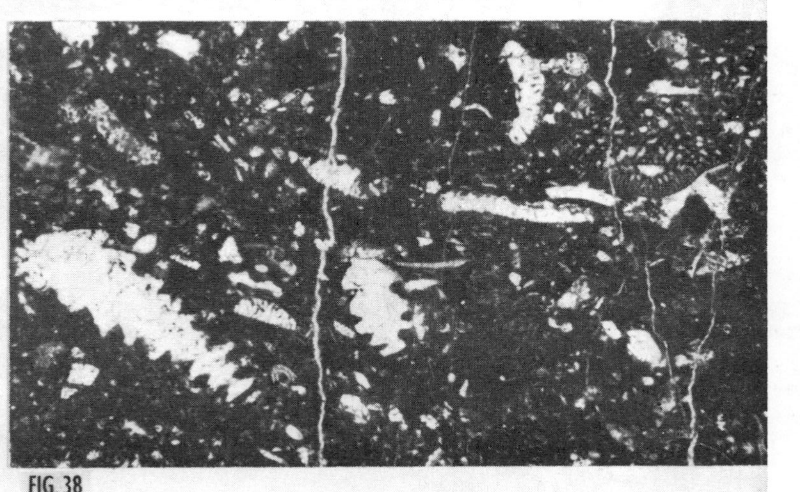

FIG. 38

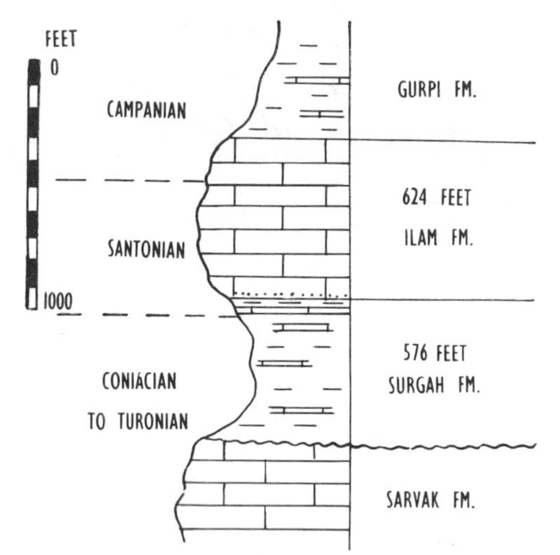

FIG.39. Type section of the Surgah and Ilam Formations. N 33°35′09″ ; E 46°19′06″

tion may be developed exclusively in one or the other of these facies or as an interdigitation of the two (Kent, Slinger, and Thomas, 1951).

The top of the Sarvak is a significant disconformity in Khuzestan and Fars Provinces. This disconformity is less apparent in Lurestan where the formation generally is in the deeper-water facies (Fig. 11). The Sarvak of Lurestan Province extends downward through the Albian because of the disappearance of the Kazhdumi Formation (Fig. 3).

Generally in coastal Fars Province only the lower part of the Sarvak is present, the top being absent because of the post-Cenomanian-Turonian disconformity (Fig. 19). Two members are recognized (Fig. 34): the Mauddud Member and the Ahmadi Member. Both of these are correlated with and named from formations of the same names in Kuwait and Iraq (Owen and Nasr, 1958).

The Mauddud Member consists of prominent-weathering, thick-bedded, gray to brown, *Orbitolina*-rich limestone. The unit ranges throughout coastal Fars from 200 to 400 feet in thickness. It conformably overlies the Kazhdumi Formation.

Low-weathering, gray to green shale and thin-bedded limestone make up the Ahmadi Member. It averages about 100-200 feet in thickness. The lower contact with the Mauddud Member is conformable. The post-Cenomanian-Turonian disconformity separates the Ahmadi Member from the overlying Ilam Formation.

The *Exogyra* marl is a synonym for the Ahmadi Member.

Fossils and age.—At the type section the basal 835 feet of nodular, marly limestone contains an abundant *"Oligostegina"-Globigerina* fauna associated with *Rotalipora* sp. and *Hedbergella* sp. The *"Oligostegina"* fauna (Figs. 5, 6) includes *Stomiosphaera conoidea* Bonet, *S. sphaerica* (Kaufmann), *Calcisphaerula innominata* Bonet, *Pithonella ovalis* (Kaufmann), and *P. trejoi* Bonet. *Ticinella* sp. becomes more common in the lower beds, and *Globigerina washitensis* Carsey occurs in the basal 200 feet of the formation.

The succeeding 360 feet of siliceous limestone contains a poor fauna of *Calcisphaerula innominata* Bonet, *Pithonella ovalis* (Kaufmann), *Globigerina* sp., and spicules. The fauna of the overlying 1,360 feet of massive, rudist-bearing limestones includes: *Praealveolina cretacea* (d'Archiac), *Ovalveolina* sp., *Nezzazata* (*Begia*) sp., *Rabanitina* sp., rare *Orbitolina* sp., *Dictyoconella* sp., and *Dicyclina* sp. In the uppermost 140 feet unit of rubbly limestones this fauna becomes more abundant, with rich occurrences of *Ovalveolina ovum* (d'Orbigny), *Cisalveolina* sp., *Cuneolina* sp., *Dicyclina* sp., *Meandropsina* sp., *Taberina* sp., and associated algal and echinoid debris (Figs. 40, 41).

Age of formation.—At the type section, the Sarvak ranges from Albian to Cenomanian and possibly to Turonian age. It is disconformably

---

→

FIG. 40.—*Praealveolina* limestone from upper Sarvak Formation of Khuzestan. Cenomanian. ×25.
FIG. 41.—Detrital rudist limestone facies of upper Sarvak Formation. Fars Province. Cenomanian. ×25.
FIG. 42.—Typical *"Oligostegina"* limestone from upper Sarvak Formation. Fars Province. Cenomanian. ×50.
FIG. 43.—Regular-bedded limestones of Ilam Formation. Type section, Lurestan.

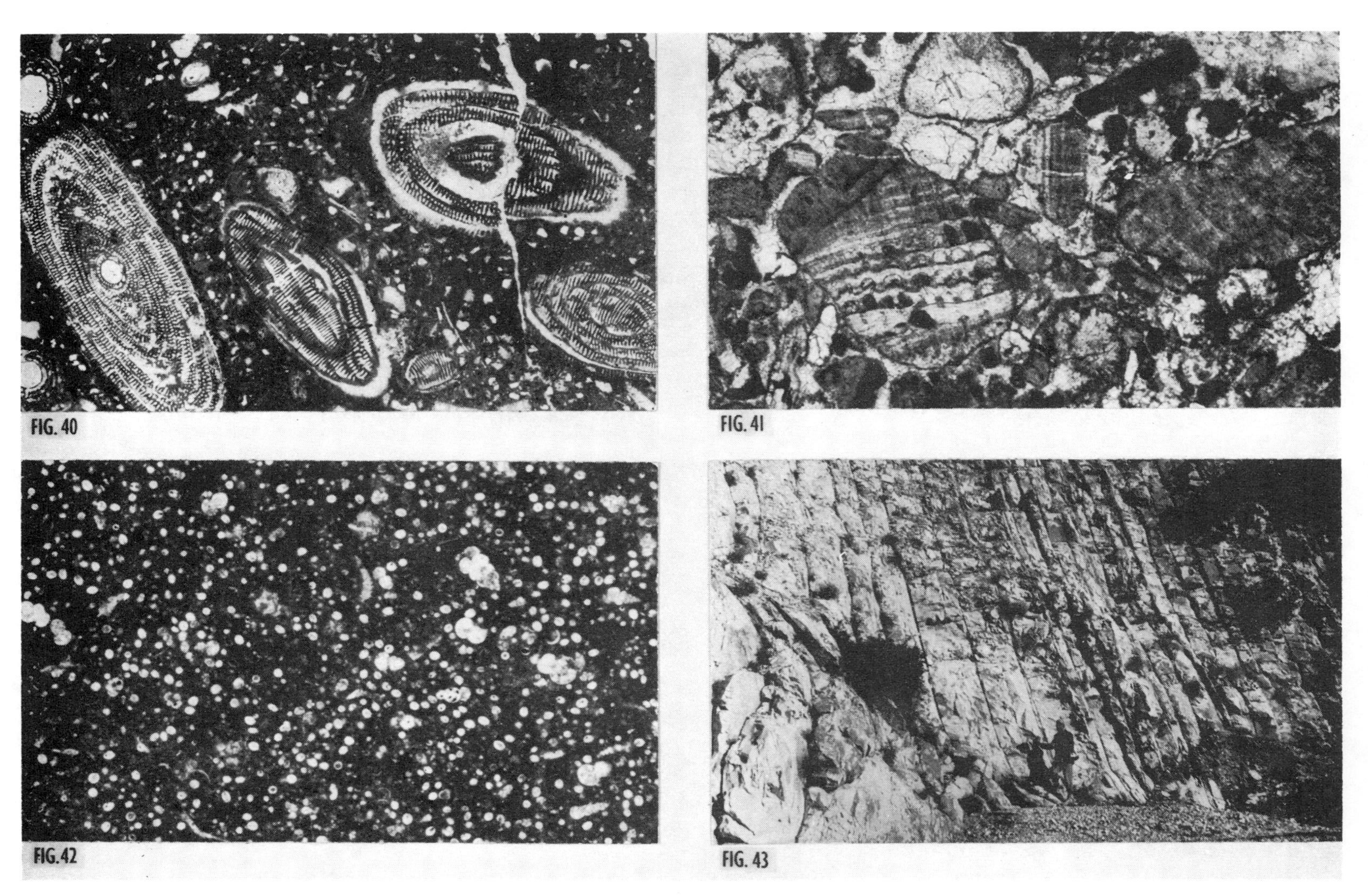

FIG. 40

FIG. 41

FIG. 42

FIG. 43

FIG. 44

FIG.45

FIG.46

FIG. 47

overlain by Maestrichtian marl of the Gurpi Formation.

Regional faunal aspects.—Considerable lateral changes in the microfaunas of the Sarvak take place in Fars Province. The basal Mauddud Member is characterized by a zone of *Trocholina* cf. *lenticularis* Henson and *Orbitolina concava* Lamarck (Fig. 38). This zone is not present in the type area, but can be correlated readily with identical faunas in the Mauddud Formation of Kuwait.

In coastal Fars, the succeeding Ahmadi Member contains oyster-bearing marly intervals, previously called the "*Exogyra* marls." The rich Cenomanian megafauna includes *Exogyra conica* (Sowerby), *E. flabellata* Goldfuss, and the echinoids *Orthopsis granularis* Cotteau, *Caenholectypus serialis* (Deshayes), and *Anorthopygus orbicularis* Grateloup. Thin limestone beds occurring at the base of the Ahmadi contain *Praealveolina* sp. and siphonate algae. "*Oligostegina*"-bearing limestone is (Fig. 42) more common in the upper levels of the Ahmadi Member. The associated microfauna includes rare *Hemicyclammina sigali* Maync and other arenaceous Foraminifera, ostracods, rotaliids, and spicules. Near the base of the member are occurrences of *Globigerina washitensis* Carsey.

In most of the coastal Fars area the Cenomanian Ahmadi Member is disconformably overlain by the shallow-water, rotaliid-bearing Ilam Formation of Santonian age. Locally, as at Kuh-e Surmeh, Cenomanian to Turonian? "*Oligostegina*" limestone with rudist banks succeeds the shaly Ahmadi Member (Fig. 3). These "*Oligostegina*" limestones, which form the upper beds of the Sarvak in parts of coastal Fars, contain a microfauna similar to that of the Rumaila Formation of southeastern Iraq (Owen and Nasr, 1958, p. 1270).

In the Khaneh Kat and Jahrum area, the Ahmadi Member and the "*Oligostegina*" limestone of the coastal Fars area pass laterally into a thick zone of alveolinid-*Nezzazata* limestone similar to that of the type Sarvak at Kuh-e Bangestan.

In Lurestan Province, the Sarvak contains an almost entirely planktonic microfauna. On the northeastern flank of Kabir Kuh, near the Iran-Iraq border, the Sarvak ranges from Albian to Turonian in age. The lower half of the formation, measuring 1,400 feet in thickness, contains a poor fauna of small globigerines, textulariids, *Lenticulina* sp., Radiolaria, and spicules. *Globigerina washitensis* Carsey occurs within the upper part of this interval. The Albian ammonites *Douvilleiceras, Mortoniceras,* and *Oxytropidoceras* also have been recorded from this part of the section.

The upper 1,000 feet of limestone at Kabir Kuh is rich in "*Oligostegina*" with associated *Rotalipora* sp., *Hedbergella* sp., and *Schackoina* sp. The Cenomanian contains the main development of the "*Oligostegina*" faunas with associated *Rotalipora appenninica* (Renz), *Stomiosphaera conoidea* Bonet, *S. sphaerica* Bonet, *Calcisphaerula innominata* Bonet, and *Pithonella ovalis* (Kaufmann). Cenomanian ammonites from this level include *Puzosia denisoni* (Stoliczka), *Sharpeiceras laticlavium* (Sharpe), *Acanthoceras* sp., and *Schloenbachia* sp. At the same locality, the Turonian is represented by the uppermost 350 feet of argillaceous limestone with *Globotruncana helvetica* Bolli. These limestones grade laterally into a nodular, shelly, ammonite-bearing limestone from which J. A. Douglas identified the following Turonian fauna: *Metoicoceras* cf. *whitei* Hyatt, *Prionotropis* sp., *Coilopoceras* sp., and *Hoplitoides* sp.

Age of formation.—Albian to Turonian? in Khuzestan and Fars Provinces; Albian to Turonian in Lurestan Province.

Surgah Formation

Synonym.—Previously the Surgah has been included within the mid-Cretaceous limestone.

Type section.—The type section is located at Tang-e Garab, 7.5 miles southwest of Ilam on the southwestern flank of the northwestern end of

---

←

Fig. 44.—Ilam outcrop near type section, northeastern flank of Kabir Kuh, Lurestan.
Fig. 45.—Mould of ammonite *Texanites* sp. Ilam Formation of Lurestan. Santonian.
Fig. 46.—Ilam Formation of Lurestan; typical *Globotruncana* assemblage from lower part of formation. Santonian. ×30.
Fig. 47.—Ilam Formation of Khuzestan and Fars; shallow-water *Rotalia* sp. (R). *Dicyclina* sp. (D) fauna. Santonian-Campanian. ×35.

Kabir Kuh. It exposes 576 feet of gray to dark gray, pyritic, low-weathering shale with subordinate yellow-weathering, fine-grained, thin-bedded limestone (Fig. 39).

Three feet of limonitic clay overlies the disconformable basal contact with the rubbly, potholed surface of the underlying Sarvak. A minor disconformity, suggested by iron nodules, siltiness, and a small amount of weathering, may exist between the Surgah and the overlying Ilam Formation.

Regional aspects.—The Surgah is well developed only in Lurestan (Fig. 19). In Khuzestan it is questionably represented in some sections by a thin shale break between the Sarvak and Ilam. It is not present in Fars Province.

Fossils and age.—The Surgah Formation contains a rich planktonic fauna in which *Globotruncana schneegansi* Sigal and *Globotruncana sigali* Reichel are the most common forms. *Globotruncana ventricosa primitiva* Dalbiez occurs in the upper part of the formation. The associated fauna includes *Globotruncana imbricata* Mornod, *G. fornicata* Plummer, *G. angusticarinata* Gandolfi, *Hedbergella* sp., and *Calcisphaerula* sp. No conclusive evidence of a disconformity at the upper contact of the Surgah with the overlying Ilam can be established on paleontological grounds.

Age of formation.—Turonian? to early Santonian.

Ilam Formation

Synonyms.—Previously the formation has been included within the mid-Cretaceous, Rudist, Hippuritic, Lashtagan, or Bangestan limestone units.

References.—See Bangestan Group.

Type section.—At Tang-e Garab on the northwestern end of Kabir Kuh the Ilam consists of 624 feet of gray, regularly bedded, fine-grained, argillaceous limestone with thin, black, fissile shale partings (Fig. 39).

The basal bed is silty with large hematite nodules and minor weathering which may indicate a disconformity between the Ilam and the underlying Surgah. The upper contact with the Gurpi is conformable.

Regional aspects.—Like the Sarvak, the Ilam Formation consists of both shallow-water and deeper-water sediments. In Lurestan the deeper-water facies prevails. The Ilam of Khuzestan and Fars Provinces, however, may exhibit either facies or both (Figs. 46, 47).

In Lurestan the Ilam overlies the Surgah, but toward the southeast it rests directly on the Sarvak (Fig. 19). Throughout the southeastern half of the Agreement Area a weathered zone is at the top of the Ilam.

Though a major disconformity separates the Ilam and Sarvak, it is difficult in some places to differentiate the units in those areas where the Surgah is absent. As the separation is largely based on the disconformity, age determination commonly plays an important part in placing the boundary.

Fossils and age.—The Ilam Formation at the type locality in Lurestan contains a rich planktonic microfauna (Figs. 46, 51) which includes: *Globotruncana concavata* (Brotzen), *G. sigali* Reichel, *G. carinata* Dalbiez, *G. elevata* (Brotzen), *G. conica* White, *G. elevata stuartiformis* Dalbiez, *Planoglobulina* sp., and *Calcisphaerula* sp. The ammonite *Texanites* has been identified in the lower part of the formation (Figs. 45, 51).

In Khuzestan and Fars the Ilam becomes a shallow-water, nodular limestone with a dominantly benthonic fauna. The fauna consists of abundant echinoid and algal debris, *Rotalia* sp., forms very similar to *Rotalia skourensis* Pfender, *Ammobaculites* sp., *Dicyclina* sp., *Valvulammina* sp., small globigerines, miliolids, and ostracods.

In coastal Fars the upper part of the Ilam contains a fauna of *Archaecyclus mid-orientalis* Eames and Smout, *Pseudedomia* cf. *complanata* Eames and Smout, *Rotalia* cf. *skourensis* Pfender, and *Dicyclina* sp. The occurrence of *Globotruncana elevata* (Brotzen) above and below this fauna indicates a Campanian age. A close correlation exists between the microfaunas of the Ilam of coastal Fars, and the Hartha, Sa'di, and Mutriba Formations of Kuwait. Eames and Smout (1955) described a similar benthonic fauna from Campanian limestone of the Umm Gudair field in Kuwait.

Age of formation.—Santonian to Campanian.

Post-Cenomanian-Turonian disconformity.—Through most of Khuzestan and Fars Provinces the top of the Sarvak Formation shows strong physical evidence of a period of general emergence which began in the Cenomanian?-Turonian and persisted locally into the Maestrichtian. This break in deposition also has been observed in

Arabia (Steineke, Bramkamp, and Sander, 1958), Kuwait (Owen and Nasr, 1958), Iraq (Dunnington, 1958), and Umm Shaif (Elder, 1963).

The magnitude of this depositional break in Iran is variable (Fig. 3). In the Bandar Abbas area, Santonian marl of the Gurpi Formation overlies the lower Cenomanian *Trocholina-Orbitolina* limestone of the Mauddud Member of the Sarvak. At Kuh-e Bangestan, Maestrichtian Gurpi marl overlies the Cenomanian-Turonian? Sarvak limestone. In most of Fars and Khuzestan, the Sarvak, in either the alveolinid-*Nezzazata* or *"Oligostegina"* facies, is overlain by the Senonian Ilam or Gurpi Formation.

No faunas of known Coniacian age have been recognized in Fars and Khuzestan, but rocks of this age may occur. The Turonian probably is present locally in the topmost levels of the Sarvak where alveolinid limestone is overlain by limestone with a *Dicyclina-Valvulammina* fauna (*e.g.*, at Kuh-e Bangestan). The Turonian ammonite *Prionotropis* has been recorded from Khuzestan, but the associated microfauna is not known.

In parts of Lurestan (*e.g.*, Kabir Kuh) the uppermost beds of the Sarvak, which are of Turonian age, are strongly iron-stained. The presence of this staining suggests the possibility of subaerial exposure before the subsequent deposition of the shales of the Surgah.

Radiolarites

The radiolarites are a thick sequence of red, green to gray chert, shale, and siliceous limestone with included masses of basic to ultrabasic igneous rocks. They crop out in a linear belt, which is coincident with a zone of thrusting and tectonic disturbance just northeast of, and parallel with, the northeastern boundary of the Agreement Area.

A Late Triassic to Late Cretaceous age has been assigned to the radiolarites (de Böeckh, Lees, and F. D. S. Richardson, 1929; Lees, 1938, 1953; Gray, 1950; The British Petroleum Company Ltd., 1956). Lately, however, work by company geologists in Fars and Lurestan, just northeast of the Agreement Area, has indicated a Late Cretaceous age for the radiolarites.

Tectonic movements within the area of radiolarite outcrop caused great quantities of radiolarite detritus to be deposited in the Agreement Area during Late Cretaceous, Paleocene, and Eocene time (Amiran and Kashkan Formations).

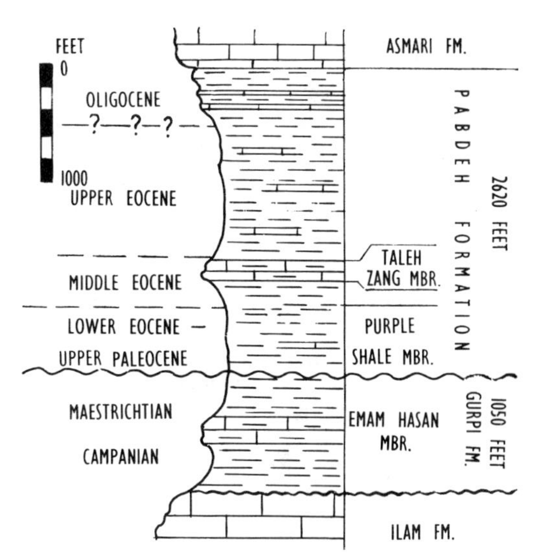

FIG. 48. Type section of the Gurpi and Pabdeh Formations. N 32°27'00"; E 49°14'00"

Chert particles, possibly derived from the radiolarites, are also common in sandstones of the Agha Jari Formation.

A formal name has not been applied to the radiolarites because their area of outcrop is outside of the Agreement Area.

Gurpi Formation

Synonyms.—The unit, previously called the *Globigerina* marl or Dezak Marl, is now divided into the Gurpi and the Pabdeh Formations. The basis for division is a disconformity with associated lithologic criteria; the disconformity generally separates the Upper Cretaceous from the Paleocene.

References.—Cotteau and Gauthier, 1895; Douvillé, 1900, 1902, 1904, 1905, 1907; Gauthier, 1902; Gregory and Currie, 1920; de Böeckh, Lees, and F. D. S. Richardson, 1929; Lees and F. D. S. Richardson, 1940; Rutsch and Schenck, 1940; Kent, Slinger, and Thomas, 1951; Lees, 1953; The British Petroleum Company Ltd., 1956; Falcon, 1958; Slinger and Crichton, 1959.

Type section.—The type section at Tang-e Pabdeh on the southwestern flank of the southeastern plunge of Kuh-e Pabdeh, just north of the Lali oil field, exposes 1,050 feet of low-lying, dark bluish gray, marine marl and shale with subordinate marly limestone (Figs. 48, 69. 71). Its low-weathering profile is broken only by the relatively prominent Emam Hasan Member.

The formation overlies the Ilam with minor disconformity marked by a one-foot ferruginous-weathering zone. The upper contact is placed at the junction of the dark gray shale of the Gurpi with the sandy, silty, purple shale of the lowermost Pabdeh.

Emam Hasan Limestone Member.—In most of Lurestan and Khuzestan a white-weathering, marly limestone, called the Emam Hasan Limestone Member, forms an important unit within the Gurpi Formation. At its type section in Tang-e Emam Hasan, on the southwestern flank of the anticline of the same name, the member consists of 365 feet of beds, 1-2 feet thick, of marly, gray limestone with interbeds of blocky, gray marl and marlstone (Fig. 49).

Lopha limestone member.—This informal member is a prominent mapping unit in Lurestan consisting of shelly limestone and marl. The name is derived from the profusion of *Lopha* shells which characterize the unit (Fig. 50).

Regional aspects.—The Gurpi is present in most of the Agreement Area. It is a transgressive shale which, together with the partly equivalent Ilam Formation, overlies the post-Cenomanian-Turonian disconformity (Fig. 3).

The upper boundary is also disconformable in Fars and parts of Khuzestan, but conformable in Lurestan. It is placed at the base of the purple shale member of the Pabdeh in the northwestern half of the Agreement Area. In the southeastern half, the upper boundary is marked by a thin conglomeratic zone or a prominent glauconite bed, beneath a zone of thin-bedded, cherty, phosphatic limestone.

The formation is best developed in Lurestan and Khuzestan from where it interfingers toward the southwest with the dominantly carbonate sequence of southwestern Iraq, Kuwait, and Saudi Arabia (Figs. 4, 9, 10). Toward the northeast of Lurestan the upper part passes into the clastic Amiran Formation. Limestone of the Tarbur grades into and takes the place of the upper part of the Gurpi in interior Fars (Figs. 3, 7, 8).

Fossils and age.—Over the entire Agreement Area the Gurpi Formation contains abundant planktonic Foraminifera rich in species of *Globotruncana*. The more important species are *Globotruncana concavata* (Brotzen), *G. sigali* Reichel, *G. carinata* Dalbiez, *G. elevata* (Brotzen), *G. conica* White, *G. calcarata* Cushman, *G. stuarti* (de Lapparent), *G. gansseri* Bolli, *G. contusa* (Cushman), and *Abathomphalus mayaroensis* (Bolli).

In Fars and Khuzestan, the basal Gurpi, with *Globotruncana concavata* (Brotzen), is of Santonian age, and overlies the Cenomanian-Turonian? Sarvak Formation disconformably. Where the Senonian Ilam Formation is developed, the overlying basal Gurpi Formation, with *Globotruncana elevata* (Brotzen), is Campanian in age (Fig. 3). *Abathomphalus mayaroensis* (Bolli), indicative of Maestrichtian age, has not been recorded from the top of the Gurpi in Fars and Khuzestan, which suggests that the uppermost Maestrichtian is missing. In this area the upper contact of the Gurpi with the overlying Pabdeh Formation is marked by a highly glauconitic marl bed of up to 2 feet in thickness. This bed indicates a hiatus which represents the topmost zone of the Maestrichtian and the lower Paleocene (see Pabdeh Formation).

In Lurestan the Gurpi ranges from Campanian to Paleocene in age. Near the Iran-Iraq border, the shallow-water, rubbly, shelly limestone known as the *Lopha* limestone member contains an abundant megafauna. This fauna includes *Lopha dichotoma* (Bayle), *Lopha dichotoma* (Bayle) var. *sollieri* (Coquand), *Alectryonia zeilleri* (Bayle), *Pycnodonta vesicularis* (Lamarck), *Sphenodiscus* sp., *Indoceras* sp., *Pachydiscus* sp., and abundant echinoids. *Monolepidorbis* sp., *Sirtina* sp., and *Orbitoides* sp. occur in the associated

→

FIG. 49.—Type section of Emam Hasan Limestone Member of Gurpi Formation, Lurestan.
FIG. 50.—Outcrop of *Lopha* limestone member of Gurpi Formation, Lurestan.
FIG. 51.—Typical plankton assemblage of Ilam Formation of Lurestan and Gurpi Formation throughout Agreement Area. *Globotruncana elevata stuartiformis* Dalbiez (G) fauna. Campanian. ×30.
FIG. 52.—Tarbur-Gurpi Formations, Kuh-e Jahrum, northeastern flank. Interior Fars.

FIG.49

FIG.50

FIG.51

FIG.52

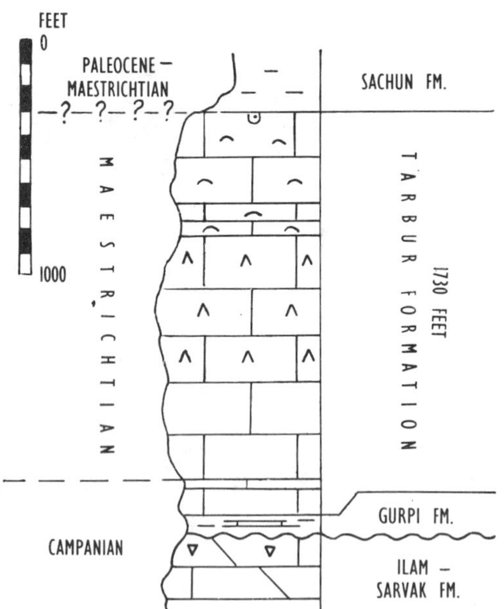

FIG. 53. Type section of the Tarbur Formation. N 29°38'01"; E 52°54'05".

microfauna. This fauna indicates that the *Lopha* limestone occurs at or near the upper Campanian-Maestrichtian boundary.

The Emam Hasan Limestone Member occurs in the uppermost levels of the Gurpi in Lurestan. It contains a rich Maestrichtian fauna which includes *Globotruncana gansseri* Bolli, *G. stuarti* (de Lapparent), and *Pseudotextularia varians* (Rzehak). The 100-200 feet of Gurpi marls directly above the Emam Hasan contains *Abathomphalus mayaroensis* (Bolli). The Cretaceous-Tertiary boundary can be established only by microfaunal evidence at a level within the marls where *Globotruncana* becomes extinct and *Globigerina daubjergensis* Brönnimann appears. The succeeding fauna of the uppermost Gurpi in Lurestan includes the Paleocene species *Globorotalia uncinata* Bolli, *Globigerina triloculinoides* Plummer, *Globorotalia pseudobulloides* (Plummer), and *Globorotalia compressa* (Plummer).

Age of formation.—Santonian to Maestrichtian in Khuzestan and Fars Provinces; Campanian to Paleocene in Lurestan Province.

Tarbur Formation

Synonyms.—Upper Cretaceous limestone, Maestrichtian limestone.

Reference.—The British Petroleum Company Ltd., 1956.

Type section.—The type section was measured at Kuh-e Gadvan, 0.75 mile north of the village of Tarbur. It is composed of 1,730 feet of massive, shelly, anhydritic limestone (Fig. 53).

The basal contact with the Gurpi Formation is conformable. The upper contact with the gray to green, ferruginous shale of the Sachun Formation is associated with ironstone nodules which may indicate a pause in sedimentation.

Regional aspects (Fig. 52).—The Tarbur is present only in interior Fars Province where it is a prominent, feature-forming unit. It may be predominantly dolomite or limestone. Toward the southwest it grades into shale of the Gurpi.

Figures 7 and 8 show the extreme thickening of the Cretaceous-Tertiary section from coastal Fars to interior Fars which is coincident with the development of the Tarbur Formation.

Fossils and age.—The Tarbur Formation contains a rich shallow-water, reefal microfauna associated with abundant rudist, mollusk, and algal detritus. *Monolepidorbis douvillei* Astre occurs in the basal 200 feet of the formation. The succeeding 1,530 feet of limestone contains the following fauna: *Omphalocyclus macroporus* (Lamarck), *Siderolites calcitrapoides* Lamarck, *Orbitella media* (d'Archiac), *Loftusia* sp., *Dictyoconella* sp., *Dicyclina* sp., and *Lepidorbitoides* sp. A similar fauna is in the Tayarat Formation of Kuwait and the upper part of the Aruma Formation of Saudi Arabia (Fig. 54).

Age of formation.—Late Campanian to Maestrichtian.

Amiran Formation

Synonym.—The "flysch."

References.—P. T. Cox, 1938; Kent, Slinger,

FIG. 54.—Tarbur Formation; detrital, reef limestone with *Omphalocyclus macroporus* (Lamarck) (O) and *Siderolites* sp. (S). Interior Fars. Maestrichtian. ×30.

FIG. 55.—Dark, somber-colored Amiran Formation. Kuh-e Pusht-e Jangal. 1 mile north of type section, Lurestan.

FIG. 56.—Type section of Kashkan and Amiran Formations. Kuh-e Amiran, Lurestan.

FIG. 54

FIG. 55

FIG. 56

and Thomas, 1951; Lees, 1953; The British Petroleum Company Ltd., 1956; Falcon, 1958.

Type section.—The type section was measured on Kuh-e Amiran, near the village of Ma'mulan, where the Kashkan River cuts through the northeastern flank of the structure. It consists of 2,960 feet of low-weathering, dark olive-brown siltstone and sandstone with local developments of chert conglomerate and shelly limestone (Figs. 55, 56, 62). Chert grains are the dominant sandstone constituent.

The basal contact with the gray marl of the Gurpi is gradational. In most localities the Amiran is overlain by the lenticular, shelly limestone of the Taleh Zang Formation. Where the Talech Zang is not present the top is placed at the junction of the dark, olive-green siltstone and sandstone of the Amiran with the red to green, conglomeratic sandstone of the Kashkan Formation.

Regional aspects.—During Late Cretaceous and Paleocene time the area of radiolarites and basic igneous rocks northeast of the Agreement Area was subjected to uplift, folding, and erosion, which provided the detritus of the Amiran Formation.

In the Agreement Arca the Amiran is present only in the northeastern part of Lurestan Province (Fig. 3).

The unit increases in time-range and thickness from southwest to northeast (Figs. 3, 11).

Fossils and age.—The Amiran Formation locally contains rich planktonic faunas of Paleocene age. The fauna includes *Globigerina daubjergensis* Brönnimann, *Globorotalia uncinata* Bolli, *G. pseudobulloides* (Plummer), *G. velascoensis* (Cushman), and *G. pseudomenardii* Bolli. Reworked Maestrichtian *Globotruncana* sp., *Omphalocyclus* sp., and *Loftusia* sp. at some places occur in the lowermost beds of the Amiran.

In central Lurestan the Amiran is of Paleocene age. Toward the northeast in the Kermanshah-Khurramabad area the basal units of the formation are Maestrichtian in age. In this area outcrops of *Omphalocyclus*-bearing limestone provided the reworked Maestrichtian faunas found in the basal Amiran on the southwest.

Age of formation.—Maestrichtian to Paleocene.

Sachun Formation

Synonym.—Infra-nummulitic gypsum.

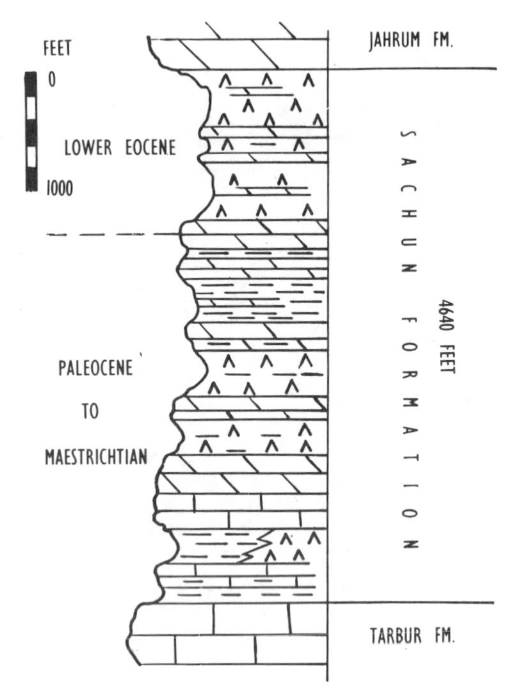

FIG.57. Type section of the Sachun Formation. N 28°33'37"; E 54°26'04".

Reference.—The British Petroleum Company Ltd., 1956.

Type section.—The type section was measured at Kuh-e Sachun, about 3 miles north of Sachun village. It is composed of 4,640 feet of gypsum, marl, and dolomite in about equal proportions (Fig. 57). The gypsum is highly lenticular, commonly interfingering with ochre-colored marl and thin-bedded dolomite along strike.

The Sachun Formation conformably and gradationally overlies the Tarbur Formation and is conformably overlain by the Jahrum Formation.

Regional aspects.—The area of development of the Sachun, which is present only in interior Fars Province (Figs. 3, 61), is coincident with the area of marked thickening of the Cretaceous-Tertiary section from coastal Fars to interior Fars (Fig. 8). The unit may have been deposited during part of the time that is represented elsewhere by the post-Cretaceous disconformity.

Lithologically the formation ranges from gypsum to ochreous-weathering marl and dolomite to silty, sandy redbeds. These changes may take place over short distances.

Fossils and age.—At the type section the basal

1,200 feet of the Sachun Formation contains a Maestrichtian microfauna similar to that of the underlying Tarbur Formation. This fauna includes *Omphalocyclus macroporus* (Lamarck), *Loftusia* sp., *Dictyoconella* sp., *Siderolites* sp., and rudist detritus. *Elphidiella multiscissurata* Smout also has been identified from this interval in nearby sections.

The succeeding 1,500-foot interval of gypsum and ochreous marl contains a non-diagnostic fauna of indeterminate algae and arenaceous Foraminifera. This interval is directly overlain by 200 feet of purple and red-stained limestone containing a Paleocene fauna of *Miscellanea* sp. and *Glomalveolina* sp.

The upper 1,740 feet of the Sachun type section is almost barren of fossils, but rare occurrences of *Opertorbitolites* sp. indicate an early Eocene age.

Age of formation.—Late Maestrichtian to early Eocene.

The position of the Cretaceous-Tertiary boundary can not be defined exactly because of lack of faunal control in the lower part of the formation.

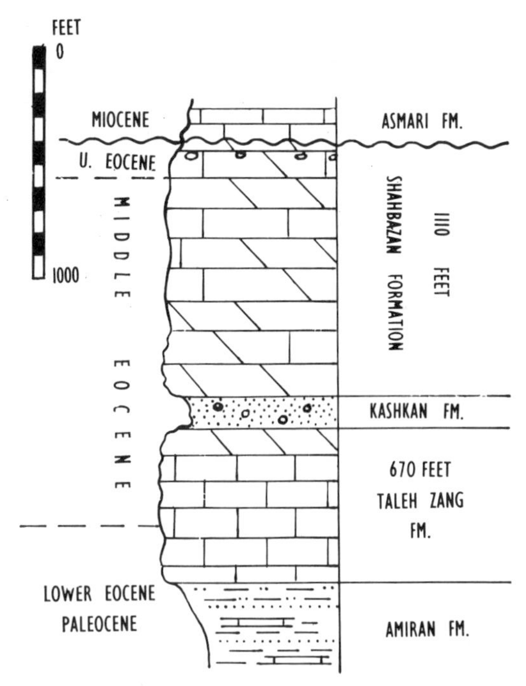

FIG. 59. Type section of the Taleh Zang and Shahbazan Formations. N 32°47′38″ ; E 48°42′00″

FIG. 58.—Type section of Taleh Zang and Shahbazan Formations, Khuzestan.

TALEH ZANG FORMATION

Synonym.—Mid-Eocene limestone.

References.—Loftus, 1855; Cotteau and Gauthier, 1895; Douvillé, 1900, 1902, 1904, 1905; Gauthier, 1902; Priem, 1908; Pilgrim, 1908; R. K. Richardson, 1924; de Böeckh, Lees, and F. D. S. Richardson, 1929; Lees and F. D. S. Richardson, 1940; Lees, 1938, 1953; Kent, Slinger, and Thomas, 1951; The British Petroleum Company Ltd., 1956; Falcon, 1958.

Type section.—At Tang-e Do, about 3 miles southeast of the Taleh Zang railroad station, the type section consists of 670 feet of medium-bedded, gray limestone with an abundant fauna of alveolinids, nummulites, assilinids, discocyclinids, and *Orbitolites* (Figs. 58, 59). The Taleh Zang conformably overlies the Amiran Formation and is overlain by red sandstone and conglomerate of the Kashkan Formation. There is some indication that a disconformity may be present at the latter boundary (see section on fossils and age, Kashkan Formation).

Regional aspects.—In northeastern Lurestan

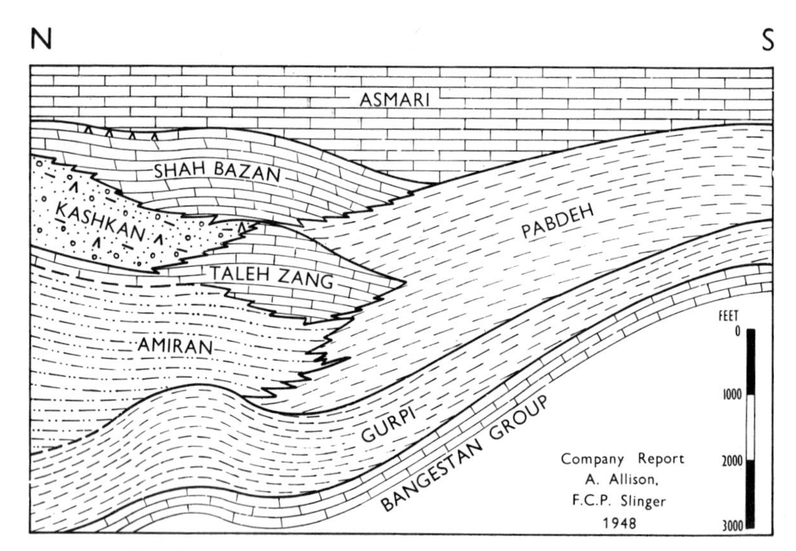

FIG. 60. Idealized N-S cross section near Lurestan-Khuzestan boundary, showing Upper Cretaceous, Paleocene, Eocene facies relationships.

two shallow-water carbonate formations ranging in age from Paleocene to Eocene, and separated by the Kashkan Formation, are present. They are lithologically similar to, and correlative with, the Jahrum Formation of Fars Province. The lower formation is called the Taleh Zang, the upper, the Shahbazan (Figs. 3, 4, 61).

The Taleh Zang ranges greatly in thickness from place to place because of its interdigitation with both the underlying Amiran and the overlying Kashkan. Allison and Slinger (1948, company report) depict the Taleh Zang as a reefal development interfingering with terrestrial redbeds of the Kashkan toward the northeast, and marl and marly limestone of the Pabdeh toward the south and southwest (Fig. 60).

The Taleh Zang extends into the marl of the Pabdeh Formation as a thin-bedded, marly, pelagic limestone and marl unit which commonly contains chert nodules, fish scales, and glauconite. Here it is termed informally the Taleh Zang member of the Pabdeh Formation.

Fossils and age.—At the type section, the Taleh Zang Formation contains a rich fauna, of which the more important species, oldest to youngest, are: *Miscellanea* sp., *Nummulites globulus* Leymerie, *Glomalveolina* sp., *Opertorbitolites* sp., *Orbitolites complanatus* Lamarck, *Nummulites* cf. *curvispira* d'Archiac and Haime, *Nummulites beaumonti* d'Archiac and Haime, and *Halkyardia* sp.

Age of formation.—Paleocene to middle Eocene at the type section.

In most of central Lurestan, where the Taleh Zang is a thin reefal limestone at the top of the Amiran, the fauna is exclusively of Paleocene age. Similar faunas are found in the basal beds of the Jahrum Formation in Fars and Khuzestan.

Kashkan Formation

Synonym.—Eocene redbeds.

References.—Loftus, 1955; de Böeckh, Lees, and F. D. S. Richardson, 1929; Kent, Slinger, and Thomas, 1951; The British Petroleum Company Ltd., 1956.

Type section.—The type section at Kuh-e Amiran was measured where the Kashkan River cuts through the northeastern flank of the anticline. It is composed of 1,125 feet of deep-red-colored siltstone, sandstone, and conglomerate which become coarser upward. The major constituent is chert (Figs. 56, 62, 64).

The lower contact with the Taleh Zang is abrupt. The upper contact exhibits a prominent weathered zone.

Regional aspects.—Deposition of the clastic Kashkan Formation resulted from orogenic movements in the area on the northeast. South-

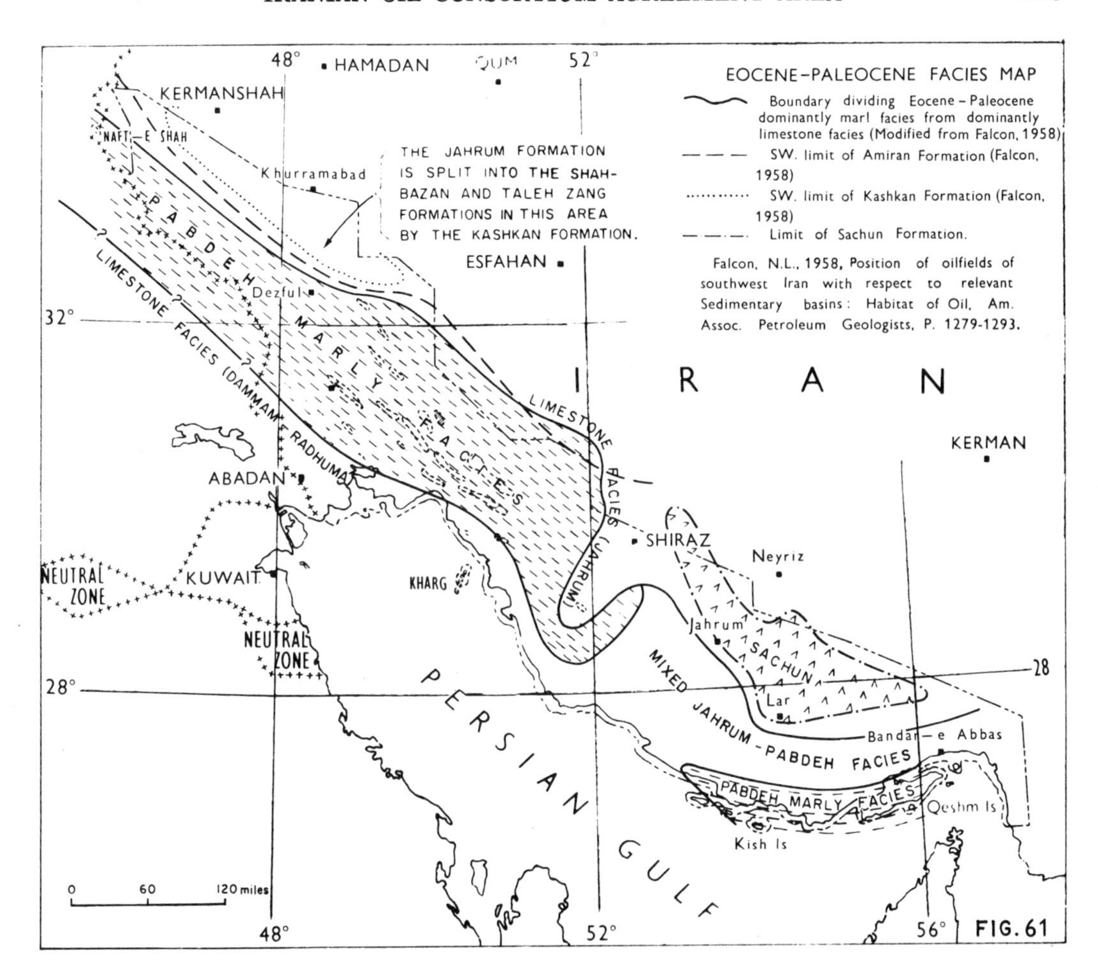

west of the type section it interfingers with the Pabdeh Formation (Fig. 11), and southeastward toward Khuzestan, it is progressively replaced by limestone of the Shahbazan and Taleh Zang (Fig. 60). The formation diminishes in both time and thickness from northeast to southwest.

Chert derived from the radiolarites is the primary constituent. The formation, by virtue of its deep red color, is a prominent and important marker throughout northeastern Lurestan.

Fossils and age.—The Kashkan Formation has yielded little faunal evidence of its age. At Tang-e Do, the type locality of the Taleh Zang and Shahbazan Formations, the non-fossiliferous Kashkan is underlain and overlain by limestone of middle Eocene age. In the Amiran area the Kashkan basal conglomerate directly overlies Paleocene Taleh Zang limestone. In this area, a hiatus may be present between the Paleocene Taleh Zang and the overlying middle? Eocene Kashkan. This hypothesis can not be supported paleontologically at present.

Age of formation.—Paleocene? to middle Eocene.

Shahbazan Formation

Synonym.—The barren limestone.

References.—See Taleh Zang Formation.

Type section.—The type section was measured at Tang-e Do, 3 miles southwest of the Taleh Zang railroad station. It consists of 1,110 feet of medium-bedded, white-weathering, porous, saccharoidal dolomite and dolomitic limestone (Figs. 58, 59). A leached brecciated zone which occurs in the upper few feet marks the disconformable contact with the overlying Asmari. The basal contact with the Kashkan is gradational.

Regional aspects.—The Shahbazan is present

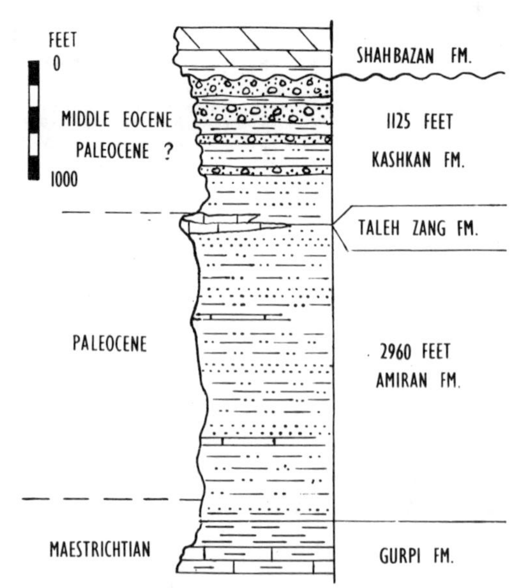

FIG. 62. Type section of the Amiran and Kashkan Formations. N 33°22′15″; E 47°58′10″

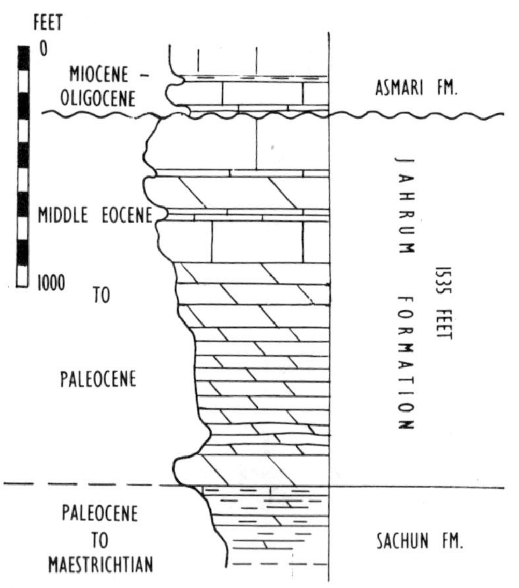

FIG. 63. Type section of the Jahrum Formation. N 28°25′53″; E 53°44′47″.

only in northeastern Lurestan (Figs. 3, 4, 61), where, together with the Asmari, it forms a prominent topographic unit. It is differentiated from the Asmari by an intervening conglomeratic leached zone and a change from limestone of the Asmari to dolomite of the Shahbazan. This boundary is commonly difficult to place, thus making it necessary to map the two formations as one unit. When the two formations can not be differentiated, the two names are hyphenated.

In southern and southwestern Lurestan the Shahbazan interfingers with marl of the Pabdeh Formation. It is progressively replaced toward the northeast by the Kashkan (Fig. 60).

Fossils and age.—Over most of Lurestan, the dolomitic limestone of the Shahbazan Formation contains few fossils. At the type section, however, rare *Nummulites* cf. *beaumonti* d'Archiac and Haime and *Alveolina oblonga* d'Orbigny have been noted near the base of the formation. The upper 100 feet of limestone, directly beneath the basal conglomerate of the lower Miocene Asmari, contains a rich late Eocene fauna of *Pellatispira* sp., *Nummulites* cf. *fabianii* (Prever), *Discocyclina* sp., and *Aktinocyclina* sp.

Age of formation.—Middle to late Eocene at the type section.

JAHRUM FORMATION

Synonyms.—Eocene limestone, Gishun Limestone.

References.—See Taleh Zang Formation.

Type section.—The type section was measured at Tang-e Ab on the northern flank of Kuh-e Jahrum (Figs. 63, 65). The basal 100 feet is massive, brown-weathering dolomite which is overlain by 535 feet of medium-bedded, relatively low-weathering dolomite. The upper 900 feet is mas-

→

FIG. 64.—Kashkan Formation; showing poorly sorted, angular nature of Kashkan components. Type section, Lurestan.

FIG. 65.—Kuh-e Jahrum, northeastern flank; showing massive Jahrum Formation overlain by Asmari. Interior Fars.

FIG. 66.—Jahrum Formation. *Orbitolites* sp. (O), *Somalina* sp. (S), *Alveolina* sp. (A). Interior Fars. Middle Eocene. ×40.

FIG. 67.—Jahrum Formation; detrital limestone with *Discocyclina* sp. (D), *Nummulites* sp. (N), and rare *Globigerina sp.* Middle Eocene. ×40.

FIG. 64

FIG. 65

FIG. 66

FIG. 67

sive, feature-forming, dolomitic limestone with abundant microfauna.

The basal contact is apparently conformable. The top of the unit is placed at a regional disconformity which is commonly marked by a low-weathering zone of rubbly limestone and solution brecciation.

Regional aspects.—The Jahrum is the lithologic and time equivalent of the Radhuma-Dammam Formations of Saudi Arabia, Kuwait, and southeastern Iraq. Its areal extent and facies relation with the Pabdeh Formation are shown in Figures 3, 7, 8, and 61.

Difficulty is commonly encountered by the field geologist in separating the Jahrum Formation from the overlying Asmari. Where this happens the two formations are mapped as one unit and the two names are hyphenated.

Fossils and age.—The Jahrum Formation yields a rich, shallow-water, benthonic microfauna. The following species, listed oldest to youngest, constitute the more important microfossils: *Miscellanea* sp., *Sakesaria* sp., *Distichoplax biserialis* (Dietrich), *Nummulites globosus* d'Archiac and Haime, *Lockhartia* sp., *Discocyclina* sp., *Opertorbitolites* sp., *Somalina* sp., *Dictyoconus* sp., *Coskinolina* sp., *Alveolina oblonga* d'Orbigny, *Linderina* sp., *Orbitolites complanatus* Lamarck, *Nummulites beaumonti* d'Archiac and Haime, and *Nummulites atacicus* Leymerie. At the type section, the Jahrum Formation is Paleocene to middle Eocene in age.

In the interior Fars area the Jahrum is no younger than middle Eocene and is disconformably overlain by the Oligo-Miocene Asmari Formation.

Evidence of the post-middle Eocene break is not seen in coastal areas of Fars where the complete Eocene is represented. The top of the Jahrum Formation contains a late Eocene fauna of *Nummulites fabianii* (Prever), *Pellatispira* sp., *Chapmanina* sp., and *Baculogypsinoides* sp.

Microfossils, similar to those of the Jahrum Formation, occur in the type section of the time-equivalent Taleh Zang and Shahbazan Formations of Lurestan (Figs. 66, 67, 68).

Age of formation.—Paleocene to late Eocene.

Pabdeh Formation

Synonyms.—The Pabdeh Formation is the Paleocene, Eocene, and Oligocene part of the unit previously called the *Globigerina* marl or Dezak Marl. Synonyms that have been applied to parts of this unit are: the Eocene blue and purple shales, Eocene green and purple marls, lower Eocene marls, upper Eocene marls, Eocene cherty limestone, mid-Eocene limestone, Eocene fissile limestone, and the Spatangid shales.

References.—R. K. Richardson, 1924; de Böeckh, Lees, and F. D. S. Richardson, 1929; Kent, Slinger, and Thomas, 1951; The British Petroleum Company Ltd., 1956; Falcon, 1958; Slinger and Crichton, 1959.

Type section.—At the Tang-e Pabdeh type section on the southeastern end of Kuh-e Pabdeh, 2,620 feet of low-weathering gray shale and thin argillaceous limestone were measured (Figs. 48, 69, 71). Two informal members are differentiated: the purple shale member at the base of the Pabdeh, and the Taleh Zang member. The former consists of silty to sandy, purple-red-gray shale. Thin-bedded argillaceous limestone containing fish scales and chert makes up the latter member (also see Taleh Zang Formation). The Taleh Zang member is considered to be the deeper-water equivalent of the Taleh Zang Formation.

The Pabdeh disconformably overlies the Gurpi Formation. The upper contact is transitional with the Asmari Formation.

Regional aspects.—The Pabdeh Formation is the shale-marl part of the multi-facies Paleocene, Eocene, and Oligocene. It is best understood by referring to Figures 3, 7, 8, 9, 10, 11, 60, and 61.

→

Fig. 68.—Jahrum Formation; detrital limestone with *Alveolina* and *Nummulites*. Interior Fars. Middle Eocene. ×35.

Fig. 69.—Kuh-e Pabdeh; Type section of Pabdeh-Gurpi Formations. Kuh-e Pabdeh, Khuzestan.

Fig. 70.—Pabdeh Formation; typical plankton assemblage with *Hantkenina* sp. Coastal Fars. Middle Eocene. ×35.

Fig. 71.—Outcrop of Sarvak-Ilam, Gurpi, Pabdeh, and Asmari Formations. Northeastern flank. Kuh-e Pabdeh, Khuzestan.

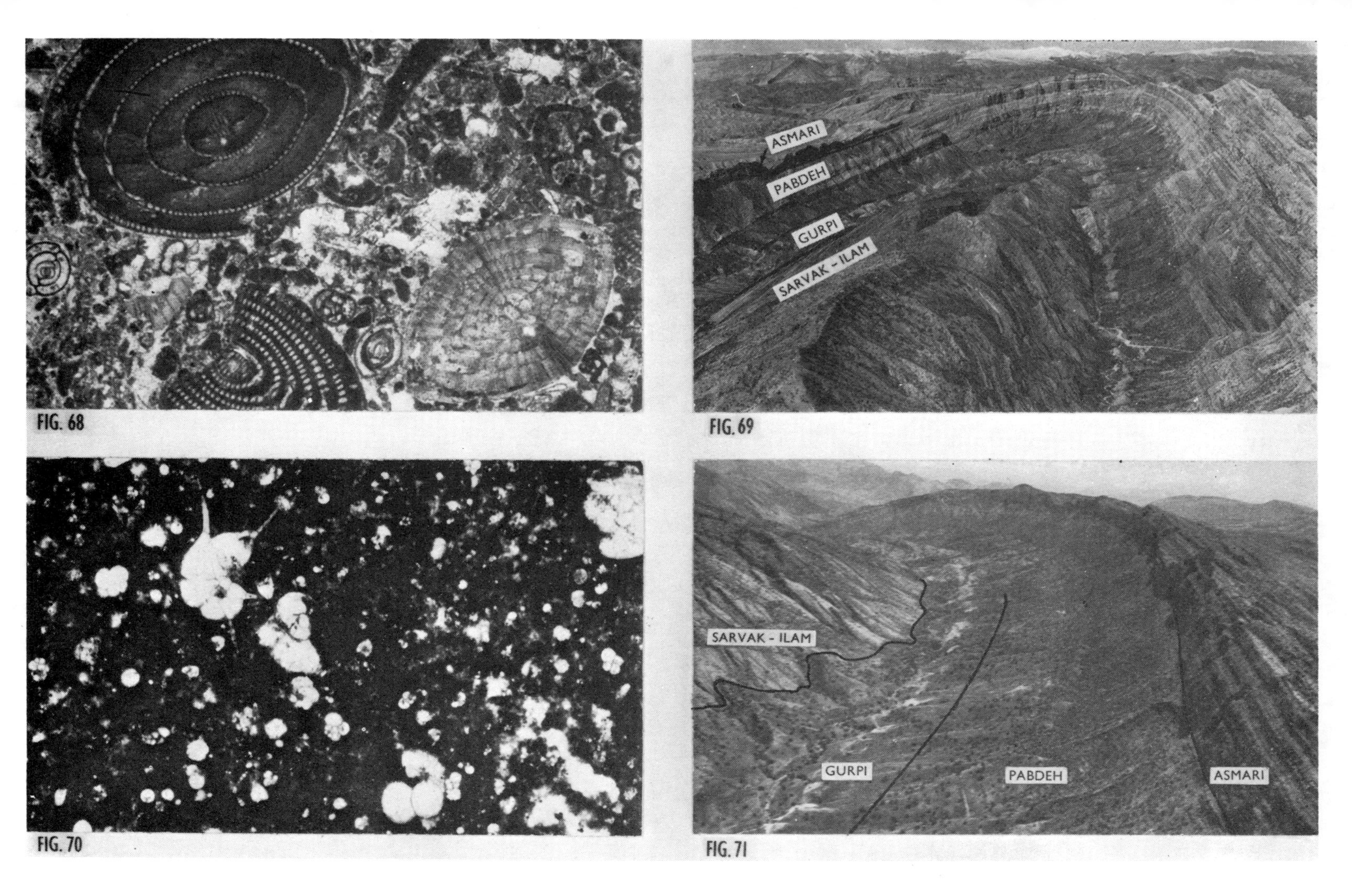

FIG. 68

FIG. 69

FIG. 70

FIG. 71

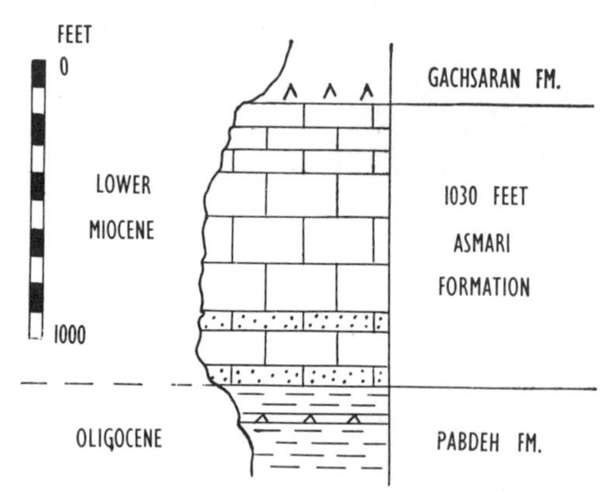

FIG. 72. Type section of the Asmari Formation.
N 31°43′09″ ; E 49°34′29″

These show the marl of the Pabdeh Formation of Lurestan and Khuzestan passing into carbonates toward Iraq, Kuwait, and Arabia, and into clastics and carbonates toward the northeast.

In Fars Province the Pabdeh is developed along only part of the coast (Fig. 61). Toward the interior it passes into a mixed Jahrum-Pabdeh facies, and then into carbonates of the Jahrum Formation. On the south and southeast toward Arabia, the marl of the Pabdeh is replaced by carbonate (Fig. 8).

Throughout Lurestan and Khuzestan the base of the Pabdeh is placed at the bottom of the purple shale member. In Fars, however, the purple shale generally is not present. The lower boundary is placed at the bottom of a cherty limestone unit where phosphate nodules, shark teeth, glauconite, and in some places pebble conglomerate with detrital bitumen, lie at the post-Cretaceous disconformity.

Fossils and age.—The Pabdeh Formation throughout the entire Agreement Area contains a rich planktonic fauna. In Khuzestan and Fars, the entire basal purple shale member, where present, is late Paleocene, with rich occurrences of *Globorotalia velascoensis* (Cushman) and *Globorotalia pseudomenardii* Bolli. This fauna overlies directly the *Globotruncana gansseri*-bearing marl of the Gurpi Formation. The glauconitic-phosphatic bed which separates the Pabdeh from the Gurpi represents a period of non-deposition from the late Maestrichtian until the end of the early Paleocene. *Abathomphalus mayaroensis* (Bolli), *Globigerina daubjergensis* Brönnimann, *Globorotalia uncinata* Bolli, and *Globorotalia compressa* (Plummer), all forms which date this missing interval, have not been recorded from Khuzestan or Fars.

In Lurestan the basal purple shale member ranges from late Paleocene to early Eocene in age. The fauna includes *Globorotalia velascoensis* (Cushman), *G. pseudomenardii* Bolli, *G. rex* Martin, *G. wilcoxensis* Cushman and Ponton, and *"Globorotalia" palmerae* Cushman and Bermudez. There is no faunal or physical evidence of a break in deposition between the basal Pabdeh and the underlying Gurpi. The planktonic faunas of the Maestrichtian and Paleocene are fully represented in Lurestan.

In all areas the succeeding Eocene units of the Pabdeh contain faunas which occur in the same zonation used by Bolli (1957, p. 159) for the Eocene Navet Formation of Trinidad. The Eocene fauna of the Pabdeh includes: *"Globorotalia" palmerae* Cushman and Bermudez, *Globorotalia aragonensis* Nuttall, *G. rex* Martin, *G. spinulosa* Cushman, *G. lehneri* Cushman and Jarvis, *Hantkenina* sp., *Truncorotaloides* sp., *Porticulasphaera* cf. *mexicana* (Cushman), *Globorotalia centralis* Cushman and Bermudez, *Globigerapsis semi-involuta* (Keijzer), *Cribrohantkenina* sp., and *Globorotalia cerro-azulensis* Cole (Fig. 70).

The planktonic faunas occurring above the last definite Eocene species are at present considered to be of Eocene-Oligocene age. Non-diagnostic *Nonion, Uvigerina,* and Textulariidae occur in this interval. *Haplophragmium slingeri* Thomas, *Zeauvigerina khuzestanica* Thomas, and the echinoids *Schizaster* and *Hemiaster* also occur. At the Pabdeh type section, the marl with *Haplophragmium* and *Zeauvigerina* is interbedded with Oligocene basal Asmari-type limestone containing *Nummulites intermedius* d'Archiac.

At Kabir Kuh in Lurestan, the uppermost Pabdeh marl beds contain a globigerinid fauna with *Globigerinoides* cf. *triloba* (Reuss). This occurrence indicates that the youngest Pabdeh is of early Miocene age in Lurestan.

Age of formation.—Late Paleocene to Oligocene in Khuzestan and Fars Provinces; late Paleocene to Miocene in Lurestan Province.

Asmari Formation

Synonyms.—Euphrates Limestone, Kalhur Limestone, Khamir Limestone.

Thomas (1950) defined the Asmari in *sensu*

lato and *sensu stricto* units. The latter was restricted essentially to the oil fields area and excluded the Euphrates Limestone of Lurestan and the Khamir Limestone of Fars. His *sensu lato* Asmari included these two lithologic equivalents as well as the *Brissopsis* shales and basal anhydrite which occur beneath the limestone body in parts of Khuzestan.

The Asmari as now defined includes the synonymous Euphrates and Khamir Limestones. In Khuzestan, the base of the formation is placed at the bottom of the limestone succession. The underlying *Brissopsis* shales and the basal anyhydrite are included in the Pabdeh Formation. Two members, the Ahwaz Sandstone Member and the Kalhur Member (previously Kalhur Anhydrite Formation of Lurestan), are included within the Asmari.

References.—Pilgrim, 1908, 1924; Busk and Mayo, 1918; R. K. Richardson, 1924; Douglas, 1927-1928; de Böeckh, Lees, and F. D. S. Richardson, 1929; Lees, 1933, 1934, 1938, 1950, 1951, 1953; Reichel, 1936-37; Henson, 1950; Kent, Slinger, and Thomas, 1951; Thomas, 1950; Ion, Elder, and Pedder, 1951; The British Petroleum Company Ltd., 1956; Falcon, 1958; Slinger and Crichton, 1959.

Type section.—The type section as designated by R. K. Richardson (company report) is at Tang-e Gel-e Tursh (Valley of Sour Earth) on the southwestern flank of the Kuh-e Asmari anticline where the formation forms the breached carapace of the structure (Fig. 73).

Thomas (1950) has described the type section in detail. It consists of 1,030 feet of resistant, feature-forming, cream to brown-weathering, well-jointed limestone with shelly intercalations (Fig. 72). Contacts with the overlying Gachsaran and underlying Pabdeh are conformable.

Only the upper and middle parts of the Asmari are present in the type section; the lower Asmari interval is represented by the laterally equivalent shale of the Pabdeh Formation.

In spite of being such a prolific oil reservoir in nearby oil fields, virtually no trace of oil remains in the exposed rock at Kuh-e Asmari and in other similar outcrops.

Ahwaz Sandstone Member.—In the Ahwaz and Mansuri oil fields the basal few hundred feet of the Asmari is a sequence of calcareous sandstone, sandy limestone, and minor shale named the Ahwaz Sandstone Member. It is correlated as the wedge-edge of the Ghar Formation of Kuwait and southeastern Iraq (Figs. 4, 9, 10).

The interval, 8,056-8,756 feet, in the Ahwaz 6 well is the type section for the Ahwaz Sandstone Member (Fig. 76). Here sandstone directly overlies the Pabdeh, and at Mansuri, just south of Ahwaz, limestone is found both above and below the Ahwaz Member (Fig. 10).

The Ahwaz Sandstone Member ranges in age from Oligocene to early Miocene. At Ahwaz well 5 the sandstone overlies Oligocene lower Asmari containing *Nummulites intermedius* d'Archiac. At Ahwaz well 6, the underlying Pabdeh marls with *Haplophragmium slingeri* Thomas are also of Oligocene age. The sandstone is succeeded by, and interfingers with, middle Asmari limestone bearing *Austrotrillina howchini* (Schlumberger) and *Peneroplis evolutus* Henson of early Miocene age.

Kalhur Member.—The Kalhur Member of the Asmari Formation was previously treated as a formation called the Kalhur Anhydrite or Kalhur Gypsum.

The type section was measured on the southern flank of Kuh-e Anaran, 2 miles southeast of where the road passing along the southwestern flank of Kuh-e Anaran crosses the Changuleh River. It is composed of three units: a basal gypsum 15 feet thick, 140 feet of marl and thin-bedded marly limestone, and 270 feet of massive gypsum (Fig. 77).

The member conformably overlies marl of the Pabdeh Formation and is conformably overlain by limestone of the Asmari. Elsewhere, the member may be present as a tongue within the Asmari Formation limestone.

The Kalhur Member is present only in southwestern Lurestan. Toward the northeast it interfingers with limestone of the middle part of the Asmari Formation (Figs. 3, 11).

The planktonic faunas of the underlying Pabdeh have a strong Miocene aspect but the details of this fauna still have to be studied. Laterally, the Kalhur is probably the time equivalent of the middle Asmari limestone. The overlying upper Asmari limestone containing *Neoalveolina melo curdica* Reichel and *Miogypsina* sp. is of definite early Miocene age. The Kalhur Member is considered to be of early Miocene age.

Regional aspects.—The Asmari serves as a reservoir for the vast proportion of the oil produced

FIG. 73

FIG. 74

FIG. 75

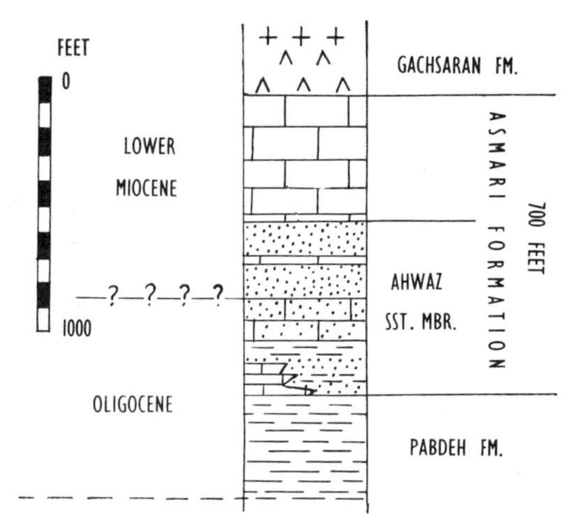

FIG.76 . Type section of the Ahwaz Sandstone Member, Asmari Formation. N 31°12'34" ; E 48°56'56".

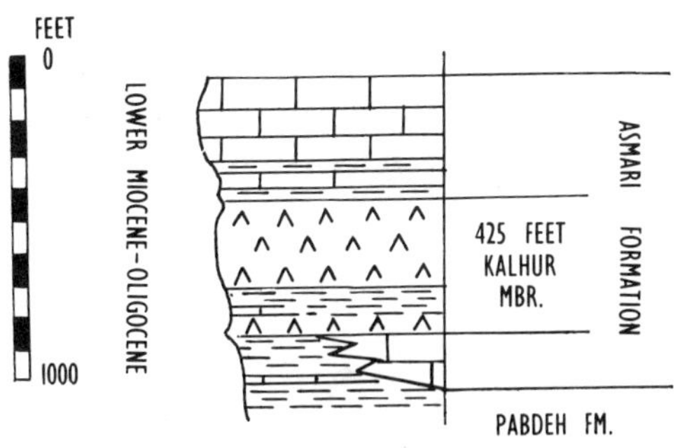

FIG.77 . Type section of the Kalhur Member, Asmari Formation. N 33°01'47" ; E 46°36'27"

in the Agreement Area. An average well in the better fields produces about 25,000 barrels of oil per day, amply justifying the formation's fame as one of the world's most prolific reservoirs. The high rate of production is attributable to a prominent system of fracturing which makes accessible large volumes of the reservoir to intersecting well bores.

The formation is present through most of the Agreement Area as a prominent limestone unit. It commonly forms the resistant carapace of the long, anticlinal mountains characteristic of the Zagros foothills area. An evaporite wedge, the Kalhur Member, replaces the limestone of the middle part of the Asmari in southwestern Lurestan (Figs. 3, 11). In southwestern Khuzestan, sandstone of the Ahwaz Member, derived from the Arabian shield, interfingers with limestone of the middle and lower parts of the Asmari (Figs. 3, 4, 9, 10). Near Bandar Abbas in southeastern Fars Province, the Asmari passes into marl of the Pabdeh toward the southeast (Fig. 7).

Southwest of the Agreement Area, where the thick Paleocene-Eocene limestones of Iraq, Kuwait, and Saudi Arabia are developed, the Asmari disappears as a result of interfingering with Arabian shield sandstones and because of the important post-Eocene disconformity (Figs. 8, 9, 10). A somewhat similar relation holds in the Agreement Area; where the Jahrum or Shahbazan Formations are well developed, the Asmari in many places is poorly represented (Figs. 7, 8, 11).

Throughout Lurestan and Khuzestan the Asmari-Gachsaran contact apparently remains at a relatively constant time horizon. In Fars Province, however, the Gachsaran progressively replaces the top of the Asmari from northwest to southeast (Fig. 3). Conversely, in Fars, the base of the Asmari generally is found at a fairly constant horizon, whereas in Lurestan and Khuzestan it is of different ages in different places (Fig. 3).

In the southeastern end of the Agreement Area the Asmari grades into marl of the Pabdeh Formation in wells on Qeshm Island (Fig. 7).

Fossils and age.—The formation is divided into lower, middle, and upper Asmari, each containing a diagnostic microfauna; these are summarized as follows.

Lower Asmari (Fig. 74).—*Nummulites intermedius* d'Archiac, *Nummulites vascus* Joly and Leymerie, *Archaias operculiniformis* Henson, *Austrotrillina paucialveolata* Grimsdale, *Peneroplis thomasi* Henson, *Eulepidina dilatata* Miche-

---

←

FIG. 73.—Type section of Asmari Formation. Kuh-e Asmari, Khuzestan.
FIG. 74.—Lower Asmari limestone. *Nummulites intermedius* d'Archiac. Khuzestanfl Oligocene. ×20.
FIG. 75.—Middle Asmari limestone. *Peneroplis evolutus* Henson (Pe), *Peneroplis thomasi* Henson (Pt), and *Austrotrillina howchini* (Schlumberger) (A). Khuzestan. Early Miocene. ×15.

lotti, and *Praerhapydionina delicata* Henson. This fauna indicates an Oligocene age for the lower Asmari, and has been correlated with the Rupelian to Chattian stages by Eames *et al.* (1962, p. 12).

Middle Asmari (Fig. 75).—*Austrotrillina howchini* (Schlumberger), *Peneroplis evolutus* Henson, *P. thomasi* Henson, *Archaias* sp., *Meandropsina anahensis* Henson, and *Miogypsina* sp.

Upper Asmari (Fig. 78)—*Neoalveolina melo curdica* Reichel, *Taberina malabarica* (Carter), *Meandropsina iranica* Henson, and *Ostrea latimarginata* Vredenberg.

The middle to upper Asmari is early Miocene in age. At the type section, the upper Asmari is represented by 750 feet of *Neoalveolina melo curdica*-bearing limestone with associated *Meandropsina iranica* Henson and *Dendritina rangi* d'Orbigny. In the same section 70 feet of sandy limestone with *Miogypsina* sp. forms the base of the upper Asmari.

The middle Asmari consists of 210 feet of shelly and miliolid limestone forming the base of the type section. The microfauna is not typical of the rich middle Asmari peneroplid-miliolid facies of other parts of Iran, *e.g.*, Gachsaran. The limestone is much recrystallized, but miliolids and *Dendritina rangi* d'Orbigny are common. A specimen of *Austrotrillina howchini* (Schlumberger) has been recorded in the upper part (Thomas, 1950, p. 39).

The rich nummulitic-lepidocycline limestone of the lower Asmari is not present at the type section. This part of the section may be represented by the alternating shale and thin limestone of the Oligocene *Brissopsis* beds which are now included in the Pabdeh Formation.

The typical microfauna of the lower Asmari is well developed at the Gachsaran oil field, 120 miles southeast of the Kuh-e Asmari type section. Thomas (1950, p. 40) recorded the following Oligocene microfauna from stages 1 and 2 (lower Asmari) at Gachsaran: *Nummulites* cf. *vascus* Joly and Leymerie, *N.* cf. *incrassatus* (de la Harpe), *Cycloclypeus, Heterostegina, Rotalia viennoti* Greig, *Eulepidina dilatata* Michelotti, *Nummulites* cf. *fichteli* Michelotti, *Archaias operculiniformis* Henson, and *Praerhapydionina delicata* Henson.

In Fars Province the upper Asmari with *Neoalveolina melo curdica* is absent, the limestone grading laterally into the Gachsaran or Razak Formations. In Lurestan the Asmari overlying the Kalhur Member is restricted to the upper part of the formation with *Neoalveolina melo curdica* Reichel.

Age of formation.—Oligocene to early Miocene.

Fars Group

In 1908, Pilgrim applied the name Fars series to a thick sequence of largely Miocene deposits cropping out in Fars Province. He divided this into a basal gypseous group, *Ostrea virleti* beds, and *Pecten vasseli* beds. Boundaries, however, were obscure. Later, near Masjed-e Suleyman, S. L. James and G. W. Halse (company report) divided the Fars into lower, middle, and upper Fars. These names have been used for many years in the oil fields area.

The term Fars is now given group status and divided into the Gachsaran Formation (formerly lower Fars), the Mishan Formation (formerly middle Fars), and the Agha Jari Formation (formerly upper Fars). The Razak Formation of interior Fars, which is a non-evaporitic redbed equivalent of the evaporitic Gachsaran Formation, is also included within the group.

References.—Loftus, 1855; Pilgrim, 1908; Harrison, 1924; R. K. Richardson, 1924; Douglas, 1928; Busk, 1929; de Böeckh, Lees, and F. D. S. Richardson, 1929; Strong, 1937; Lees, 1938, 1953; Lees and F. D. S. Richardson, 1940; O'Brien, 1948, 1953, 1957; Thomas, 1948, 1950; Ion, Elder, and Pedder, 1951; Kent, Slinger, and Thomas, 1951; The British Petroleum Company

»»→

Fig. 78.—Upper Asmari limestone. *Neoalveolina melo curdica* Reichel (N). Khuzestan. Early Miocene. ×60.
Fig. 79.—Gachsaran Formation, bryozoan-algal, detrital limestone with *Peneroplis farsensis* Henson. Coastal Fars. Early Miocene. ×35.
Fig. 80.—Outcrop of Gachsaran, Mishan-Agha Jari, and Bakhtyari Formations. Khuzestan.
Fig. 81.—Gachsaran, Mishan-Agha Jari, and Bakhtyari outcrop north of Masjed-e Suleyman. Bakhtyari type section in right-center of photograph.

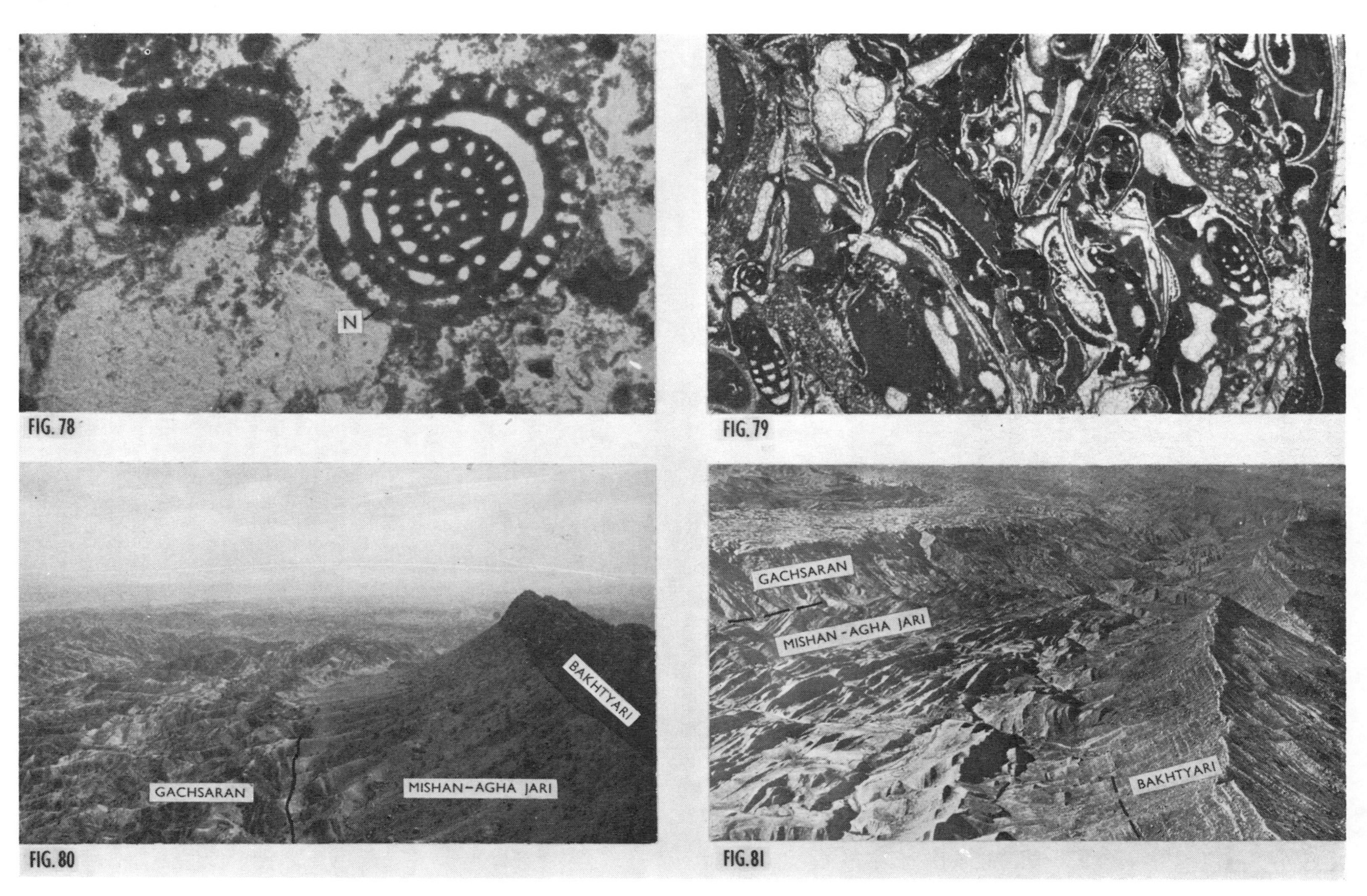

FIG. 78

FIG. 79

FIG. 80

FIG. 81

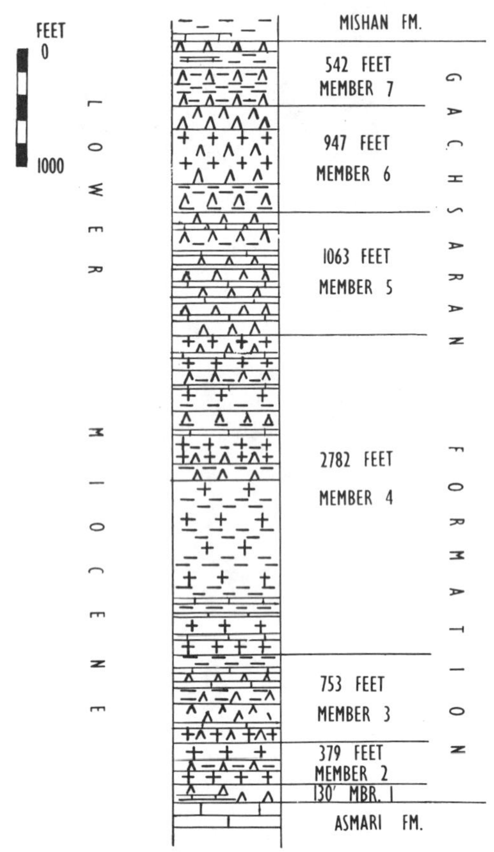

FIG. 82. Type section of the Gachsaran Formation.

Ltd., 1956; Falcon, 1958; Slinger and Crichton, 1959.

Gachsaran Formation

Synonym.—Lower Fars.

References.—See Fars Group.

In Fars Province the Gachsaran is divided into the Chehel, Champeh, and Mol Members. Synonyms for the Chehel Member are gypsum series, basal gypsum beds, and Chel Anhydrite. Synonyms for the Champeh Member are *Gypsina* limestone and lower limestone. The Mol Member was previously called the redbeds.

Type section.—A complete section of the Gachsaran Formation is found in very few wells or surface outcrops because of its mobile and saliferous nature (O'Brien, 1948). In 1946, F. D. S. Richardson, by virtue of a detailed study of the sequence near the Gachsaran oil field, was the first worker successfully to reduce to order the complexities of the formation. This was supplemented in 1948 by F. C. P. Slinger who, while resident geologist at the Agha Jari oil field, worked out a sequence of key beds of the formation. This has served for many years as a guide to drilling progress within this complex, variable evaporitic sequence. Micropetrographic work by Strong (1937) has also aided in defining correlative units of the Gachsaran Formation.

The type section is a composite of intervals compiled by S. E. Watson from wells of the Gachsaran oil field (1960, company report). It should be borne in mind, however, that the Gachsaran Formation is highly subject to solution and also responsive to differential pressures exerted upon it; consequently, one seldom finds elsewhere a sequence that is similar to the type section (Fig. 82).

In the oil fields area the Gachsaran has been divided into seven informal members. Members 6 and 7 are the same as stages 2 and 3 referred to in previous publications. Members 1 through 5 are subdivisions of stage 1.

The type section members are described in ascending order.

Member 1 (GS 25, 7,875-7,745 feet).—The "Cap Rock" conformably overlies the Asmari. It consists of 130 feet of alternating thick-bedded anhydrite and thin-bedded limestone associated with bituminous shale. It is an important unit separating high pressures in the Gachsaran Formation from the lower pressures of the Asmari.

Member 2 (GS 21, 7,713-7,334 feet).—Member 2 is made up of thick salt units with intervening anhydrite and thin limestone.

Member 3 (GS 27, 4,947-4,194 feet).—Member 3 consists of thick anhydrite beds with subordinate salt in the lower half, and alternating anhydrite, thin-bedded limestone, and marl in the upper half.

Member 4 (GS 21, 5,175-2,393 feet).—Thick salt beds alternate with intervening gray marl, limestone, or anhydrite.

Member 5 (GS 20, 3,913-2,850 feet).—Member 5 is composed of marly anhydrite alternating with limestone and red to gray marl.

Member 6 (GS 18, 4,490-3,553 feet).—The basal 340 feet consists of alternating anhydrite, red marl, and limestone. The middle part is alternating salt and anhydrite, and the upper 200 feet is anhydrite with minor red to gray marl.

Member 7 (GS 14, 3,068-2,610 feet).—Member 7 is alternating anhydrite and gray marl or marly limestone. It is conformably overlain by the Mishan Formation.

Regional aspects (Figs. 80, 81).—The Gachsaran Formation which forms the all-important seal over Asmari oil accumulations was deposited in a northwest-southeast-trending asymmetric trough. The steep side of the trough was toward the northeast and the axis trended through the Gachsaran, Masjed-e Suleyman, and Lali oil fields (O'Brien, 1948; Falcon, 1958). Figures 8-10 are cross-sectional views of part of this basin showing the marked northeastward thickening of the formation, approximately coinciding with the southwestern border of the Agreement Area.

In Khuzestan the Gachsaran-Asmari boundary is sharp, and generally considered to be isochronous. The base of the Gachsaran in Fars Province, however, progressively replaces the upper part of the Asmari from northwest to southeast (Fig. 3).

Toward the interior of Fars Province the evaporites of the Gachsaran pass into the redbeds and thin beds of limestone of the Razak Formation (Figs. 7, 8).

Three members, the Chehel, Champeh, and Mol, are differentiated within the Gachsaran Formation of Fars Province. These were originally designated as formations by L. E. T. Parker (1959, company report). Their type section (Fig. 83) is Tang-e Chehel, in the central part of the southern flank of Kuh-e Gachestan, also called Kuh-e Namaki, about 40 miles northeast of the Fars port of Lengeh, and 2 miles north of the village of Rustami.

The Chehel Member is a prominent anhydrite or gypsum unit, 970 feet thick, which makes up the basal part of the Gachsaran Formation in southeastern Fars (Fig. 83). From the type section the Chehel becomes thinner west and north, and thickens east and southeast.

In its type section the Champeh Member consists of 360 feet of gray to red, gypseous marl, chalky-gypseous limestone or dolomite, and nodular to massive gypsum (Fig. 83). It is characteristically higher-weathering than either the underlying Chehel Member or overlying Mol Member. Limestone and dolomite within the Champeh are correlatives of the upper part of the Asmari Formation in Khuzestan.

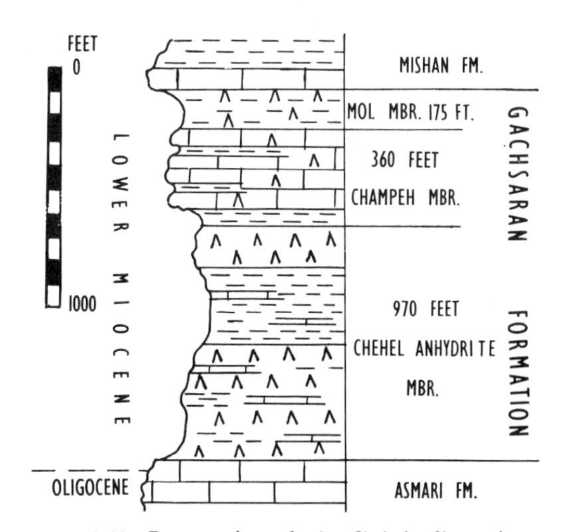

FIG. 83. Type section of the Chehel, Champeh and Mol Members of the Gachsaran Formation, Fars Province. N 26°54'04"; E 54°05'22"

The Mol Member is a prominent marker unit in Fars Province because of its predominantly red color. It consists of red to gray, gypseous marl and thin, gypseous limestone. In the type section it is 175 feet thick, which is an average thickness for southeastern Fars Province (Fig. 83).

Fossils and age.—In Khuzestan the Gachsaran Formation contains a predominantly brackish-water microfauna of ostracods, bryozoans, and rotaliids. Thin beds of limestone contain swarms of small miliolids and chilostomellids with associated *Peneroplis farsensis* Henson, *Dendritina rangi* d'Orbigny, rare *Miogypsina* sp., and charophyta (Fig. 79). Similar microfossils occur in the Gachsaran Formation of Lurestan.

The microfaunas of the Chehel, Champeh, and Mol Members of the coastal Fars area are summarized as follows.

Chehel Member.—Thin beds of limestone within the gypsum sequence contain rich occurrences of polymorphinids, miliolids, and small peneroplids. Associated marl beds contain ostracods, radiolarians, and small globigerinids. *Austrotrillina howchini* (Schlumberger) and *Peneroplis evolutus* Henson have been recorded from marl which occurs locally at the base of the Chehel Member.

Champeh Member.—The limestone and marl of the Champeh contain the following microfauna: *Taberina malabarica* (Carter), *Sphaerogypsina* sp., *Peneroplis farsensis* Henson, *Dendritina rangi* d'Orbigny, *Miogypsina* sp., *Flosculinella* sp., *Neoalevolina (Borelis) melo* (Fichtel and Moll) with associated *Elphidium* sp., rotaliids, and bryozoans. The succeeding Mol Member contains a similar, but less rich, fauna.

In interior Fars, limestone in the lower part of the Gachsaran Formation contains an early Miocene microfauna which includes: *Neoalveolina melo curdica* Reichel, *Taberina malabarica* (Carter), *Meandropsina anahensis* Henson, and *Peneroplis farsensis* Henson. This fauna is also typical of the upper Asmari limestone. Basal Gachsaran limestone and marl of interior Fars yield a poor peneroplid-miliolid fauna which includes *Peneroplis evolutus* Henson and *Austrotrillina howchini* (Schlumberger).

The megafauna of the Gachsaran Formation includes *Ostrea latimarginata* Vredenburg, *Antigona granosa* (Sowerby), *Clementia papyracea* (Gray), and *Ostrea praevirleti* (Douglas).

Throughout Khuzestan and Lurestan the Gachsaran Formation overlies the upper Asmari and is late early Miocene in age. In interior Fars the lower part of the formation is the time equivalent of the upper Asmari and probably part of the middle Asmari of Khuzestan.

Age of formation.—Early Miocene.

Razak Formation

Synonyms.—Previously the Razak has been included within the lower Fars or described as the central Iran facies (company report).

Type section.—The type section is located on the northern flank of Kuh-e Jahrum, 23 miles southeast of the city of Jahrum, and 0.5 mile southeast of the village of Chah Tiz. The succession consists of 2,540 feet of low-weathering, silty, red to green to gray marl with subordinate silty limestone (Fig. 84). It forms a broad, low, multicolored outcrop area between the resistant Asmari-Jahrum and the Guri Limestone Member of the Mishan Formation.

Both lower and upper contacts are gradational and conformable.

Regional aspects.—The Razak interfingers toward the south and southwest with the Gachsaran Formation (Figs. 3, 7, 85). This interfingering approximately coincides with the southwestern limit of the Sachun and Tarbur Formations as well as with the thick development of the Guri Limestone Member of the Mishan Formation. The Razak is considered to be in part equivalent to the upper red formation of central Iran.

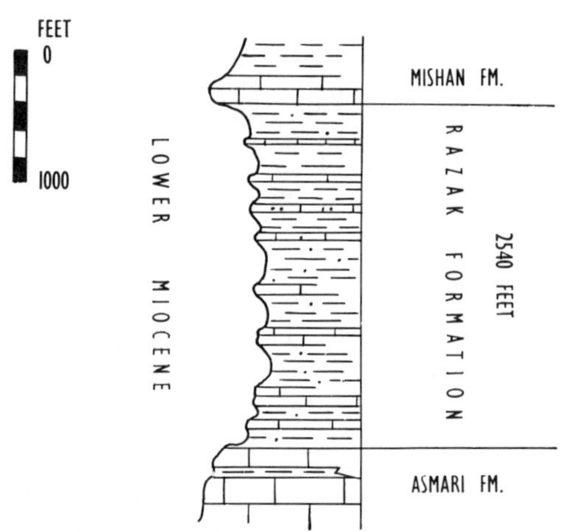

FIG. 84. Type section of the Razak Formation. N 28°33'00"; E 53°50'39"

Fossils and age.—The Razak Formation contains a fauna similar to that of the laterally equivalent Gachsaran Formation. The microfossils from the marl include ostracods, rotaliids, *Elphidium* sp., and charophyta. The thin silty limestone contains the following microfossils: *Miogypsina* sp., *Taberina* sp., *Neoalveolina (Borelis) melo* (Fichtel and Moll), *Dendritina rangi* d'Orbigny, and *Tubucellaria* sp. This fauna is also found in the Champeh Member of the Gachsaran Formation. *Flosculinella* cf. *bontangensis* (Rutten) has been recorded from a thin limestone 480 feet above the base of the formation. Rare planktonic Foraminifera, which occur near the base of the Razak, include: *Globigerinoides* cf. *rubra* (d'Orbigny), *G. bispherica* Todd, *G. triloba triloba* (Reuss), *G. triloba immatura* LeRoy, and *G. triloba sacculifera* (Brady). Reworked Oligocene fossils are common in the base of the formation in the interior Fars area.

Age of formation.—Early Miocene.

Mishan Formation

Synonyms.—The formation previously called middle Fars is now termed the Mishan Formation. Its usage is also extended into Fars Province, thus replacing the names argillaceous group and Anguru Marl.

References.—See Fars Group.

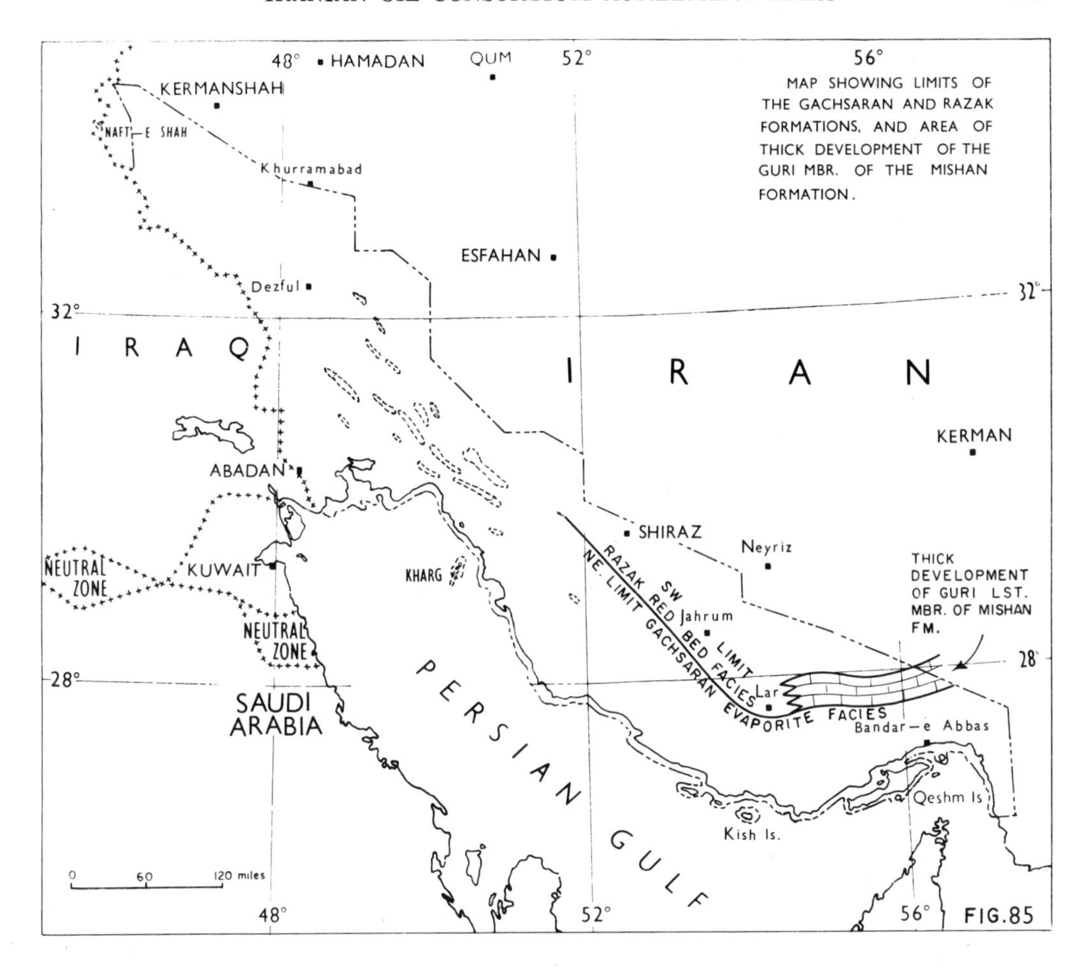

Type section.—The type section was measured along the "Golden Staircase" road on the southwestern flank of the Gachsaran oil field where approximately 2,330 feet of Mishan is present (Fig. 86). It consists of low-weathering gray marl and ridge-forming ribs of shelly limestone with abundant microfauna (Fig. 87).

The basal 200 feet is made up of alternating shelly, "worm bed" limestone and marl which may change abruptly along strike to massive reef limestone (Fig. 88).

The sharp basal contact with the gypsum of the Gachsaran Formation is accompanied by a minor amount of ferruginous staining. The upper contact is gradational.

Guri Limestone Member.—Through most of the Agreement Area the basal part of the Mishan is a limestone facies named here the Guri Limestone Member. Previously this unit has been called the *Operculina* limestone or the Guri Formation (company reports).

The Guri Member type section is just east of the salt plug on Kuh-e Gach, 18 miles southeast of the city of Lar and 11 miles east-southeast of the village of Nimeh. It consists of 370 feet of cream, hard, feature-forming, brown-weathering, fossiliferous limestone with thin bands of marlstone.

Both the upper contact with the marl of the Mishan and the lower contact with the gypseous red marl of the Gachsaran are sharp and conformable.

In an east-west-trending belt in the central part of the southeastern end of Fars Province (Figs. 7, 85) the Guri Member changes to a massive reef development, in many places exceeding 2,000 feet in thickness. This buildup is approximately coincident with the southern and southwestern

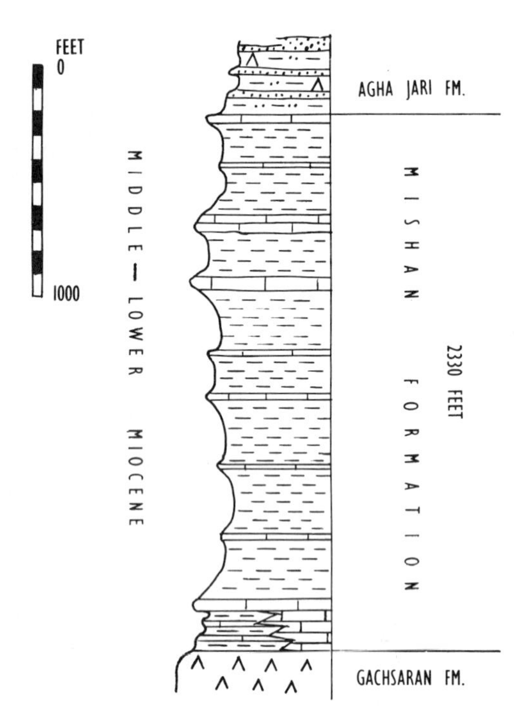

FIG. 86. Type section of the Mishan Formation.
N 30°13'08" ; E 50°45'34"

limits of the Sachun, Tarbur, and Razak Formations.

Regional aspects.—The marine conditions in which Mishan sedimentation took place existed for a shorter length of time in the northwest than the southeast. Consequently, the formation increases toward the southeast in time span and thickness (Fig. 3). From the oil fields area toward Lurestan, southeastern Iraq, Kuwait, and Saudi Arabia the Mishan is replaced by clastics of the Agha Jari Formation (Figs. 8-10). The formation therefore appears to consist of the deposits of a linear, southeast-to-northwest-trending, marine transgressive sea situated dominantly within what is now the Agreement Area. Erosion in the northeastern part of the Agreement Area has obscured the position of the northeastern limit of Mishan deposition.

Fossils and age.—The Mishan Formation abounds in rich occurrences of shallow-water marine faunas which include: *Ostrea virleti* (Deshayes), *Ostrea digitata* Eichwald, *Chlamys senatoria* (Gmelin), *Chlamys pusio* (Linné), *Lithophaga lithophaga* (Linné), *Clementia papyracea* (Gray), *Antigona granosa* (Sowerby), *Echinolampas jacquemonti* (d'Archiac and Haime), *Turritella* sp., *Conus* sp., *Natica* sp., and *Anadara* sp.

The rubbly, reef limestones of the Mishan, in particular the basal Guri Limestone Member, contain abundant Foraminifera (Fig. 89). The identified microfauna includes *Operculina complanata* Defrance, *Nephrolepidina* sp., *Miogypsina* sp., *Flosculinella bontangensis* (Rutten), *Dendritina* cf. *rangi* d'Orbigny, *Taberina malabarica* (Carter), and *Neoalveolina (Borelis) melo* (Fichtel and Moll). The bryozoan *Tubucellaria* and many forms of algae also occur.

Planktonic Foraminifera have been identified from a few marl zones in Fars Province: *Globigerinoides triloba triloba* (Reuss), *G. triloba sacculifera* (Brady), *G.* cf. *rubra* (d'Orbigny), *Orbulina universa* d'Orbigny, *Hastigerina* cf. *aequilateralis* (Brady), and *Sphaeroidinellopsis* sp.

Age of formation.—Early to middle Miocene.

Agha Jari Formation

Synonyms.—The name Agha Jari Formation replaces the term upper Fars. The transition beds of southwestern Khuzestan and Lurestan, which referred to transitional sediments between the argillaceous middle Fars and clastic upper Fars, are now included in the Agha Jari Formation. The Lahbari Member of the Agha Jari Formation is a term now applied to deposits which are essentially the same as those heretofore called lower Bakhtyari.

References.—See Fars Group.

Type section (Figs. 93, 94).—The type section was measured along the road crossing the central part of the Agha Jari oil field (Fig. 91). It consists of 9,730 feet of feature-forming, brown to gray, calcareous sandstone and low-weathering,

---

→

FIG. 87.—Light-colored Mishan marl with ribs of resistant, shelly, rubby limestone. Near Gachsaran.
FIG. 88.—Surface of "worm bed" limestone in basal Mishan Formation. Type section, near Gachsaran.
FIG. 89.—Mishan Formation, basal Guri Limestone Member showing typical assemblage of ***Operculina*** **(O),** ***Nephrolepidina*** **(N), and** ***Miogypsina*** **(M). Interior Fars. Early Miocene. ×35.**
FIG. 90.—Agha Jari oil field frontal thrust. Mishan Formation thrust over Agha Jari.

FIG. 87

FIG.88

FIG. 89

FIG.90

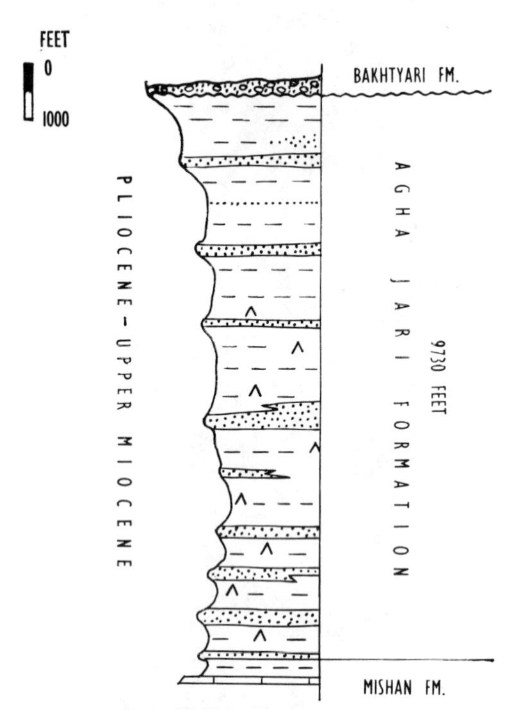

FIG. 91. Type section of the Agha Jari Formation. N 30°48′42″ ; E 49°47′29″

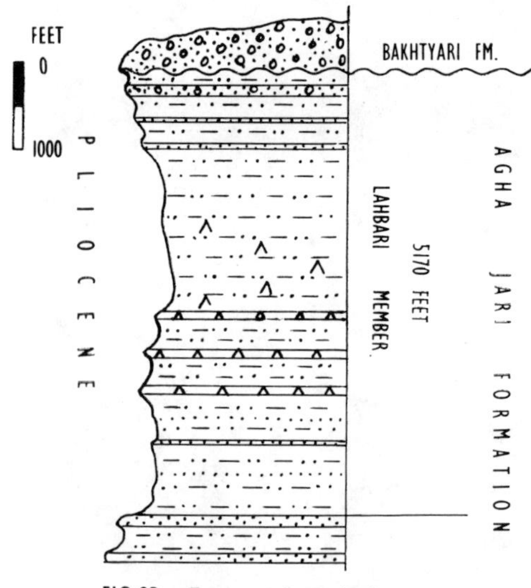

FIG. 92. Type section of the Lahbari Member, Agha Jari Formation. N 31°28′23″ ; E 49°37′33″

gypsum-veined, red marl and siltstone. Chert is one of the most common sandstone constituents. Local developments of thin, discontinuous, gypsum beds are common.

The Agha Jari transitionally overlies the Mishan Formation and is unconformably overlain by conglomerate of the Bakhtyari. The basal contact is defined as the top of the highest gray, marine marl of the Mishan Formation.

Lahbari Member of Agha Jari Formation.—The Lahbari Member refers to a rock unit that is essentially the same as the one previously called the lower Bakhtyari. Different criteria are used, however, in defining the Lahbari Member from those used to define the lower Bakhtyari.

The latter was a formation differentiated from the underlying upper Fars on the basis of the lowest pebble conglomerate bed. The Lahbari is separated from the main body of the Agha Jari on the bases of its low-weathering profile, buff color, and less consolidated nature, in sharp contrast to the rest of the Agha Jari, with its red color and well-cemented, prominent-weathering sandstone.

The type section of the Lahbari Member was measured at Tang-e Tukab on the northeastern flank of the Haft Kel anticline, 6 miles northeast of Haft Kel along the road to Malamir. The name is derived from the Lahbari syncline which separates the Masjed-e Suleyman and Kuh-e Asmari structures from the Haft Kel and Naft Safid anticlines. The type succession consists of 5,170 feet of low-lying, gypsum-veined, buff-colored siltstone, silty marl, and subordinate sandstone and gypsum beds (Fig. 92). The grade of deposit becomes coarser upward.

The Lahbari overlies the more prominent-weathering part of the Agha Jari and is uncon-

---

FIG. 93.—Massive, cross-bedded sandstone. Agha Jari Formation. Type section at Agha Jari oil field.
FIG. 94.—Outcrop of Agha Jari Formation silty marl at Agha Jari oil field.
FIG. 95.—Type section of Lahbari Member of Agha Jari Formation, overlain unconformably by Bakhtyari Formation. North of Haft Kel.
FIG. 96—Type section of Bakhtyari, showing Bakhtyari overlying Agha Jari Formation. Godar Landar water pumping station in foreground. North of Masjed-e Suleyman.

FIG.93

FIG.94

FIG.95

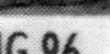

FIG.96

formably overlain by the Bakhtyari Formation (Fig. 95).

The Lahbari Member is a diachronous unit which intergrades both laterally and vertically with the main body of the Agha Jari Formation. It is developed in parts of the Khuzestan plains area and possibly in southwesternmost Lurestan and Fars Provinces. Toward the northeast it interfingers with, and becomes indistinguishable from, the rest of the Agha Jari.

Regional aspects (Figs. 80, 81, 90).—The Agha Jari Formation ranges from about 2,000 to 10,000 feet in thickness. In Lurestan and Khuzestan it is an estuarine and lacustrine deposit, but is partly marine in Fars Province. It is extremely diachronous throughout the Agreement Area, becoming younger from northwest to southeast and northeast to southwest.

The lower contact is customarily selected as the top of the first prominent gray marine marl of the Mishan; however, it is preferable in some places, particularly in Fars, to place the boundary at the base of the lowest significant red marl.

The clastics of the Agha Jari were derived from the northeast. Although erosion has obscured the picture, it is probable that the thickest accumulation was in a northwest-southeast trough trending through the center of what is now the Agreement Area.

Toward Kuwait and Arabia the Agha Jari thins and merges with the Dibdibba and Hofuf Formations. The area of provenance for the latter two formations probably was the Arabian shield.

Fossils and age.—The Agha Jari Formation contains a predominantly brackish- to fresh-water fauna with a few marine intercalations in the basal part of the unit. The microfauna is composed mainly of *Elphidium-Rotalia-Nonion* assemblages associated with numerous ostracods and charophytes. The microfauna includes: *Elphidium hauerinum* (Cushman), *Rotalia beccarii* (Linné), *R. beccarii* (Linné) var. *dentatus* Parker and Jones, *R. stachi* Asano, *R. takanabensis* (Ishizaki), *Nonion incisum* (Cushman), and the ostracods *Trachyleberis exanathemata* (Ulrich and Bassler), *Cytheridea* sp., and *Bairdiocypsis* sp.

The megafauna of the Agha Jari includes *Crassostrea gryphoides* (Schlotheim) var. *cuneata* Douglas, *Temnopleurus iranicus* Douglas, *T. toreumaticus* (Leske), *Chlamys* sp., and crustacean remains.

At Qeshm Island and on the adjoining mainland, the uppermost siltstone and sandstone of the Agha Jari Formation contain a rich marine megafauna: *Pecten vasseli* Fuchs, *Chlamys prototranquebarica* (Vredenburg), *Minnivola pascoei* (Cox), *M. ishtakensis* Eames and Cox, *Indoplacuna iranica* (Vredenburg), *Ostrea protoimbricata* Vredenburg, and *Temnopleurus latidunensis* Clegg. A Pliocene age has been assigned to this fauna.

The Lahbari Member of the Agha Jari Formation contains a shallow, brackish- to fresh-water microfauna consisting of *Rotalia, Elphidium,* and *Eponides* with associated ostracods and charophytes. Harrison (1932, company report) records remains of the Pliocene horse *Hipparion* from the Lahbari of Khuzestan.

Age of formation.—Late Miocene to Pliocene.

Bakhtyari Formation

The name Bakhtyari is derived from the Bakhtyari tribe of Khuzestan which has played a prominent role in the history of Iranian oil operations. It was first applied to chert and limestone conglomerate and sandstone unconformably overlying the Fars sediments of Lurestan, Khuzestan, and the Bakhtyari Mountains (Pilgrim, 1908). Later Anglo-Persian Oil Company geologists extended its usage to include underlying sandstone, siltstone, and thin conglomerate which they called lower Bakhtyari.

Bakhtyari is now restricted to the rock unit originally designated by Pilgrim. It is also extended to include homotaxial conglomerates of Fars Province.

Synonyms.—Upper Bakhtyari and Dardam Conglomerate.

References.—Loftus, 1855; Pilgrim, 1908; Busk and Mayo, 1918, 1929; de Böeckh, Lees, and F. D. S. Richardson, 1929; Lees, 1953; Ion, Elder, and Pedder, 1951; The British Petroleum Company Ltd., 1956; Falcon, 1958; Slinger and Crichton, 1959.

Type section (Figs. 81, 96).—The type section was measured at Godar Landar just north of Masjed-e Suleyman where the Karun River emerges from a gorge cut through the resistant Bakhtyari. Approximately 1,800 feet is present in the type succession (Fig. 97). The lower third consists of alternating massive, resistant conglomerate, and relatively low-weathering, strongly lensing, conglomeratic sandstone and gritstone. The

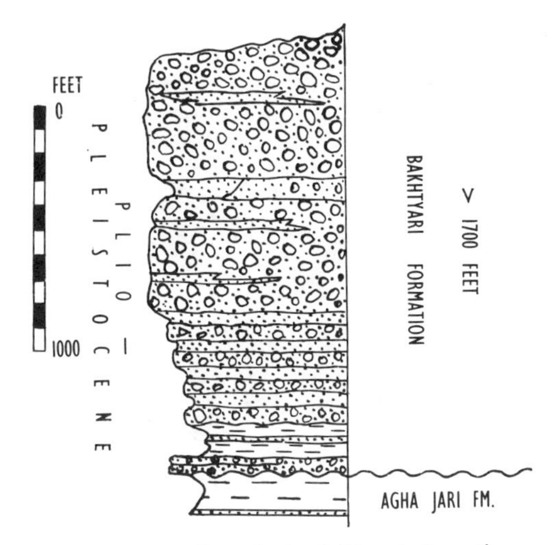

FIG.97. Type section of the Bakhtyari Formation. N 30°01'57"; E 49°23'59"

upper two-thirds consist of massive, cliff-forming conglomerate with a thin central unit of less resistant gritstone.

Constituents are rounded pebbles and cobbles of Oligocene, Eocene, and Cretaceous limestones and cherts firmly held in a matrix of sand, grit, and calcareous cement (Fig. 98). The chert generally is dark ferruginous brown. Conglomeratic material from the Fars Group is notably absent.

The basal contact with the Agha Jari Formation was obscured at the type locality, but could be seen nearby where an angular unconformity is present.

Regional aspects (Figs. 80, 81, 95).—Following the Mio-Pliocene uplift of the Zagros Mountains, great quantitie? of coarse Bakhtyari clastics were shed into the foothills region which now makes up the Agre' .nent Area. The growth of the foothill structures was concurrent with deposition, resulting in extreme thickness variations from a few feet to 8,000 feet (O'Brien, 1948; Lees, 1953). These were the last allochthonous sediments deposited in the trough between the Arabian shield and the high Zagros Mountains. Post-Bakhtyari sedimentation has been local in nature.

In coastal Fars the Bakhtyari is represented by a thin, transgressive, conglomeratic limestone, which rests with angular unconformity upon Agha Jari sediments.

Fossils and age.—No diagnostic fossils have been found in the Bakhtyari conglomerates. The formation is considered to be late Pliocene or younger in age.

FIG. 98.—Typical Bakhtyari pebble conglomerate. Khuzestan.

REFERENCES

American Commission on Stratigraphic Nomenclature, 1961, Code of stratigraphic nomenclature: Am. Assoc. Petroleum Geologists Bull., v. 45, no. 5, p. 645–656.

Arkell, W. J., 1956, Jurassic geology of the world: Oliver and Boyd Ltd., Edinburgh, 806 p.

Banner, F. T., and Wood, G. V., 1964, Lower Cretaceous—Upper Jurassic stratigraphy of Umm Shaif field, Abu Dhabi marine areas, Trucial Coast, Arabia: Am. Assoc. Petroleum Geologists Bull., v. 48, no. 2, p. 191–206.

Böeckh, H. de, Lees, G. M., and Richardson, F. D. S., 1929, Contribution to the stratigraphy and tectonics of the Iranian ranges, p. 58–177, *in* The structure of Asia, ed., J. W. Gregory: Methuen, London.

Bolli, H. M., 1957, Planktonic Foraminifera from the Eocene Navet and San Fernando Formations of Trinidad, B.W.I: U.S. Natl. Mus. Bull. 215, p. 165–172.

British Petroleum Co. Ltd., 1956, Geological maps and sections of southwest Persia: Proc. 20th Sess. Internatl. Geol. Cong., Mexico.

Busk, H. G., 1929, Earth flexures: Cambridge Univ. Press, 106 p.

——— and Mayo, H. T., 1918, Some notes on the

geology of the Persian oilfields: Jour. Inst. Petroleum Technology, v. 5, no 17, p. 3–26.
Cotteau, G., and Gauthier, V., 1895, Echinides fossiles, *in* J. de Morgan, Mission scientifique en Perse: t. 3, 2e pte., p. 5–142, Paris.
Cox, L. R., 1936, Fossil mollusca from southern Persia and Bahrein: Mem. Geol. Survey India, Paleontologia Indica, n.s., v. 22, mem. 2, 69 p.
Cox, P. T., 1938, The genus *Loftusia* in southwestern Iran: Eclogae Geol. Helv., Bd. 30, p. 431–450.
Currie, E. D., 1921, Fossils from western Persia: Geol. Mag., v. 68.
Douglas, J. A., 1927-1928, Contributions to Persian palaeontology: I, Fauna of Fars series; II, *Kuphus arenarius;* III, Fauna of Mio-Pliocene series: Anglo-Persian Oil Co., Holywell Press, Oxford.
Douvillé, H., 1900, Les explorations géologiques de M. de Morgan en Perse: Cong. Internatl. Géol., Paris, v. 8.
——— 1902, Sur les analogies de faunes fossiles de la Perse avec celles de l'Europe et de l'Afrique: Bull. Soc. Géol. France, ser. 4, t. 2.
——— 1904, Les explorations de M. de Morgan en Perse: Bull. Soc. Géol. France, ser. 4, t. 4, p. 539–553.
——— 1905, Les découvertes paléontologiques de M. de Morgan en Perse: C. R. Acad. Sci., Paris, v. 140.
——— 1907, Mollusques fossiles, *in* J. de Morgan, Mission scientifique en Perse: t. 3, 4e pte., Paris.
——— 1910, Etude sur les Rudistes, Rudistes de Sicile, d'Algerie, d'Egypte, du Liban et de la Perse: Mém. Soc. Géol. France, Paléont., t. 18, mém. 41, p. 1–84.
Dunnington, H. V., 1958, Oil in northern Iraq, *in* Habitat of oil: Am. Assoc. Petroleum Geologists, p. 1194–1251.
Eames, F. E., and Smout, A. H., 1955, Complanate alveolinids and associated Foraminifera from the Upper Cretaceous of the Middle East: Ann. and Mag. Nat. History, v. 8, no. 91, p. 505–512.
——— Banner, F. T., Blow, W. H., and Clarke, W. J., 1962, Fundamentals of mid-Tertiary stratigraphical correlation: Cambridge Univ. Press, 163 p.
Elder, S., 1963, Umm Shaif oilfield. History of exploration and development: Jour. Inst. Petroleum Technology, v. 49, no. 478, p. 308–314.
Falcon, N. L., 1958, Position of oilfields of southwest Iran with respect to relevant sedimentary basins, *in* Habitat of oil: Am. Assoc. Petroleum Geologists, p. 1279–1293.
Gauthier, V., 1902, in J. de Morgan, Mission scientifique en Perse: t. 3, 3e pte., Echinides, supplement.
Gray, K. W., 1950, A tectonic window in southwestern Iran: Quart. Jour. Geol. Soc. London, v. 105, pt. 2, p. 189–224.
Gregory, J. W., and Currie, E. D., 1920, Echinoidea from western Persia: Geol. Mag. v. 57, p. 500–503.
Harrison, J. V., 1924, The gypsum deposits of southwestern Persia: Econ. Geology, London, v. 19, no. 3, p. 259–274.
Henson, F. R. S., 1950, Cretaceous and Tertiary reef formations and associated sediments in Middle East: Am. Assoc. Petroleum Geologists Bull., v. 34, no. 2, p. 215–238.
——— 1951, Observations on the geology and petroleum occurrences of the Middle East: Proc. 3d World Petroleum Cong., The Hague, sect. 1, p. 118–140.
Hudson, R. G. S., and Chatton, M., 1959, The Musandam Limestone (Jurassic to Lower Cretaceous) of Oman, Arabia, *in* Notes et memoires sur le Moyen-Orient: v. 7, p. 69–93, Mus. Natl. D'Histoire Naturelle, Paris.
International Subcommission on Stratigraphic Terminology, 1961, Stratigraphic classification and terminology: Rept. 21st Sess., Norden, 1960, pt. 25, ed., H. D. Hedberg, Det Berlingske Bogtrykkeri, Copenhagen.
Ion, D. C., Elder, S., and Pedder, A. E., 1951, The Agha Jari Oilfield: Proc. 3d World Petroleum Cong., The Hague, sect. 1, p. 162–186.
Kent, P. E., Slinger, F. C. P., and Thomas, A. N., 1951, Stratigraphical exploration surveys in southwest Persia: Proc. 3d World Petroleum Cong., The Hague, sect. 1, p. 141–161.
Lees, G. M., 1933, The reservoir rocks of Persian oil fields: Am. Assoc. Petroleum Geologists Bull., v. 17, no. 3, p. 229–240.
——— 1934, The source rocks of Persian oil: Proc. 1st World Petroleum Cong., London, sect. 1, p. 3–5.
——— 1938, The geology of the oilfield belt of Iran and Iraq, *in* Science of petroleum: Oxford Univ. Press, v. 1, 1st ed., p. 140–148.
——— 1950, Some structural and stratigraphic aspects of the oilfields of the Middle East: Rept. 18th Internatl. Geol. Cong., London, pt. 6, p. 26–33.
——— 1951, The oilfields of the Middle East: Proc. 3d World Petroleum Cong., The Hague, General volume.
——— 1953, Persia, *in* Science of petroleum, v. 6, pt. 1, p. 73–83.
——— and Richardson, F. D. S., 1940, The geology of the oil-field belt of S.W. Iran and Iraq: Geol. Mag., v. 77, no. 3, p. 227–252.
Loftus, W. K., 1855, On the geology of the Turko-Persian frontier and of the districts adjoining: Quart. Jour. Geol. Soc. London, v. 11, p. 247–344.
O'Brien, C. A. E., 1948, Tectonic problems of the oilfield belt of southwest Iran: Rept. 18th Internatl. Geol. Cong., London, pt. 6, p. 45–58.
——— 1953, Salztektonik in Sudpersien: Z. Deutsch Geol. Gesel., Bd. 105 (4), Hanover, 1955.
——— 1957, Salt diapirism in south Persia: Geol. en Mijnbouw, 19e jaarg.
Owen, R. M. S., and Nasr, S. N., 1958, Stratigraphy of the Kuwait-Basra area, *in* Habitat of oil: Am. Assoc. Petroleum Geologists, p. 1252–1278.
Pilgrim, G. E., 1904, Cretaceous fossils from Persia: Rec. Geol. Survey India, v. 31.
——— 1908, The geology of the Persian Gulf and the adjoining portions of Persia and Arabia: Mem. Geol. Survey India, v. 34, pt. 4, p. 1–177.
——— 1924, The geology of parts of the Persian Provinces of Fars, Kirman and Laristan: Mem. Geol. Survey India, v. 48, pt. 2, 111 p.
Priem, F., 1908, Poissons fossiles de Perse: Ann. Hist. Nat., Délégation en Perse, t. 1, Paléontologic.
Redmond, C. D., 1964, The foraminiferal family Pfenderinidae in the Jurassic of Saudi Arabia: Micropaleontology, v. 10, no. 2, p. 251–263.
Reichel, M., 1936–1937, Etude sur les alveolines: Soc. Pal. Suisse, Mém. 57 and 59.
Richardson, R. K., 1924, The geology and oil measures of south-west Persia: Jour. Inst. Petroleum Technology, v. 10, no. 43, p. 256–283.
Rutsch, R., and Schenck, H. G., 1940, Upper Cretaceous pelecypods of the *Venericardia beaumonti*

group from Iran (abs.): Geol. Soc. America Bull., v. 51, p. 1976.

Slinger, F. C. P., and Crichton, J. G., 1959, The geology and development of the Gachsaran field, southwest Iran: Proc. 5th World Petroleum Cong., New York, sect. 1, paper 18, p. 349–375.

Steineke, M., and Bramkamp, R. A., 1952, Mesozoic rocks of eastern Saudi Arabia (abs.): Am. Assoc. Petroleum Geologists Bull., v. 36, no. 5, p. 909.

——— Bramkamp, R. A., and Sander, N. J., 1958, Stratigraphic relations of Arabian Jurassic oil, *in* Habitat of oil: Am. Assoc. Petroleum Geologists, p. 1294–1329.

Strong, M. W., 1937, Micropetrographic methods as an aid to the stratigraphy of chemical deposits: Proc. 2d World Petroleum Cong., Paris, t. 1, sec. 1, p. 395–399.

Thomas, A. N., 1950, The Asmari Limestone of southwest Iran: Rept. 18th Internatl. Geol. Cong., London, pt. 6, p. 35–55.

Van Bellen, R. C., Dunnington, H. V., Wetzel, R., and Morton, D. M., 1959, Lexique stratigraphique international: v. III, Asie, fasc. 10a, Iraq, 333 p.

Vredenberg, E. W., 1908, Occurrence of the genus *Orbitolina* in India and Persia: Geol. Survey India Rec., v. 36.

Asmari Oil Fields of Iran[1]

CEDRIC E. HULL[2] and HARRY R. WARMAN[2]
London, England

Abstract The Oligo-Miocene Asmari oil fields of Iran are truly giants, most of them having recoverable reserves greater than 1 billion bbl each, and many having much more. The fields are close together in a region of relatively constant stratigraphy and structure, and have a common genetic history. The individual accumulations occupy very large rock volumes in large-amplitude folds and, although the Asmari reservoir is poor in porosity and matrix permeability, very high production rates are possible because of extensive reservoir fracturing. These rates can be maintained for very long periods because of the great vertical extent of the oil columns. The Asmari fields are prime examples of anticlinal traps and of the effect of fracturing on reservoir performance.

The Asmari reservoir is limestone of Oligo-Miocene age and consists mostly of shallow-water but nonreefal carbonate rocks, with a significant sandstone member in the northwest part of the area. The Asmari is the uppermost wholly marine unit in a shelf-carbonate sequence interspersed with shale which was deposited, with only minor interruptions, from Carboniferous through Oligo-Miocene time. At the end of the time of Asmari deposition, increasing tectonic instability caused more varied sedimentation; this phase of instability terminated in a strong orogeny (Zagros orogeny) which formed the enormous anticlinal traps in this thick sedimentary sequence.

Introduction

The Oligo-Miocene Asmari oil fields of Iran are within the sedimentary basin which is between the Arabian shield and the Central Iranian plateau and which contains a significant proportion of the world's oil reserves. The general distribution pattern of fields in this area is varied (Fig. 1). Production is from reservoirs ranging in age from Middle Jurassic to Miocene, both carbonate rocks and sandstone. All of the fields discovered are primarily structurally controlled, but the structural style changes from the Arabian side of the Persian Gulf, where gentle folds have been growing since Mesozoic time, to Iran, where the folding of the Asmari fields is very pronounced. The Zagros orogeny which formed the Asmari structures took place during very late Tertiary time.

The size and prolific production of the Asmari fields are the results of a favorable combination of stratigraphic and structural conditions. Within the basin very large quantities of oil have been generated in sediments of different ages at least from the Middle Jurassic to the Miocene. The age of the Asmari oil is not firmly established. There is good reason to believe that some of it originated in the uppermost Cretaceous and Eocene shaly beds underlying the Asmari. However, some of it may be older.

The scale of oil accumulation in the Asmari can be gauged by the concentration within an area less than 200 mi long and 60 mi wide of 17 fields, a quarter of which have more than 5 billion bbl of recoverable reserves each. Many of the rest have recoverable reserves in excess of 1 billion bbl.

Stratigraphy

A generalized depiction of the total sedimentary fill of the basin is shown on the right side of Figure 2.

The sedimentary section of principal interest to the petroleum industry begins with transgressive sandstone of late Carboniferous age. This unit spread over a basement complex about which little is known in Iran, except that it includes lower Paleozoic sedimentary rocks. This sandstone grades upward into Permian shallow-water limestone up to 3,000 ft thick. After the Permian, a succession of depositional cycles was laid down. Among them are some regional unconformities and regressive phases, marked by minor angular unconformities and by the deposition of terrigenous clastics and evaporites. Folding movements were small, however, and relatively quiet deposition of mainly marine sediments continued until early Miocene time. Carbonate deposition predominated in much of the area throughout this time.

The part of the sedimentary section most relevant to consideration of the Asmari oil fields is from the base of the Bangestan Group up to the Gachsaran Formation (Fig. 2, left).

For the purpose of this paper, the Bangestan is considered to be one carbonate unit 2,500–3,000 ft thick. It is developed in two interleaved facies, one neritic detrital packstone to wackestone and the other dense, fine-grained,

[1] Read before the 53d Annual Meeting of the Association, Oklahoma City, Oklahoma, April 25, 1968. Manuscript received, August 26, 1968; accepted, November 1, 1968.

[2] British Petroleum Co. Ltd.

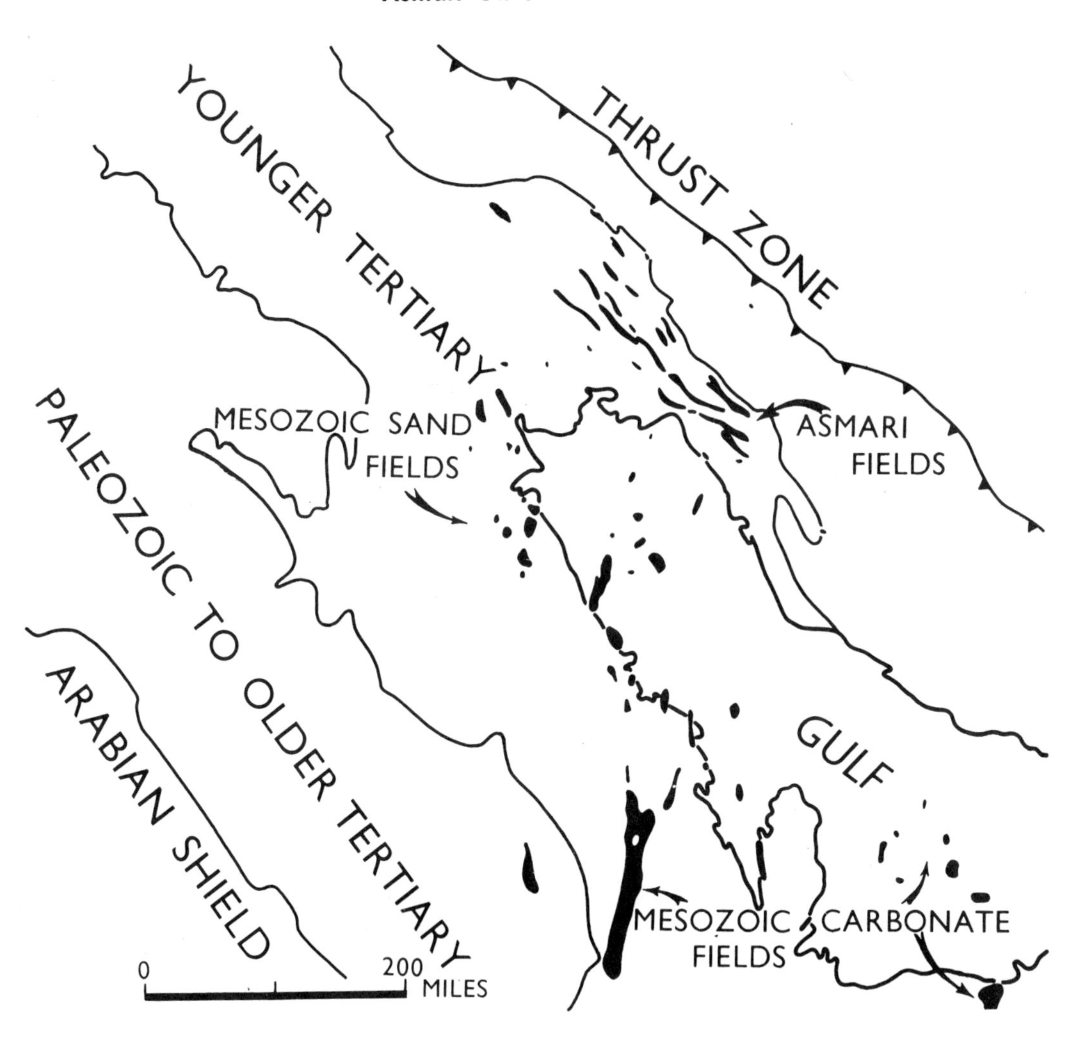

FIG. 1.—General distribution pattern of oil fields in Persian Gulf area.

calcite mudstone containing abundant *Oligostegina* (indicating a Cretaceous age).

Above the Bangestan is an argillaceous sequence about 2,500–3,000 ft thick, ranging in age from Santonian (Late Cretaceous) to Oligocene. The detailed stratigraphy of this sequence is complex, but the dominant rock type is dark-colored variably calcareous marine shale which has a fairly high organic content. The upper part of this unit grades into limestone, particularly in the southwestern part of the area of the Asmari oil fields.

The Asmari Limestone, which unconformably overlies shaly beds and their carbonate equivalents, of varied ages, is a major independent depositional cycle. Considering the magnitude of later tectonic events, there is strikingly little angularity between the Asmari and underlying beds. In places, both at outcrop and in wells, the basal contact is not easy to identify.

The Asmari Limestone ranges in age from Oligocene (with a rich microfauna including *Nummulites intermedius* d'Archiac, and *Nummulites vascus* Joly and Leymerie) to early Miocene. The upper Asmari has a rich microfauna including *Miogypsina* at the base and *Neoalveolina melo curdica* at the top. Contrary to some earlier reports, the Asmari shows no features related to reef development and any terms relating it to reef deposition are misleading. It consists of varied packstone and wackestone. The ratio of micrite to grain develop-

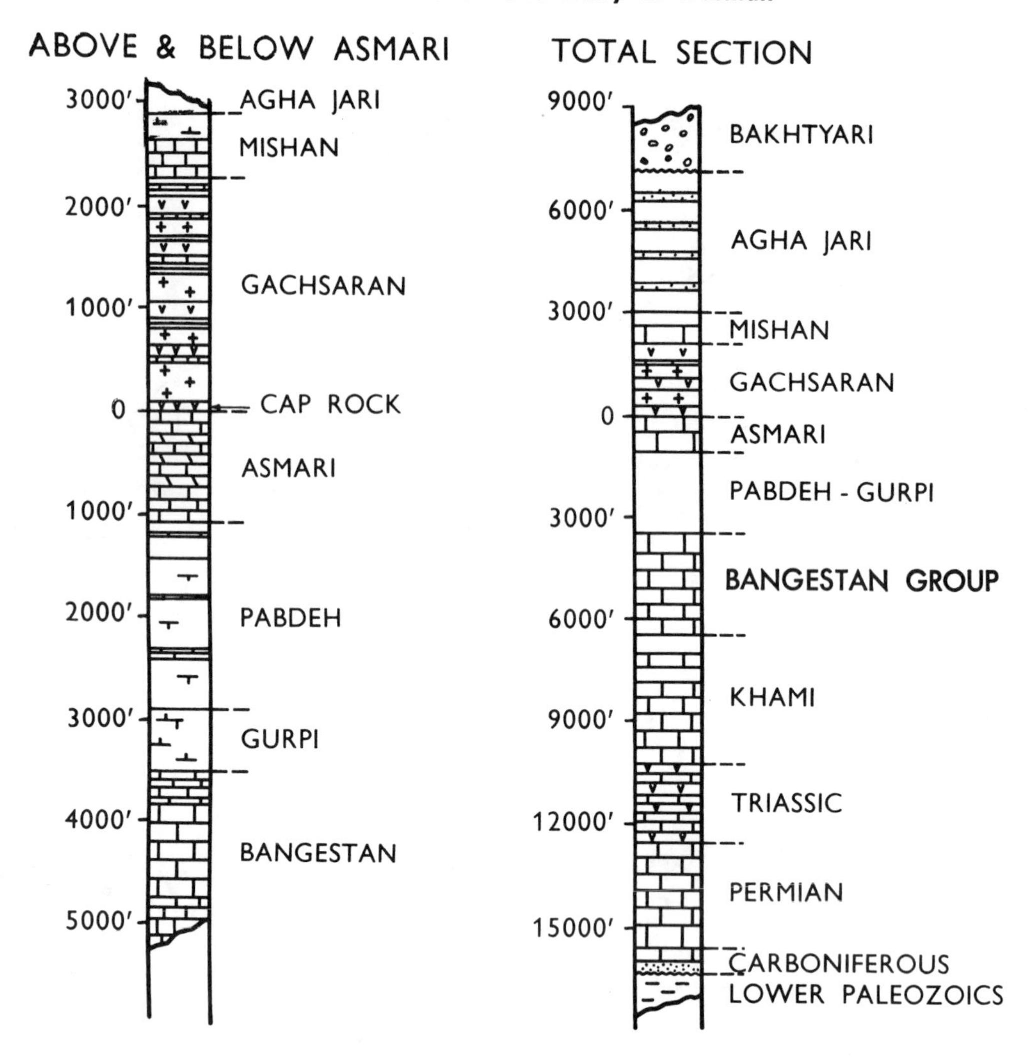

FIG. 2.—Stratigraphic section for southwest Iran. Column on right is a general depiction of total sedimentary fill of the basin. Left column shows part of section most relevant to Asmari oil fields.

ment differs in broad belts related to the overall basin shape rather than to reef developments. Diagenesis similarly affected broad belts, producing dolomitization and anhydritization. The formation of dolomite is particularly important with regard to the reservoir characteristics and in some of the fields provides the only signficant porosity.

In the more northwesterly fields, such as Ahwaz, sandstone forms an appreciable part of the Asmari reservoir. The sandstone, like some Mesozoic sandstone on the west side of the Persian Gulf, is an extensive sheet that spread out from the Arabian shield. At Ahwaz the sandstone forms 50 percent of the total thickness of the Asmari but it thins markedly northeast and southeast. In most of the known Asmari fields, sandstone does not constitute a significant part of the reservoir.

In the study area the Asmari ranges in thickness from 1,000 to 1,500 ft.

The usual convention in Iran is to divide the carbonate rocks of the Asmari into net pay and nonproductive rock. The division is made somewhat arbitrarily, depending on porosity, permeability, and water saturation.

The rock classified as nonproductive forms a high percentage of the total Asmari thickness, ranging throughout the study area from 25 to 75 percent. Its porosity is less than 5 percent and permeability everywhere is less than 1 md, generally much less.

The net pay carbonate rocks have average porosity normally ranging in any one field from 9 to 14 percent and permeability averaging 10 md, but in a few places exceeding 20 md.

Where significant quantities of sandstone are present, the reservoir properties are much better; porosity ranges from 20 to 30 percent and permeability values commonly are large.

An interesting feature is the apparent absence of fracturing in the sandstone, and the reservoirs with mixed sandstone and carbonate rocks have different production-behavior patterns. Some of the sandstone is uncemented and the high production rates obtained by the wells producing wholly from carbonate rocks are not possible because of the entry of sand into the borehole, and associated sand-cutting of well fittings.

Above the Asmari is a fairly thick succession composed of four formations. The lowest is the Gachsaran which consists of alternate anhydrite, salt, gray marl, and a few thin limestone beds. This unit shows considerable mobility and its thickness is difficult to estimate, though generally it ranges from 2,000 to 4,000 ft. The most important member of the Gachsaran is the "Cap Rock," a unit of constant overall character in all the Asmari fields. It consists of very compact, nodular anhydrite interbedded with thin shale beds which in one or two places are bituminous. There are also thin limestone beds with a marine microfauna of few species but abundant individuals. The "Cap Rock" forms a permeability and pressure barrier of great importance.

Below the "Cap Rock," 7,000 ft of limestone and shale (of Asmari and pre-Asmari strata) are fractured so intensively that in many places there is complete fluid connection. Above the "Cap Rock," the pressure regime in the Gachsaran, particularly in the limestone beds, has a wide range. Local pressure anomalies presumably are due largely to compaction and squeezing of the plastic mobile evaporites. These pressures are in many places nearly equal to the gross rock overburden pressure, and the mud problems associated with this part of the section are one of the main concerns in drilling in Iran.

Above the Gachsaran are marine shale and limestone of the Mishan Formation of middle Miocene age. The Mishan is well developed southeast of the Asmari oil-field belt where it is more than 2,000 ft thick, but wedges out near the northeast margin of the belt. The marine Mishan Formation passes gradationally upward into a thick redbed sequence named the Agha Jari Formation, which consists of 5,000–10,000 ft of interbedded red shale and sandstone devoid of marine fauna. The Agha Jari is overlain with marked unconformity by the Bakhtyari Formation, which consists of sandstone and well-cemented, coarse conglomerate, the pebbles and boulders of which consist mainly of Asmari and older limestone derived from the rising fold mountains of the Zagros.

Structure

The overall structure of the basin is shown in Figure 3, which is a very diagrammatic cross section from the Arabian shield to the zone of major thrusting along the northeast margin of the Zagros Mountains. The tectonic details of the whole Zagros system are very complicated. In general, the fold belt can be considered to consist of two main elements. The more northeasterly element, which forms the main chains of the Zagros Mountains, consists of relatively simple, large-amplitude folds exposing the main reservoir groups of the Asmari, Bangestan, and Khami. Southeast of this group of folds is a foothill zone in which folds of similar type and amplitude underlie a cover of younger Tertiary strata. The Asmari oil fields are in this foothill zone.

The amplitude of the folding is varied, but in several places the elevation difference between the crest of a fold and the bottom of the adjacent syncline only 5-10 mi away is more than 20,000 ft. It can be seen in Figure 4 that on Kuh-e-Khami the extrapolated elevation of the Asmari top is nearly 20,000 ft above sea level, whereas in the syncline on the southwest —only 12 mi away—it is 10,000 ft below sea level.

Critical closures on the anticlines are on the plunging ends of the structures. Because of a northwest regional tilt in the area of the main producing fields, the critical closure is on the southeast ends of the structures. The vertical difference between this spill point and the crest maximum, on most of the larger structures, is between 4,000 and 7,000 ft.

A concept of the scale of folding can be seen on the generalized cross section through the Gachsaran, Garangan, and Bibi Hakimeh structures (Fig. 4).

Two interpretations of the structural data are shown. Both seismic and drilling control

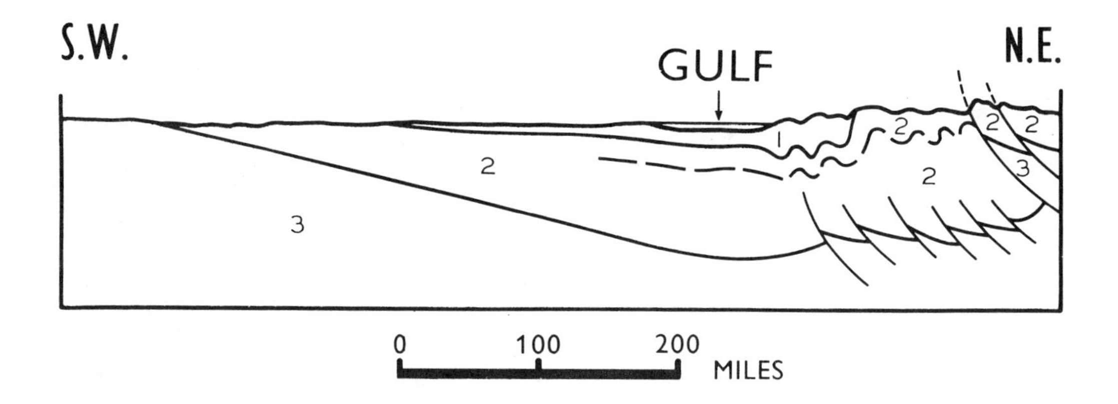

FIG. 3.—Diagrammatic section from Arabian shield to zone of major thrusting along southwest margin of Zargros Mountains.

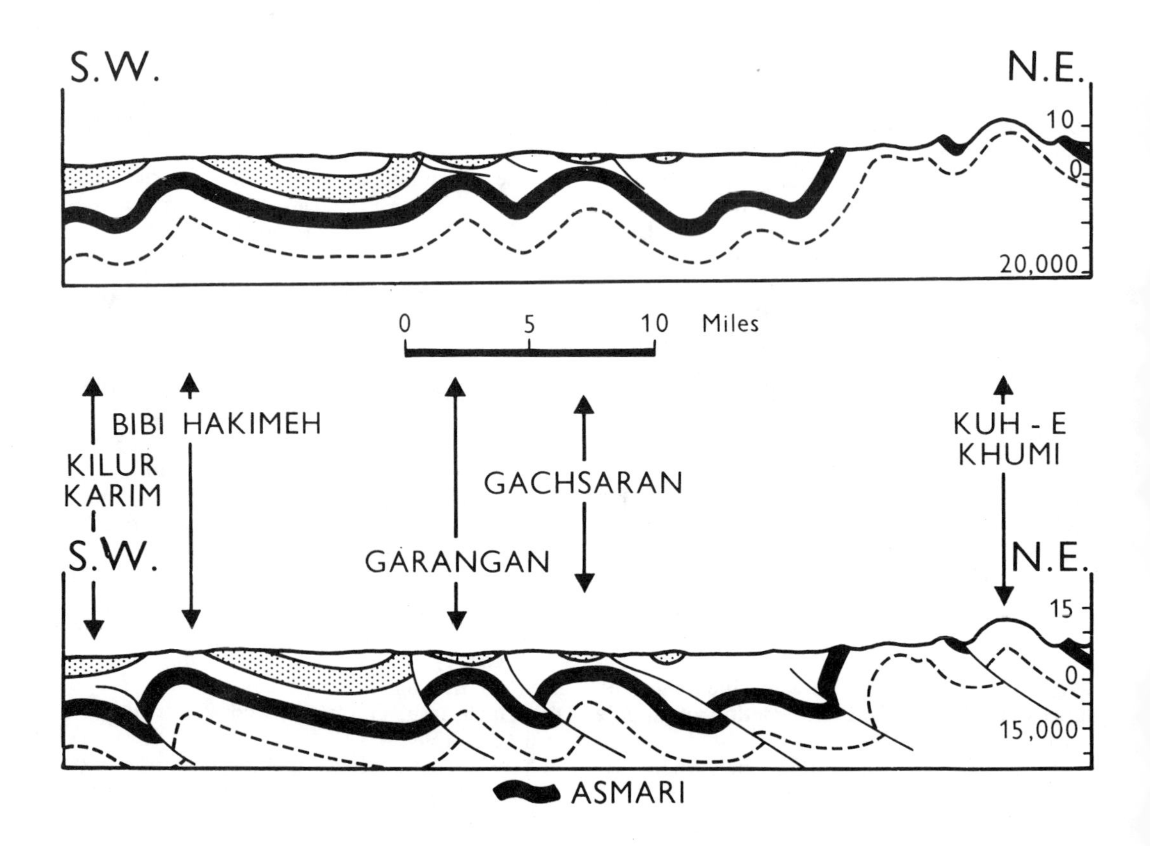

FIG. 5.—Kabir Kuh, one of the larger typical Zagros folds. Shows 100 mi of its total 130-mi length. Core is formed by limestones of Bangestan Group, overlain by soft Gurpi and Pabdeh Formations. Flanking flatirons are Asmari limestone.

are restricted mainly to the gentler parts (the tops) of the folds. The structure can be drawn either with or without thrusting in the deeper beds, according to one's preference. Thrusting is known to affect the contorted Gachsaran Formation and also the more competent overlying beds. Because of the deformation of the Gachsaran, it is not possible to draw simple sections showing the trace of faults affecting both the upper and lower sets of competent formations.

Gachsaran field has a single 6,000-ft vertical column of hydrocarbons. Bibi Hakimeh is another major field, with a 3,000-ft column of hydrocarbons.

Figure 4 also shows the disharmonic relation between the simpler folds of the Asmari and older beds and the surface structures. The two are separated by the mobile evaporite complex of the Gachsaran Formation.

A good example of the problems of defining the details of the Asmari fields is the relation of two of the largest fields—Marun and Agha Jari. These fields overlap *en échelon* and, in the region of the tight syncline between their plunging ends, the two oil accumulations are horizontally 3,000 ft apart but the original oil-water contacts differ vertically by about 5,000 ft.

Figures 5–7 are oblique aerial photographs

FIG. 4.—Sections through Asmari oil fields. **Upper** is drawn without thrust faults; **lower** is drawn with thrust faults.

FIG. 6.—Kuh-e-Kialan, an Asmari fold adjacent to Kabir Kuh. Fold plunges away for 16 mi to pass beneath the Saidmarreh landslide. Scale is shown by 1,000-ft thickness of Asmari and Eocene limestone beds.

that illustrate typical Zagros folds. The first (Fig. 5) is of Kabir Kuh, which is one of the larger folds; of its total length of 130 mi, more than 100 mi can be seen on the photograph. The core of the fold, formed by limestones of the Bangestan Group, is overlain by the soft, easily eroded Gurpi and Pabdeh Formations. The Asmari forms the prominent flanking flatirons.

Figure 6 shows an adjacent Asmari fold called Kuh-e-Kialan. The scale is shown by the 1,000-ft thickness of the Asmari and Eocene limestone beds. The Asmari fold can be seen to plunge away for 16 mi to the place where it passes beneath the Saidmarreh landslide (Harrison and Falcon, 1937). This remarkable landslide consists of approximately 65 mi^3 of the Asmari which once flanked Kabir Kuh, and which slid off and travelled 15 mi across country as one mass of ground and jumbled blocks of limestone.

Figure 7 shows folds somewhat farther back in the fold-belt system, where the folds are crowded together. The syncline is isoclinal; upsidedown Asmari strata overlie Asmari strata in normal position. The more easily eroded Pabdeh and Gurpi Formations are clearly differentiated and the visible core is limestone of the Bangestan Group.

Figure 8 shows Asmari Mountain, a typical but somewhat smaller Asmari anticline. The carapace of the mountain consists of Asmari Limestone. As can be seen from the hollow areas or holes on the southwest flank, once the more resistant limestone is breached, erosion soon cuts downward into the underlying Pabdeh shale. Surrounding the Asmari Limestone is the moderately eroded outcrop of contorted Gachsaran Formation.

Oil Accumulations

The crude oil in typical Asmari fields generally fills the large structures to the spill point. The oil is asphaltic in general type, and gravity values in the belt of main fields range from about 30 to 38° API. The more northerly fields have the higher gravity values. On the southwest the smaller accumulations contain heavier oils, with gravity values as low as 20°; in Kuh-e-Mund, at the southeast extremity of the known Asmari fields, there is a large accumulation of 11° API oil. Sulfur content ranges from slightly less than 1 to about 3.5 percent.

Associated gas is rather unsystematically varied in composition. Some of the fields have original free gas caps, ranging from very large to small; others nearby have undersaturated oil. Some crudes are undersaturated originally by as much as 4,000 psi.

Generally the oil in any particular accumulation is of very constant composition. This homogeneity is interpreted to be a result of the extremely good fracture connection throughout the reservoir that allows mixing by convection currents which prevent gravity segregation. there is no general evidence in the Asmari of the formation of tar mats or similar heavy basal segregations.

Pressure connection *via* the fissure systems is good, and on sustained production the movement of the gas-oil and oil-water interfaces is kept relatively constant. Pressure responses to production are essentially immediate across distances of tens of miles.

There is a regional pressure gradient through the Asmari water system, with an excess above hydrostatic pressure near the mountains and approximate hydrostatic equilibrium near the

Fig. 7.—Folds farther back in fold-belt system. Syncline is isoclinal; upsidedown Asmari overlies Asmari in normal position. Softer Padbeh and Gurpi Formations are clearly differentiated, and core is limestone of Bangestan Group.

FIG. 8.—Asmari Mountain. Carapace of mountain consists of Asmari limestone, and contorted outcrops of the Gachsaran Formation surround the Asmari.

Persian Gulf. This gradient causes differences in the levels of the oil-water interfaces on the two flanks of oil fields—as much as 500 ft in some of the structures.

An interesting feature of some of the oil accumulations is the good connection through fractures across the 2,000-3,000-ft interval of shale and marl separating the Asmari from the Bangestan. Where there is such connection the oils are essentially identical in composition and type, and fluid interfaces in the two separate formations appear to move in unison as a result of production from the Asmari. In places the Asmari and Bangestan have separate oil accumulations which differ in gravity values and other properties. In such places, the two units act as separate reservoirs, and there is no pressure response or fluid exchange as a result of production from either.

Despite the generally poor porosity and matrix permeability of the limestone, production rates from the Asmari are high because of the fractures. The Agha Jari field consistently produces approximately 1 million bbl/day from about 40 producing wells. This average production rate is common in many fields and exceptional wells are capable of producing more than 80,000 bbl/day, each. Because of the great vertical extent of the oil columns it is possible for wells to sustain these rates for many years. For example, more than 15 Asmari wells have each produced more than 100 million bbl.

Productivity indices (PI) differ in all fields, from a minimum of 2 bbl/day/psi at Lali to a maximum of 7,800 bbl/day/psi at Gachsaran. Many wells in different fields have a PI in excess of 1,000 bbl/day/psi.

Cumulative production to date (April 1968) was slightly more than 8.25 billion bbl.

CONCLUSION

Much has become known from the many

years of work on the detailed stratigraphy and structure of the region, but there is much to be learned, particularly about the origin, migration, and accumulation of the oil. The writers believe that, eventually, this prolific oil-producing region will provide a valuable contribution to the general understanding of oil-field formation.

In conclusion the writers wish to pay tribute to the foresight and gambler's instinct that led D'Arcy and the other early industry pioneers to prospect for oil in Iran, and to persist despite years of setbacks. The writers also pay homage to the early workers in the field who endured, a long way from home, hardships from a terrain and summer climate that have few equals for severity.

Selected References

British Petroleum Co. Ltd., 1956, Geological maps and sections of southwest Persia: 20th Internat. Geol. Cong. Proc., Mexico.

Falcon, N. L., 1958, Position of oilfields of southwest Iran with respect to relevant sedimentary basins, *in* Habitat of oil: Am. Assoc. Petroleum Geologists, p. 1279–1293.

Harrison, J. V., and N. L. Falcon, 1936, Gravity collapse structures and mountain ranges as exemplified in southwestern Iran: Geol. Soc. London Quart. Jour., no. 365, v. 92, pt. 1, p. 91–102.

——— and ——— 1937, The Saidmarreh landslip, southwest Iran: Geog. Jour., v. 89, no. 1, p. 42–47.

James, G. A., and J. G. Wynd, 1965, Stratigraphic nomenclature of Iranian Oil Consortium Agreement Area: Am. Assoc. Petroleum Geologists Bull., v. 49, no. 12, p. 2182–2245.

Kent, P. E., F. G. P. Slinger, and A. N. Thomas, 1951, Stratigraphical exploration surveys in southwest Persia: 3d World Petroleum Cong. Proc., The Hague, sec. 1, p. 141–161.

Lees, G. M., 1938, The geology of the oilfield belt of Iran and Iraq, *in* Science of petroleum, ed. 1: Oxford Univ. Press, v. 1, p. 140–148.

——— 1953, Persia, *in* Science of petroleum: Oxford Univ. Press, v. 6, pt. 1, p. 73–83.

O'Brien, C. A. E., 1948, Tectonic problems of the oilfield belt of southwest Iran: 18th Internat. Geol. Cong., London, pt. 6, p. 45–58.

Slinger, F. C. P., and J. G. Crichton, 1959, The geology and development of the Gachsaran field, southwest Iran: 5th World Petroleum Cong. Proc., New York, sec. 1, p. 349–375.

STRATIGRAPHIC RELATIONS OF ARABIAN JURASSIC OIL[1]

MAX STEINEKE,[2] R. A. BRAMKAMP,[3] AND N. J. SANDER[3]

New York, N.Y.

ABSTRACT

Jurassic rocks over much of the central and northeastern parts of Saudi Arabia constitute a major cycle of carbonate deposition closing with evaporites. Lower, Middle, and lower Upper Jurassic rocks apparently represent shallow-water shelf deposits in the west, passing to the east into rocks presumably originally lime muds deposited in water of unknown but greater depth. Two of the units (Tuwaiq Mountain and Hanifa) prominent in the shallow-water sequence of central Arabia seem to be lacking in the east. It is reasonable to assume that oil accumulations in these rocks (Dhruma and Jubaila formations) of coastal Arabia, originated in and near the rocks in which the oil is now found.

The later Jurassic Arab formation (middle or upper Upper Jurassic) shows increased paleogeographic differentiation. During the time of lower Arab deposition an extensive evaporite-depositing lagoon covered much of the western part of the area in which sediments of this unit survive. This lagoon was apparently bounded to the east by a bar, or group of bars, composed predominantly of cleanly washed calcarenite. Farther east are granular, fine-grained limestones, no doubt originally silt sized, with minor calcarenite (generally with a lime mud matrix). Still farther, lithographic limestones which must have been laid down as essentially pure lime mud, become prominent. In the area of the oil fields, the upper Arab evaporites periodically expand from west to east and the general cycle finally closed with the Hith anhydrite apparently blanketing the whole area.

The largest Arabian Jurassic oil accumulations occur in the lowest member of the Arab ("D member") in the area where the dominant rocks are clean-washed calcarenites, and other more or less porous fine-grained carbonate rocks, apparently as a belt between the evaporite-depositing lagoon on the west and a lime-mud-depositing area on the east. Oil is generally present in the upper Jubaila in this same general area when porosity is sufficient.

Oil of the upper Jubaila and "D member" of the Arab may have originated either 1. In the general stratigraphic units in which it is now found, or 2. In the lime muds of the underlying Jubaila from which it was forced upward during compaction. Oil in higher members of the Arab formation probably was generated in the members in which it is now found.

The preferred hypothesis for the geological history of upper Jubaila and Arab oil accumulations is that permeability barriers of stratigraphic type surrounding the main calcarenite lenses and other bodies of porous rock maintained the oil in the general area until the gentle folding of Middle Cretaceous to Eocene times, the main period of growth of the structures, concentrated the oil into the fields as they are now known.

Definitions of the main sedimentary rock units currently used in Saudi Arabia are included.

INTRODUCTION

The relatively complete Saudi Arabian sequence of late Paleozoic and Mesozoic rocks is exposed in central Arabia as a great curved outcrop belt formed of beds dipping at low rate off the Arabian shield. The great structural stability of the area, the gentle homoclinal east dip, the arid climate, the prevalence of erosional over depositional topographic surfaces, and the tendency for certain units to form

[1] The material of this paper was originally read in two parts under titles—"Mesozoic Rocks of Eastern Saudi Arabia," by Max Steineke and R. A. Bramkamp, before the Association at Los Angeles, March 25, 1952; and "Stratigraphic Relations of Arabian Jurassic Oil," by R. A. Bramkamp and N. J. Sander, before the Association at New York, March 30, 1955. The late Max Steineke has been retained as senior author because of his extensive contributions to the stratigraphy presented, and his role in the 1952 paper. Published by permission of the Arabian American Oil Company. Manuscript received, September 26, 1955.

Many geologists have had parts in the study of the geology of Saudi Arabia during the 22 years of exploration work of the Arabian American Oil Company. It is not possible to give credit here for even the more important individual contributions, but the large amounts of excellent field and laboratory work by these men form the basis for the summary which is presented here.

[2] Deceased.

[3] Arabian American Oil Company. Dr. Sander is not now (1958) with Aramco, and resides in Paris, France.

bold escarpments result in nearly continuous clean exposures of some of the rock units for distances along strike from 500 to nearly 1,000 kilometers. Jurassic exposures are especially good. Abundant fossils at many levels allow accurate dating of a number of horizons.

The homoclinal outcrop belt carries another consequence, however. This is that, although strike information on lateral changes in the central Arabian rocks is excellent, lithologic variation normal to the strike must mainly be derived from a few drilled wells, the number of which is negligible in terms of the great area involved. To the east a substantial body of information on equivalent rocks is available only in the area of intensive oil search near the western shore of the Persian Gulf. Paleogeographic and lithofacies interpretations must be pieced together from the outcrops in Nejd and the records of distantly scattered oil wells and wildcats near the Persian Gulf. It is obvious that any current discussion of regional geological features will have many serious weaknesses.

Although some drastic lateral lithologic changes take place over short distances, most such changes are gradual. The information already gathered leaves little doubt that the basic lithofacies patterns are developed on a grand scale and that overall simplicity is a characteristic of most of them. For this reason, the first steps toward a synthesis of some of the more obvious features are attempted here in spite of the widely scattered distribution of the various pieces of information.

GEOLOGICAL WORK IN SAUDI ARABIA

References to the geology of central and eastern Arabia prior to oil search in the area are scattered and are mainly limited to descriptions of the terrain; by far the most notable are those of Philby (1922, 1928, 1933, and 1939). Several of his fossil collections gave an initial glimpse of the basic rock sequence of central Arabia, with which Lamare (1936) and von Wissmann, *et al.* (1943), made rather sound surmises as to the broad geology of the area. These two papers give relatively complete references to prior literature on Arabia and their citations will not be duplicated in this paper.

Systematic surface geological field work of the Arabian American Oil Company started from Jubail October 22, 1933, and extensive reconnaissance was continued through the pre-war years. Most of the rock units were recognized and named by Max Steineke in 1937 in unpublished reports. Much detailed stratigraphic information on the surface exposures of Paleozoic and Mesozoic sedimentary rocks of Nejd was gathered in early post-war years.

In connection with oil search, the eastern part of the area, almost completely blanketed by Miocene-Pliocene rocks and wind-blown sand, has been attacked systematically by the use of shallow drilled wells for structural and stratigraphic purposes, gravity-magnetic mapping, and seismic methods.

Deeper drilling in the Eastern Province started on Dammam Dome April 30, 1935. Following disappointing results from initially attractive shows in the Middle Cretaceous, gas was found in the Arab Zone of Dammam Dome on December 31, 1937; the first test yielding oil was on March 4, 1938.

Information on rocks down to and including the Arab is good in the areas of development where oil in it has resulted in extensive drilling to this unit. In Saudi Arabia only 12 wells near the Persian Gulf have penetrated substantially below the Arab and, of these, only 6 have been drilled completely through the Jurassic into older rocks.

GENERAL STRATIGRAPHY[4]

In the work done in Saudi Arabia by the Arabian American Oil Company the number of formally named rock units has been held to a minimum following a policy introduced by the senior author early in Arabian work. It was felt that to christen each small recognizable lithologic unit would confuse rather than simplify the problem of presenting the Arabian rock sequence, and that names of smaller rock units would be of use only to the handful of geologists actually working in the country. For routine surface and subsurface work, informal names falling outside of the rules of stratigraphic nomenclature have been used.

The nearly complete and well-exposed surface sequence of Central Arabia has allowed the use of surface sections to serve for the definitions of all but one of the formational names, thus avoiding some of the considerable difficulties which often arise in units defined on the basis of well data. A large proportion of the names were first used in company reports by Steineke, many of them first in 1937. When it has been possible better to place formational limits at horizons of maximum geological significance as determined by later work, and where the type sections could be located on the best possible exposures of the particular units, strict priority within unpublished Company reports has not been observed.

Table I lists the main units in current use. The definitions of these units and the locations of their type sections are summarized in the following pages. Figure 1 gives the approximate locations of the various type localities.

SAQ SANDSTONE (CAMBRIAN?)

TYPE SECTION

Location.—Along a traverse extending from the base of Jebel Hanadir (26°27.6'N., 43°39.8'E.) 28.3 km S. 40½° W. to Jebel Saq (26°15.9'N., 43°18.9'E.), and from the latter 20.5 km S. 86° W. to a point on the top of the crystalline basement (26°12.9'N.,43°07.0'E.).

Thickness.—More than 600 meters (calculated).

Lithology.—Brown- to black-weathering, white to buff, massive, commonly cross-bedded, medium-to coarse-grained sandstone. Locally considerable red color is present. A few thin red silty members occur locally.

Limits.—The base is at the unconformable contact with underlying crystalline basement. The top is at the contact of relatively massive sandstones of the Saq with overlying shales and sandstones with *Didymograptus* cf. *bifidus* (Hall) of the basal Tabuk formation.

FOSSILS AND AGE

"*Cruziana*"-like tracks are present in the upper part, but no diagnostic fossils have been found.

Rocks of Cambrian age, dated by fossils, occur in the Aqaba-Dead Sea area. The latest discussion of these is given by Quennell (1951, pp. 49-98). Actual tracing of Saq laterally into the area of the fossiliferous Cambrian has not yet been

[4] By Max Steineke and R. A. Bramkamp.

TABLE I. SAUDI ARABIAN ROCK UNITS

ROCK UNITS	MAIN ROCK TYPES	THICKNESS (TYPE SECTION)	AGE
Hofuf formation	Sandy marl and limestone; local quartz gravel at base	95m	Miocene or Pliocene
Dam formation	Shale, marl and limestone, with chert	90m	Probably Middle Miocene
Hadrukh formation	Sandstone, shale, marl & chert	84m	?Miocene
Unconformity			
Dammam formation	Limestone, dolomite, clay and marl	28m	Lower and Middle Eocene
Rus formation	Anhydrite, marl, shale and limestone	56m	Lower Eocene
Umm er Radhuma formation	Mainly limestone and dolomite	About 229m	Paleocene and Lower Eocene
Aruma formation	Limestone, with dolomite and shale	144m	Late Upper Cretaceous (Campanian? and Maestrichtian)
Unconformity			
Wasia formation	Sandstone and shale with subordinate limestone	42m	Early Upper Cretaceous (Cenomanian)
Unconformity			
Biyadh sandstone	Sandstone, with shale	About 270m	Lower Cretaceous
Buwaib formation	Limestone, with subordinate shale and sandstone	34m	Lower Cretaceous
Unconformity			
Yamama formation	Calcarenite and fine-grained limestone	58m	Lower Cretaceous (Neocomian)
Sulaiy limestone	Limestone, with a basal calcarenite unit	About 180m	Lower Cretaceous?
Disconformity			
Riyadh group			
Hith anhydrite	Anhydrite	71m	Probably Upper Jurassic
Arab formation	Limestone, dolomite and anhydrite	127m	Late Upper Jurassic
Jubaila limestone	Limestone	About 110m	Upper Jurassic (Kimmeridgian)
Hanifa formation	Limestone	101m	Upper Jurassic (?Oxfordian)
Tuwaiq Mountain Limestone	Limestone, mainly coral-bearing	215m	Upper Jurassic (Callovian)
Unconformity			
Dhruma formation	Limestone, and shale	383m	Middle Jurassic (Bajocian - Bathonian)
Marrat formation	Limestone, dolomite and red shale	111m	Lower Jurassic (Toarcian)
Unconformity			
Minjur sandstone	Sandstone, with varicolored shale	315m	Triassic or Jurassic
Jilh formation	Sandstone, shale and limestone	About 326m	Middle Triassic
Sudair shale	Mainly red shale	116m	Permian or Triassic
Khuff limestone	Limestone, with shale and marl	235m	Permian (Probably Upper)
Unconformity			
Jauf formation	Limestone and shale	276m	Silurian and Lower Devonian
Tawil sandstone	Sandstone	Over 200m	Silurian
Tabuk formation	Sandstone and shale	About 725m	Ordovician and Silurian
Saq sandstone	Sandstone	Over 600m	Presumably Cambrian
Unconformity			
Basement complex			

accomplished. However, there can be little doubt that the Saq represents the approximate lateral equivalent of Quennell's combined Quweira, Ram, and Um Sahm series as mapped by him in the general vicinity of the Saudi Arab-Jordan

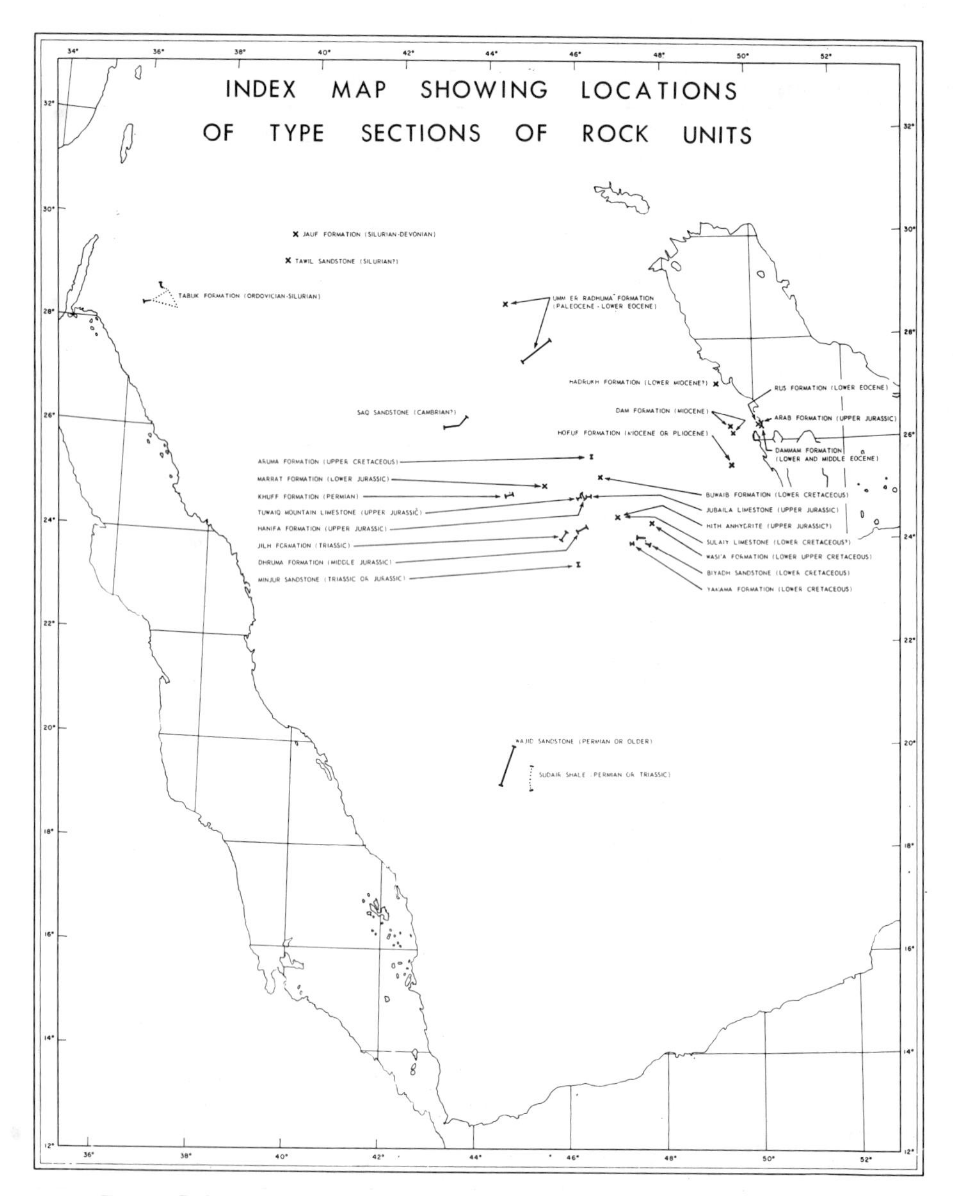

FIG. 1.—Index map showing locations of type sections of rock units, Saudi Arabia.

boundary. These, in turn, are considered by him to include beds equivalent to the fossiliferous Cambrian.

TABUK FORMATION (ORDOVICIAN-SILURIAN)

TYPE SECTION

Location.—The basal 104 meters is exposed on a rough line between 33 km S. 83–3/4° W. and 15.5 km S. 84° W. of Tabuk (28°21.1′N., 36°14.3′E., to 28°22.3′N., 36°24.9′E.). East of this lies a gravel-covered plain lacking exposures, but the beds concealed under this (about 210

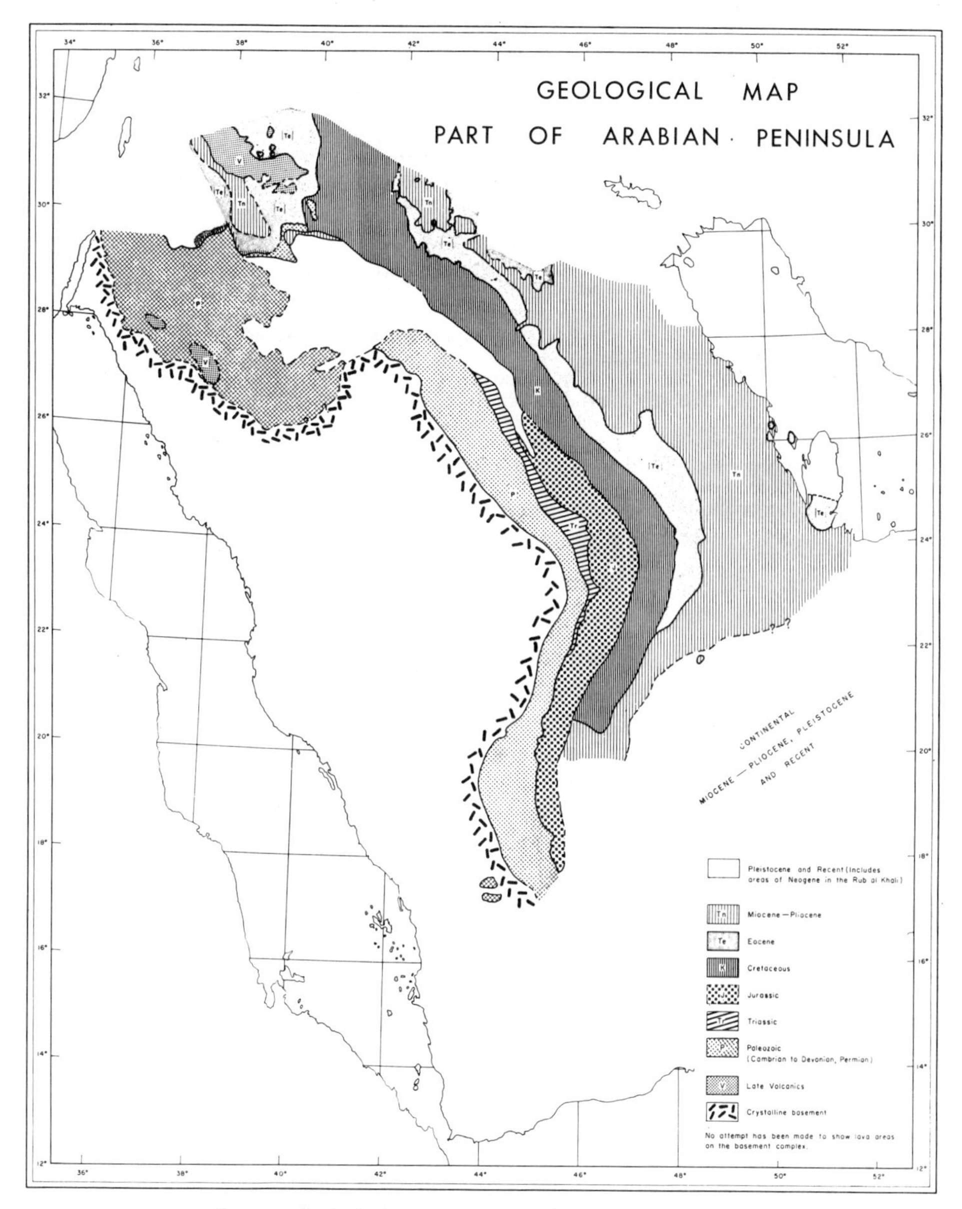

FIG. 2.—Geological map of part of the Arabian Peninsula.

meters) were studied in a supplemental type section about 50 km S. 82° E. of Tabuk (28°19.5′N., 37°0.4′E.) where a minor structural disturbance prevents completely filling the gap, but there is good reason to believe that only a few meters at the top remained unmeasured. East of the gravel-covered plain the succeeding 47 meters were found 24 km N. 45½° E. of Tabuk (28°32.8′N., 36°45.0′E.), and the succeeding and uppermost 411 meters of the unit were measured from a point 5.5 km S. 3° E. of Jebel Sharaura (38°42.0′N., 36°36.1′E.) extending to its top (38°45.0′N., 36°35.0′E.). Supplemental data on the upper 30 meters were obtained at 28°53.0′N., 36°59.6′E.

Thickness.—Slightly more than 725 meters. This thickness of beds was measured by hand-level, except for the one, probably small, gap of unknown lithology and thickness mentioned above.

Lithology.—Pastel and light-gray mottled, fine sandstone; many horizons with abundant tubular molds (? worm tubes); (127 m).

Gray, olive, brown, and red-brown fine silty sandstone and micaceous shaly siltstone (67 m).

Greenish-gray and gray slightly micaceous fine sandstone and micaceous shaly siltstone, with subordinate micaceous silty shale (61 m).

Green to greenish-gray, micaceous, shaly siltstone, and purple and green in part micaceous shale, with subordinate silty fine sandstone; *Climacograptus?* sp. in lower 20 meters; Orthoceratid cephalopod 35 meters above base (88 m).

Gray, purple, and green shale, in part silty and micaceous, with layers of light gray platy fine sandstone; at several horizons, *Climacograptus?* (47 m).

Gap of unknown, but probably small thickness.

Gray fine sandstone, in part micaceous, and green micaceous silty shale (105 m).

Green, purple, and gray shale, with subordinate fine sandstone and siltstone, mainly poorly exposed; *Diplograptus?* sp. in lower 7 meters (71 m).

Black-weathering, massive, medium-to coarse-grained quartz sandstone (a possible disconformity at the base); (± 35 m).

Brown to gray fine-to medium-grained quartz sandstone, in part cross bedded, with silty micaceous, fine sandstone interbedded in upper 25 meters; (92 m).

Purple and gray, calcareous, silty shale, with a few thin irregular impure limestone layers. *Didymograptus* cf. *bifidus* (Hall); (12 m).

Limits.—The base is at the sharp contact between *Didymograptus*-bearing shales and the underlying massive coarse sandstones of the Saq.

The top is an apparent disconformity between the pastel-colored fine sands of the upper Tabuk and the overlying locally ironstone-bearing coarse sandstone at the base of the Tawil.

FOSSILS AND AGE

Didymograptus cf. *bifidus* (Hall) in the basal Tabuk is taken to date its lower part as Lower Ordovician. *Diplograptus?* sp., possibly upper Ordovician, occurs near the middle of the formation. Probable *Climacograptus* and several other graptolites serve as the basis for a Silurian assignment for beds about 375 meters below the top. Graptolites listed by Blankenhorn, (1914, p. 13-14) and Stanislas-Meunier (1891, p. 759-770) were no doubt collected from the Tabuk formation.

In southern Jordan Quennell (1951, p. 99 and geological map) shows beds which must be equivalent to the Tabuk within what he has mapped as "Kurnub sandstone," for example, the general area of Muduwarra. The name "Kurnub" should not be used for these Paleozoic rocks as the "Kurnub" in its type area is Cretaceous.

Tawil Sandstone (Silurian?)

TYPE SECTION

Location.—The northern face of Jebel Tawil near Jebel Quwairah (about 29°20′N., 39°30′E.).

Thickness.—200 meters (minimum). The full thickness has not been accurately measured.

Lithology.—Black- to dark brown-weathering, massive, medium- to coarse-grained quartz sandstone, in part with large-scale cross bedding. Locally red shaly intervals are present in the lower portion. Ironstone is developed as platy and concretionary masses at a number of horizons, particularly in the lower part.

Limits.—The base is at the contact of the pastel, varicolored fine sands of the upper member of the Tabuk formation with the overlying sandstone of the Tawil. This contact shows apparent channeling and therefore seems to be disconformable.

The top is the apparently conformable contact of massive Tawil sandstone with the shales and limestone of the basal member of the Jauf formation.

FOSSILS AND AGE

"*Cruziana*"-like tracks are present. The Tawil is underlain by Silurian rocks and overlain by probable Silurian.

Jauf Formation (Silurian-Devonian)

TYPE SECTION

Location.—In Sha'ib Gharisa 24½ km N. 69° W. of the fort at Jauf (29°53.5'N., 39°37.8'E.).
Thickness.—276 meters.
Lithology.—Thin beds of cream to tan, thin-bedded, hard, finely crystalline limestone (locally coral-bearing) separated by gray-green silty shale layers, with subordinate interbedded sandstone; (108 m).
Banded red and gray, silty shale with minor beds of partly micaceous, fine sandstone and siltstone common in the upper part and few in the lower part; (113 m).
Gray thin-bedded, saccharoidal to finely crystalline limestone (upper 20 meters) which locally includes common reef mounds 1 to 5 meters high, grading down into grayish-green, silty shale with subordinate maroon shale and silty limestone (lower 35 meters); (55 m).
Limits.—The base is at the contact of the lower Jauf shales and silty limestones with the underlying sandstones of the Tawil.
The top is at the contact of the uppermost fossiliferous limestone of the Jauf with overlying presumably Cretaceous gray, tan and greenish shales, silts and sandstones.

FOSSILS AND AGE

Rather meager faunas have been collected at various levels in the Jauf formation. Brachiopods in the upper part, including *Anathyris* and *Rensselaeria,* have been tentatively determined as Lower Devonian by G. A. Cooper. The fauna of the lower Jauf is suspected to be Silurian.

Wajid Sandstone (Permian or Older)

TYPE SECTION

Location.—The lower portion is in the series of hills and mesas of Jebel Wajid (approximately 19°10' to 19°20'N., 44°25' to 44°35'E.). The upper portion is in the vicinity of Beni Ruhaiya (19°50'N., 44°46'E.).
Thickness.—Calculated to be greater than 300 meters.
Lithology.—Mainly relatively poorly cemented sandstone of variable coarseness, at certain levels conglomeratic, with interbedded varicolored sandy shales, and thin, black to brown, ironstone-bearing horizons. In some areas (Beni Khatma and the Mundafan) two horizons near the middle include granitic and metamorphic rock erratics reaching 2 meters in diameter. Locally a few thin, irregular, pale gray and pastel, dense, dolomite lenses are present, particularly in the upper part; (150 m).
Brown heavily bedded, partly cross bedded, poorly to moderately cemented, medium- to coarse-grained, slightly ferruginous quartz sandstone; (150 m).
Limits.—The base is at the unconformable contact of the sandstone with metamorphic and plutonic rocks of the basement complex. The basement surface was topographically irregular.
The top (subject to later revision) is at the probable disconformity between massive sandstones below and the dolomite, shale, and limestone of the basal Khuff above.

FOSSILS AND AGE

No fossils have been found. As the Wajid underlies the Khuff, which is Permian, it is presumed to be Permian or older.

Khuff Formation (Permian)

TYPE SECTION

Location.—Near the Riyadh-Jeddah road, between a point 5.3 km N. 08° W. of Ain Khuff (24°58.6'N., 44°41.8'E.), and a point to the southwest, 13.5 km S. 73° W. of Ain Khuff (24°53.2'N., 44°32.8'E.).
Thickness.—About 235 meters.
Lithology.—Alternating hard and soft, light-colored, chalky and dense limestones, commonly with thin hard shelly limestone layers. In some places considerable dolomitization is present. Cream-colored marls and red and green gypsiferous clays occur in moderate amounts in the

upper half of the unit, and in thin beds sporadically through the remainder. Generally a basal 10-(or more) meter unit of sand and varicolored shale is present.

Limits.—The base is at the unconformable contact of the lowermost sandy phase of the Khuff with underlying Cambrian? Saq sandstone in the type section. Elsewhere it rests on other lower Paleozoic units, or on basement.

The top is at the level where limestone and dolomite of the Khuff changes to red and green gypsiferous shales with a few thin sandstone beds, of the Sudair. Apparently this is a conformable contact.

FOSSILS AND AGE

Small gastropods and lamellibranchs are abundant in a few layers, generally with a single species dominating on a given bedding plane. *Foordiceras transitorium* (Waagen)? and *Coelogastroceras* aff. *mexicanum* (Girty), identified by W. M. Furnish, occur in the middle and lower Khuff. *Derbyia* sp. and *Circulopecten* sp. occur at several horizons. Wood identified by L. H. Daugherty[5] as *Dadoxylon* sp. is present in the lower part. Fusulinids have not been found.

The Khuff is considered Permian, probably upper.

Sudair Shale (Permian or Triassic)

TYPE SECTION

Location.—The total section was obtained by combining parts of three measured sequences at 1. Khashm Abu er Rumdha (19°38.5′N., 45°08.3′E.), 2. Khashm Ghudhaiy (19°17.8′N., 45°06.2′E.), and 3. Khashm es Sudair (19°11.7′N., 45°06.0′E.).

Thickness.—116 meters.

Lithology.—Brick- to dark-red, massive shale, with occasional pale greenish-white layers, many of which are silty. A pale greenish-white unit several meters thick, in which calcareous siltstones and silty, impure limestones are predominant, occurs 50 meters above the base. The lower part includes a number of thin beds of soft, fine sandstone.

Limits.—The base is tentatively taken at the top of a platy sandstone unit which is apparently equivalent to the top of the Khuff in its type section.

The top is at the contact of the red shales of the Sudair with the overlying transgressive sandstone and limestone of the Upper Jurassic. North of the type section the Sudair is overlain by silty shale and fine sandstone of the Jilh, apparently conformably.

FOSSILS AND AGE

No fossils have been found in the Sudair. On the basis of its stratigraphic position it is considered Permian or Triassic.

Jilh Formation (Triassic)

TYPE SECTION

Location.—Across the Jilh el 'Ishahr (a low escarpment near Nafud Qunaifidha) between 24°03.7′N.,45°45.5′E., and 24°12.9′N., 45°52.2′E.

Thickness.—About 326 meters.

Lithology.—The following units are found in this formation at the type section.

UPPER JILH

Fine- to medium-grained quartz sandstone, and green to purple silts and shales; several prominent layers of yellow to golden-brown hard, slabby, in part sandy, dense, finely crystalline limestone; capped by a hard layer of sandy oölite and oölitic sand. Rare molds of marine fossils including *Myophoria* sp.(67 m).

Gray to buff and white fine to coarse quartz sandstone, locally with numerous small spherical concretions, mainly massive and strongly cross bedded in upper part, with much interbedded green, olive, yellow, and red silt and shale in the lower half; several thin ironstone layers, occasionally showing molds of fossil wood; two thin limestone layers just below the middle. Fossil wood molds only;(118 m).

[5] Letter of March 16, 1953.

LOWER JILH

Limestones, varying from gray chalky to yellow-brown slabby dense, with interbedded fine quartz sandstone and marl. Rare poorly-preserved marine fossils;(36 m).

Green shale, in part gypsiferous, with a few thin, hard limestone layers. Rare poorly preserved marine fossils and a few amphibian? bone fragments;(25 m).

Gray to buff and reddish, fine- to medium-grained quartz sandstone, locally ferruginous, in places with many small spherical concretions, interbedded with an about equal amount of light- to dark-green shale. The basal portion is poorly exposed, but probably mainly sandstone;(80 m).

Limits.—The base is at the contact between soft sandstone and green shale of the Jilh above, and red shales of the Sudair below.

The top is at the contact of the resistant brown oölitic quartz sandstone of the uppermost Jilh with buff-colored, massive, commonly cross bedded sandstone of the overlying Minjur.

FOSSILS AND AGE

Between 45 and 55 meters below the top of the upper member, fragments and nuclei of ammonites, *Myophoria* sp., etc., considered by Arkell (1952, p. 249) to be Middle Triassic, were collected near the Marrat-Buraida road 11.8 kilometers from er Rukaiya (30.7 km N. 76½° E. of Buraida). *Myophoria* sp. and other poorly preserved lamellibranchs have been found at a number of localities over a wide area.

REFERENCE

Bramkamp and Steineke in Arkell (1952, p. 250–51).

Minjur Sandstone (Triassic or Jurassic)

TYPE SECTION

Location.—From the east end of the dip slope of the top of the Jilh formation (23°33′N., 46°08′E.) extending up to the base of the Marrat in the face of Khashm Minjur (23°35′N., 46°10.4′E.).

Thickness.—315 meters.

Lithology.—Buff, massive, commonly cross bedded, fine- to coarse-grained quartz sandstone, a few layers of which contain abundant small quartz pebbles, locally calcareous, weathering to give small spherical concretionary masses; several irregular zones of red, purple, and blue-gray varicolored shale, sandy shale and shaly sand; at several levels much black to brown ironstones as thin platy layers and concretionary masses, locally containing molds of fossil wood.

Limits.—The base is at the contact of the buff-colored, massive, commonly cross-bedded sandstone of the lowermost Minjur with the underlying resistant brown oölitic quartz sandstone with marine fossils of the Jilh.

The top is at the unconformable contact of this sandstone with limestone, dolomite, and calcareous sandstone of the Marrat.

FOSSILS AND AGE

Only fossil wood has been found. On the basis of its stratigraphic position the Minjur is considered to be Upper Triassic or Lower Jurassic.

REFERENCE

Bramkamp and Steineke in Arkell (1952, p. 249).

Marrat Formation (Lower Jurassic)

TYPE SECTION

Location.—In Jebel Kumait, a hill immediately northwest of the town of Marrat (25°04.5′N., 45°28.6′E.) and in the escarpment 5 km S. 70° W. of Marrat. Limited supplemental data on uppermost beds were obtained near 25°01.5′N., 45°33′E.

Thickness.—111 meters.

Lithology.—The formation is divided into the following units—

UPPER MARRAT

Golden brown- to tan-weathering, light gray, dense limestone with several beds of oölite; small amounts of tan clay shale near base; *Hildaites* and *Nejdia*. (21 m).

MIDDLE MARRAT

Dark brick red, in part silty shale, with a few thin, light-green layers; a zone of ripple-marked, silty, fine, quartz sandstone in the upper part; one thin layer of limestone like that of the upper Marrat 2 m below top. (56 m).

LOWER MARRAT

Yellowish tan to gray, compact limestone and dolomite of varying hardness, with subordinate tan, sandy shale and buff, fine- to medium-grained calcareous quartz sandstone in the lower part; locally a soft, chalky fossiliferous zone at the top.

Bouleiceras nitescens Thevenin, *B. arabicum* Arkell, *B. elegans* Arkell, *B. marraticum* Arkell, *Protogrammoceras madagascariense* (Thev.), and *Pecten ambongoensis* Thev., at or near top. (34 m).

Limits.—The base is at the contact of limestone, dolomite, and calcareous sandstone of the lower Marrat with the varicolored sandstones of the Minjur below.

The top is at the contact between the golden-weathering dense gray limestone of the upper Marrat with olive-green clay shales, and occasionally gypsum, of the lower Dhruma.

FOSSILS AND AGE

Fossils are listed above under the members in which they occur. *The Bouleiceras* fauna of the lower Marrat is considered by Arkell (1952, p. 293–94) to be lower Toarcian, and the *Nejdia* fauna of the upper Marrat is placed by him as early upper Toarcian.

REFERENCE

Bramkamp and Steineke in Arkell (1952, p. 248–49).

Dhruma Formation (Middle Jurassic)

TYPE SECTION

Location.—A series of successive partial measured sequences at and between Khashm Dhibi (24°12.4′N., 46°07.5′E.) and Khashm Madhurd (24°19.0′N., 46°19.6′E.).

Thickness.—383 meters.

Lithology.—The lithology of the Dhruma formation falls into three main parts, under which further subdivision on the basis of fauna is made—

UPPER DHRUMA (86 M)

Olive, calcareous, clay shale interbedded with white chalky limestone. *Gryphaea costellata* Douv. in lowest 20 m. (65 m).

Cream limestone with subordinate shale. "*Terebratula*" cf. *superstes* Douv., *Eudesia cardium* (Lam.), *E. cardioides* Douv., *Gryphaea costellata* Douv., *Eligmus rollandi* Douv. and *E. rollandi* var. *jabbockensis* Cox, *Mactromya aequalis* Ag., *Pholadomya lirata* (Sowerby), *Ph. aubryi* Douv., and *Homomya inornata* (Sow.). (21 m).

MIDDLE DHRUMA (170 M)

Cream to tan limestone, mainly soft, with subordinate marl and shale, and several prominent brown oölite beds. "*Dhrumaites* Zone." *Dhrumaites cardioceratoides* Arkell in upper part. (56 m).

Cream soft limestone capped by a bed of hard oölite. "*Micromphalites* Zone." *Micromphalites elegans* Arkell, *M. vertebralis* Arkell, *M.* cf. *busqueti* (de Gross.), *M. pustuliferus* (Douv.), *Daghanirhynchia* cf. *daghaniensis* Muir-Wood, *Gryphaea costellata* Douv., *Lopha solitaria* (Sow.), *Eligmus rollandi* Douv., *Bakevellia waltoni* (Lycett), *Pholadomya lirata* (Sow.), and *Homomya* cf. *gibbosa* (Sow.). (33 m.).

Cream compact limestone with several thin calcarenite beds in upper part. "*Tulites* Zone." *Tulites arabicus* Arkell, *T. tuwaiqensis* Arkell, *T. erymnoides* Arkell, *Arcomytilus somaliensis* Cox, *Modiolus* (*Inoperna*) *plicatus* (Sow.). *Eligmus rollandi* Douv., *Chlamys curvivarians* Dietrich, *Pholadomya lirata* (Sow.), and *Homomya* cf. *gibbosa* (Sow.). (37 m).

Cream to tan soft limestone, commonly calcarenitic. "*Thambites* Zone." *Thambites planus* Arkell, *Eudesia cardium* (Lamarck), *Eligmus rollandi* Douv., *E. rollandi* var. *jabbokensis* Cox, *E. polytypus* (Eudes-Deslongchamps), *Chlamys curvivarians* Dietrich, and *Homomya* cf. *gibbosa* (Sow.). (44 m).

LOWER DHRUMA (127 M)

Cream dense lithographic limestone, passing down into softer partly calcarenitic limestone with interbedded shale. "*Ermoceras* Zone." *Ermoceras* aff. *mogharense* Douv., *E. reineckeoides* Arkell, *E. coronatoides* (Douv.), *E. magnificum* Arkell, *E. elegans* (Douv.), *E. aulacostephanus* Arkell, *E. runcinatum* Arkell, *E. splendens* Arkell, *E. strigatum* Arkell, *Thamboceras mirabile* Arkell, *Stephanoceras arabicum* Arkell, *Eudesia cardium* (Lam.), *Eligmus rollandi* Douv., *E. polytypus* (Eudes-Deslongchamps), and *Lopha solitaria* (Sow.). (38 m).

Olive-green and golden-brown, slightly gypsiferous shale, with a few thin chalky limestone layers. Fossils scarce; none identified. (22 m).

Interbedded white and cream, mainly chalky, limestone and golden-brown, slightly gypsiferous clay shale; several thin hard limestone layers in lower part. Fossils rare. (36m).

Green to olive, in part varicolored clay shale, with several thin layers of calcarenite and fine sandstone in the upper part; in some places several meters of bedded gypsum at the base. "*Dorsetensia* Zone." *Dorsetensia arabica* Arkell. (31 m).

Limits.—The base is at the contact of the interbedded shale and soft limestone or gypsum of the lower Dhruma with the limestone of the upper Marrat.

The top is at the contact of the olive-green clay shales of the upper Dhruma with the soft, cream-colored basal Tuwaiq Mountain limestone.

FOSSILS AND AGE

Arkell (1952, p. 293, 295–96) gives the following age references for the Dhruma.

	Ammonite Fauna	*European stage*
Upper Dhruma	None	Upper Bathonian?
Middle Dhruma	*Dhrumaites* fauna *Micromphalites* fauna *Tulites* fauna	Middle (and upper?) Bathonian
	Thambites fauna	Lower Bathonian?
Lower Dhruma	*Ermoceras* fauna	Early-upper or late-middle Bajocian
	Dorsetensia fauna	Middle Bajocian

REFERENCE

Bramkamp and Steineke in Arkell (1952, p. 247–48).

TUWAIQ MOUNTAIN LIMESTONE (UPPER JURASSIC)

TYPE SECTION

Location.—Along the Riyadh-Jeddah road in Haisiyan Pass between 24°51′N., 46°05′E., and 24°55′N., 46°12′E.

Thickness.—215 meters (calculated).

Lithology.—The upper 175 meters of the Tuwaiq Mountain limestone is a massive cliff-forming unit composed of cream dense microcrystalline lithographic limestone, and less dense but similar limestone with a few calcarenite layers at various levels. The lower 35-40 meters is soft marly limestone with some interbedded calcarenite. Fine organic detritus is locally abundant in the lower unit. Heads of colonial corals in position of growth occur at many levels.

Limits.—The base is at the contact of soft calcarenitic limestone with the underlying olive-green clays of the Dhruma.

The top is at the contact of massive coral-bearing limestone with the overlying marls and soft limestones of the Hanifa formation.

FOSSILS AND AGE

Fossils are common in the lower marly part of the formation. They include *Musculus somaliensis* Cox, *Eligmus rollandi* Douville and *E. rollandi* var. *jabbokensis* Cox, *Lopha solitaria* (Sowerby), *Gryphaea balli* (Stefanini), *Exogyra nana* (Sow.), *Chlamys curvivarians* Dietrich, *Ch. macfadyeni* Cox, *Ceratomya* cf.

plicata (Ag.), *Ceromyopsis arabica* Cox, *Pholadomya aubryi* Douville, and *Homomya inornata* Sow. Ammonites including *Erymnoceras philbyi* Arkell, *E.* cf. *jarryi* (R. Douville), *E.* aff. *triplicatum* (Till), and *Pachyceras* cf. *schloenbachi* (Roman) have been determined by Arkell (1952, p. 298) who considers the lower Tuwaiq Mountain to be middle Callovian (*coronatum* zone), or lower Upper Jurassic. Ammonites collected by Philby (Cox, 1933, p. 383) were from the same general stratigraphic level.

REFERENCE

Bramkamp and Steineke in Arkell (1952, p. 247).

Hanifa Formation (Upper Jurassic)

TYPE SECTION

Location.—Along the north side of Wadi Hanifa for about 8 kilometers from 24°57.4′N., 46°12.8′E. to 24°55.2′N., 46°17.2′E.

Thickness.—101 meters.

Lithology.—The Hanifa formation is made up of cream to tan relatively soft chalky limestone with minor interbedded marl and tan clay shale; several prominent brown oölite units in middle and upper portions, with a particularly prominent one at the top. Colonial corals are common at several horizons in the middle and upper parts, commonly in position of growth. Reef structures have not yet been seen.

Limits.—The base is at the contact between massive, coral-bearing Tuwaiq Mountain limestone below and marls and soft limestone of the Hanifa above.

The top is at the contact of the brown coral-bearing oölite beds of the uppermost Hanifa with the very hard dense cream to tan limestone of the Jubaila formation.

FOSSILS AND AGE

Colonial corals are common at several horizons in the middle and upper parts. Some of the more common fossils are *Somalirhynchia africana* var. *mesoloba* Muir-Wood, *Modiolus imbricatus* (J. Sow.), *Lopha solitaria* (Sow.), *Gryphaea balli* (Stefanini), *Exogyra nana* (Sow.), *Chlamys macfadyeni* Cox, *Mactromya aequalis* Ag., *Ceromyopsis somaliensis* Weir, and *Pholadomya protei* (Brongniart). Only one poorly preserved ammonite has been found. Arkell (1952, p. 299) compares this with an Oxfordian species. *Somalirhynchia africana* Weir and var. *mesoloba* Muir-Wood also suggests Oxfordian age.

REFERENCE

Bramkamp and Steineke in Arkell (1952, p. 245–47).

Jubaila Limestone (Upper Jurassic)

TYPE SECTION

Location.—Along Wadi Hanifa from a point near the town of Jubaila (24°53.2′N., 46°26.7′E.) to a point about 12 kilometers west (24°53.8′N., 46°19.6′E.). Supplemental information on the upper 25 meters was collected in Wadi Hanifa between Jubaila and Riyadh.

Thickness.—About 110 meters.

Lithology.—Cream to tan, hard, dense, fine-grained limestone, mostly microcrystalline, containing a number of hard massive, partly conglomeratic, brown calcarenite layers, and several thin softer calcarenite lenses. Dolomite layers are generally present in the upper part.

Limits.—The base is at the contact of the cream to tan dense Jubaila limestone with underlying brown oölite of the Hanifa. The contact a prominent color change in the type section.

The top is at the change from dense limestone below, to softer dolomite and limestone of the Riyadh formation above. The upper contact is locally obscured by solution alteration on the old upland surface of Jebel Tuwaiq.

FOSSILS AND AGE

Fossils are rare. They include *Ceromyopsis somaliensis* Weir, *Pholadomya protei* Brongniart, *Perisphinctes jubailensis* Arkell, and *P.* aff. *progeron* von Ammon. Arkell (1952, p. 299) places the ammonites which occur in the lower part of the Jubaila as lower Kimmeridgian.

REFERENCE

Bramkamp and Steineke in Arkell (1952, p. 245)

Arab Formation (Upper Jurassic)

TYPE SECTION

Location.—4,500 to 4,918 feet depth in Dammam No. 7 (oil well) (26°19.1'N., 50°07.6'E.). Cutting samples are available from "A" and "B members" and diamond cores of "C" and "D members."

Thickness.—127.5 meters.

Lithology.—The subdivisions are as follows—

"A Member".—Various types of fine-grained limestone and calcarenite with minor anhydrite near the top. *"Diceras"* sp. (16.8 m).
Anhydrite with subordinate dolomite. (4.6 m).

"B Member".—Generally strongly recrystallized calcarenite (probably oölite), and finely granular dolomite. (6.1 m).
Anhydrite and minor dolomite and limestone. (12.8 m).

"C Member".—Various types of fine-grained limestone, and calcarenite and minor dolomite. *"Diceras"* sp. and *Nerinea* sp. (28.7 m).
Anhydrite and minor dolomite and limestone. (12.5 m).

"D Member".—Fine-grained limestone, in part partially dolomitized (dolomite rhombs in a fine-grained $CaCO_3$ matrix), with interbedded calcarenite (mainly fine grained) generally with an original fine-grained matrix; subordinate dolomite. *"Valvulinella" jurassica* Henson and *Nerinea* sp. (46.0 m).

Limits.—The base is at the contact of the basal calcarenitic rocks of the Arab with underlying dense limestone of the upper Jubaila. Correlation by sequence matching, mainly from well to well, has been used to carry the correlation of the top of the Jubaila from its type area east to Dammam No. 7. This correlation is probably accurate within a few meters.

The top is the contact of essentially continuous anhydrite above with the underlying mainly limestone unit of the Arab. The thin transitional zone from limestone to anhydrite has been included in the Arab.

FOSSILS AND AGE

The fauna found so far in the Arab has not proved diagnostic. Except for *Diceras,* identifiable forms range down into the Jubaila below. On continuity of sedimentation and fauna the Arab is considered safely placed as Upper Jurassic (Kimmeridgian or younger). The highest occurrence of *"Valvulinella" jurassica* Henson so far detected in Saudi Arabia is in the "D member" of the Arab.

Hith Anhydrite (Upper Jurassic?)

TYPE SECTION

Location.—In Dahl Hith, a solution pit in the escarpment bounding the west side of the Jubail plateau, about 32 kilometers southeast of Riyadh (24°29.1'N., 47°00.3'E.).

Thickness.—71.2 meters.

Lithology.—Bluish-gray, massive, compact anhydrite with occasional thin calcareous partings. Locally the anhydrite is spotted with white hydrated areas.

Limits.—The tentative base is at the top of the limestone and anhydrite-breccia unit in the bottom of Dahl Hith.

The top is at the disconformable contact of the anhydrite with the overlying fragmental limestone of the basal Sulaiy.

FOSSILS AND AGE

No fossils have been found. The Hith is presumed to be Upper Jurassic (Kimmeridgian or younger) because it seems to close the depositional cycle starting with the Arab formation of apparent late Jurassic age.

Sulaiy Limestone (Lower Cretaceous?)

TYPE LOCALITY

Location.—In the cliff face above Dahl Hith (24°29.1'N., 47°00.3'E.) about 32 kilometers southeast of Riyadh. The name is taken from Wadi es Sulaiy (immediately to the west).

Thickness.—About 180 meters.

Lithology.—Alternating layers of varying hardness of light grayish tan, more or less rubbly, compact, fine-grained limestone with occasional thin layers of fragmental limestone, oölite, and recrystallized coquina; partings with abundant oyster and other fossil fragments. Locally there is limited dolomitization. (157 m).

Fragmental limestone interbedded with compact limestone. Several breccia layers occur near, and at, the base. Scattered quartz sand grains are also present near the base. Some of the brecciation is taken to suggest that beds of anhydrite since removed by solution may once have been present. (22.7 m).

Limits.—The base is at the disconformable contact with the underlying Hith anhydrite.

The top is at the contact of the relatively dense massive Sulaiy limestone with overlying fragmental limestones of the Yamama.

FOSSILS AND AGE

Arenaceous Foraminifera including a *Pseudocyclammina* apparently the same as one in the Yamama, *Aporrhais* sp., and *Ostrea* sp., occur mainly in the lower part. The Sulaiy has been tentatively considered Lower Cretaceous as its arenaceous foraminiferal fauna resembles that of the overlying Yamama whose echinoids have been determined as Cretaceous. This unit is regarded as Jurassic by some geologists of the Iraq Petroleum Company, Ltd.

Yamama Formation (Lower Cretaceous)

TYPE SECTION

Location.—In the Qusa'ia upland, 14.7 km S. 35¼° E. of Ain edh Dhila' (24°00.4'N., 47°15.7'E. to 24°00.4'N., 47°20.9'E.). The name is taken from the town of Yamama, 21 kilometers north of this.

Thickness.—58 meters.

Lithology.—Tan and brown calcarenite (part oölite), alternating with cream to light tan, soft, compact, fine-grained limestone; the base formed mainly of calcarenite commonly includes scattered heads of coral. Several horizons near the top carry a molluscan fauna and a few echinoids.

Limits.—The base is at the contact of coral-bearing, coarsely fragmental limestone above, with moderately bedded compact, nonfragmental limestone below.

The top is at the contact of fragmental limestone and oölite of the Yamama below, with overlying mainly nonfragmental limestone of various types.

FOSSILS AND AGE

Echinoids from the upper part have been identified by J. Roger,[6] and include *Pygurus rostratus* Agassiz and *Trematopygus* cf. *grasi* d'Orb., both of which he regards as Neocomian. A fauna of small molluscs has not yet been studied sufficiently

[6] Personal communication to N. J. Sander, 1952.

to furnish pertinent conclusions. Foraminifera include a *Pseudocyclammina* probably the same as one found in the basal Sulaiy. On the basis of the echinoids the Yamama is considered to be Neocomian.

Buwaib Formation (Lower Cretaceous)

TYPE SECTION

Location.—In the lowest step of the escarpment below and about one kilometer west of Khashm Buwaib (25°15.1'N., 46°38.7'E.).

Thickness.—33.8 meters.

Lithology.—(Part concealed). Buff, soft, fine- to medium-grained quartz sandstone with a thin bed of brown hard argillaceous limestone at the top. (8.6 m).

Cream, soft, fine-grained limestone, in part with common quartz sand grains; a thin bed of buff, well-sorted, fine-grained quartz sandstone near the middle; a thin bed of brown, hard shelly limestone at the top. *Cyclammina greigi* Henson and other arenaceous Foraminifera. (10.5 m).

Cream, fine-grained limestone, in part argillaceous and sandy; several thin layers of brown, hard, shelly, argillaceous limestone in middle and upper part; thin partings of olive clay shale near the base. (5.0 m).

Cream, rubbly weathering, soft, fine-grained limestone, in upper part with shelly, foraminiferal calcarenite layers with *Cyclammina greigi* Henson; a thin gray clay shale bed near the top (9.7 m).

Limits.—The base is at the apparently unconformable contact with underlying limestone and calcarenite of presumed lower Sulaiy limestone.

The top is at the top of highest limestone bed below the sandstone of the overlying Biyadh.

FOSSILS AND AGE

In the Kharj area the upper part of the basal unit carries a fair molluscan fauna (not yet determined). Ammonites collected from the basal 5 meters were determined by C. W. Wright as an undescribed genus showing affinities with *Knemiceras* and *Hypengonoceras* (Wright, 1952, p. 220, footnote 23). None of these fossils so far has served to indicate the age of the Buwaib.

On the basis of its stratigraphic position between presumed Neocomian (Lower Cretaceous), based on echinoids, below, and the Biyadh formation which is dated as Aptian-Albian, above, the Buwaib is probably Aptian, but could be late Neocomian.

Biyadh Sandstone (Lower Cretaceous)

TYPE SECTION

Location.—Along a traverse from the base of continuous sandstones near Ain Thulaima (24°06.5'N., 47°24.0'E.) to a point 24.2 km S. 85° E. (24°05.0'N., 47°38.1'E.); and from a point at the same stratigraphic level as the preceding 22.5 km S. to a point 33.4 km S. 64° E. of Ain Thulaima (23°58.5'N., 47°41.6'E.).

Thickness.—About 270 meters (mainly calculated).

Lithology.—Coarse- to medium-grained quartz sandstone, partly cross bedded, with quartz pebble-bearing zones at many levels. (61 m).

Red, green, and tan, in part silty, shale with minor tan marl; thin layers of tan to brown, strongly crystalline ferruginous dolomite at middle and top. (10 m).

Poorly exposed, probably mainly sandstone and red shale. (40 m).

Poorly exposed, probably medium-grained sandstone and subordinate shale, locally ferruginous, in part with quartz pebbles. (63 m).

Brown to buff medium- to coarse-grained quartz sandstone, in part cross bedded, with subordinate red shale; several horizons strongly ferruginous, weathering to give chips of black sandy ironstone. (96 m).

Limits.—The base is at the top of highest of the thin marine limestone beds of the uppermost Buwaib.

The top is at the contact of coarse quartz sand with quartz pebbles and the overlying ferruginous silty sandstones of the basal Wasi'a formation.

FOSSILS AND AGE

Fossil wood, including *Laurinoxylon* sp., identified by L. H. Daugherty[7] has been found in the upper part.

Poor marine molluscs are present 61-71 meters below the top, but nothing diagnostic has been found.

In the oil wells of the Eastern Province, beds equivalent to the Biyadh contain *Cyclammina greigi* Henson in the lower part, and *Choffatella decipiens* Schlumberger and *Orbitolina* cf. *discoidea* Gras in the upper. The unit is therefore regarded as Aptian-Albian (upper Lower Cretaceous).

WASI'A FORMATION (LOWER UPPER CRETACEOUS)

TYPE SECTION

Location.—In the lower slope of Khashm Wasi'a, extending 1.35 km S. 50° E. to low hills near 'Ain Wasi'a (24°22.6'N., 47°45.8'E.).

Thickness.—42.0 meters.

Lithology.—Brown- and yellow-weathering, fine-grained to very fine-grained quartz sandstone, in part silty, with scattered quartz pebbles. Pronounced cross bedding is generally well developed except in uppermost part. Fossil wood is present as ferruginous molds. (19.2 m).

Red and purple silty, sandy shale; some green clay shale; common thin brown to black ironstone partings; a 0.1 m reddish-brown sandy silty granular dolomite with calcite-filled vugs at the base. Elsewhere nearby similar dolomite layers occur at higher levels. At one locality just east of 'Ain Wasi'a a local lens of cream, soft, nodular, fine-grained limestone up to 2 meters in thickness, containing rare ammonites, extends for about 200 meters laterally. *Neolobites vibrayeanus* (d'Orbigny), cidaroid, and lamellibranchs. (8.1 m).

Buff and red banded, fine-grained sandstone and siltstone, showing small-scale cross bedding. (5.3 m).

Brown-weathering, green, sandy silty shale. (6.4 m).

Gray, green, and red, silty fine-grained sandstone and coarse siltstone, relatively thin-bedded. (3.0 m).

Limits.—The base is at the contact of varicolored fine-grained sandstone and siltstone of the Wasi'a with underlying light-colored, locally pebbly, coarse sandstone of the Biyadh. The contact is an apparent unconformity.

The top is at the presumably disconformable contact of Wasi'a sandstones with overlying carbonate rocks of the Aruma, the base of which is generally 1-2 meters of brown coarsely crystalline dolomitic? limestone with calcite-filled vugs.

FOSSILS AND AGE

Neolobites vibrayeanus (d'Orbigny) determined by C. W. Wright[8] is considered to date the Wasi'a as Cenomanian (lower Upper Cretaceous).

ARUMA FORMATION (UPPER CRETACEOUS)

TYPE SECTION

Location.—A compiled full sequence based on parts of several measured sections in the Aruma plateau or cuesta (El 'Arma), between a point 2.6 km N. 41° E. of Khashm Ruwaiqib (25°38.6'N., 46°26.2'E.) and a promontory 15.4 km N. 51½° E. of the first point (25°42.7'N., 46°26.2'E.)

Thickness.—144 meters.

Lithology.—Yellowish-tan to yellowish-brown dolomitic and calcareous shale, argillaceous dolomite, and olive shale, with minor interbedded limestone and dolomite. Small Foraminifera, *Omphalocyclus* sp., *Cyclolites* near *medlicotti* Noetling, Rudistids, and *Austrosphenodiscus*. (33 m).

Blue-gray to cream, massive dolomite and limestone. (28 m).

Cream-colored, soft, chalky, fine-grained limestone, with minor interbedded olive to olive-

[7] Letter of March 16, 1953

[8] Letter of February 17, 1951

green calcareous shale in the lower part. Small Foraminifera, *Globotruncana, Lepidorbitoides, Omphalocyclus* sp., Rudistids (varied fauna), and *Sphenodiscus* sp. (43 m).
Cream-colored, massive, nodular, chalky, fine-grained limestone with abundant small fragmental organic remains. Small Foraminifera, *Meandropsina* sp., *Orbitoides (s.s.)* sp., echinoids, and *Ostrea dichotoma* Bayle. (40 m).

Limits.—The base is at contact of Aruma carbonate rocks with underlying clastic sediments of the Wasi'a formation.

The top is at the contact of tan shale and dolomitic shale of the upper Aruma with overlying gray granular crystalline dolomite with *Lockhartia* of the basal Umm er Radhuma (Paleocene-lower Eocene).

FOSSILS AND AGE

The Aruma is generally abundantly fossiliferous and some of the fossils are listed under the members in which they occur. Ammonites and nautiloids are currently being studied by A. K. Miller and W. M. Furnish (paper in preparation). Early collections were examined by C. W. Wright.[9] On the basis of its ammonites most of the Aruma is apparently Maestrichtian, a conclusion with which fossils other than ammonites are consistent.

Umm Er Radhuma Formation (Paleocene-Lower Eocene)

TYPE SECTION

Location.—Near Umm er Radhuma wells (28°41'N., 44°41'E.). Because of difficulties in working out the full sequence near the type locality, the standard section for the unit is that in Wadi al Batin between points 149.8 km and 70.4 km southwest of Hafar al Batin (between 27°32'N., 44°52'E., and 27°59'N., 45°29'E.).

Thickness.—About 229 meters.

Lithology.—Gray and cream-colored limestone, dolomitic limestone, and dolomite with molds of fossil fragments and poor Foraminifera at several levels; silicification takes place at several horizons. (140 m).
Light colored, relatively soft, part chalky limestone. (74 m).
Light tan to gray, hard, dolomitic limestone and dolomite. (15 m).

Limits.—The base is at the contact of the *Lockhartia*-bearing basal dolomite and dolomitic limestone with underlying Aruma tan dolomitic shale.

The top is not exposed in the type sequence because of overlapping Miocene-Pliocene rocks, but it is taken at the upper limit of dolomites with *Lockhartia tipperi* (Davies) where they underlie soft dolomitic limestone of the Rus.

FOSSILS AND AGE

In general, the Umm er Radhuma is rather poorly fossiliferous in its type sequence but in eastern Nejd and the area of the Hasa oil fields Foraminifera are abundant and varied.

Lower Umm er Radhuma faunas are characterized by great abundance of large rotaloids including such species as *Lockhartia haimei* (Davies), *L. diversa* Smout, *L. conditi* (Nuttall), and *Daviesina khatiyahi* Smout.

The similarly rich middle Umm er Radhuma fauna includes such species as *Kathina delseota* Smout, *Lockhartia haimei* (Davies), *Sakesaria cotteri* Davies, *Miscellanea miscella* (d'Archiac & Haime), *Saudia discoidea* Henson, etc.

The upper Umm er Radhuma fauna is typified by *Nummulites lahirii* Davies, *N. globula* Leymerie, *Lockhartia tipperi* Davies, *Rotalia trochidiformis* (Lamarck), *Alveolina* cf. *subpyrenaica* Leymerie, *Flosculina globosa* (Leymerie), and others.

The larger of these Foraminifera have been described by Sander (1952) and Smout (1954). They agree that most of the Umm er Radhuma is Paleocene, but that the upper part is lower Eocene. Also excellent faunas of small Foraminifera

[9] Letter of February 17, 1951

are present at many levels in the Umm er Radhuma, but nothing on them has reached publication.

RUS FORMATION (LOWER EOCENE)

TYPE SECTION

Location.—On the southeast flank of Dammam Dome below Jebel Umm er Rus (26°19.5′N., 50°10.0′E.).

Thickness.—56 meters.

Lithology.—The lithology and thickness of the Rus formation are variable, with most of the variation occurring in the middle one of the three units described below. Thicknesses are those of the type section.

White, soft, chalky porous limestone, with one or more calcarenite beds at the top. (3.6 m).

Light-colored marls with local irregular masses of crystalline gypsum and occasional thin harder limestone beds; geodal quartz at several levels. In other areas this unit is highly variable, including as common equivalents—(a) white, compact, finely-crystalline anhydrite with interbedded green shales and minor amounts of dolomitic limestone, or (b) gray marls with coarsely crystalline calcite and interbedded shale and limestone. (31.8 m).

Gray to buff compact crystalline limestone commonly partly dolomitized, with minor amounts of soft limestone made porous by the leaching of small organic remains. Quartz geodes occur rarely in the lower part, and are typical of the uppermost part. (21.0 m).

Limits.—The base is at the contact of dolomite containing *Lockhartia tipperi* Davies of the upper Umm er Radhuma, with overlying light-colored dolomitic limestone commonly with leached indeterminate molds of small molluscs of the basal Rus.

The top is at the contact of light-colored calcarenite layers of the upper Rus, with overlying thin-bedded impure limestone and shale of the basal Dammam formation.

FOSSILS AND AGE

Diagnostic fossils have not been found in the Rus, but it is underlain and overlain by rocks considered lower Eocene.

DAMMAM FORMATION (LOWER AND MIDDLE EOCENE)

TYPE SECTION

Location.—In the rim-rock of Dammam Dome, 7.2 km N. 85° W. of Jebel Umm er Rus (26°17.3′N., 50°07.7′E.).

Thickness.—About 28 meters.

Lithology.—Light colored, chalky, porous, generally dolomitic limestone with abundant molds and casts of molluscs and minute organic remains, locally silicified near the top. *Orbitolites complanata* Lamarck, *Dictyoconus egyptiensis* (Chapman), and many small molluscs. (9 m).

Dolomitic marl or clay (generally abundant but discrete dolomite rhombs in a clay or marl matrix), generally light-colored, commonly cream or orange. Darker grays and browns also occur, normally where the rocks have a higher clay content. (6 m).

Crystalline dolomite or limestone above, becoming soft marly limestone below. *Nummulites staminea* Nuttall, *N. discorbina* Schlotheim var. *major* Rozlozsnik, *N. somaliensis* Nuttall & Brighton, *N. beaumonti* d'Archiac & Haime, *Dictyoconoides kohaticus* (Davies), *Dictyoconus indicus* Davies, *D. egyptiensis* (Chapman), *Linderina buranensis* Nuttall & Brighton, and *Euphenax jamaicensis* (Trechmann). (9.2 m).

Marl and calcareous clay, generally gray or blue in subsurface but brown in outcrop, with one or two light-tan foraminiferal limestone layers characterized by *Alveolina subpyrenaica* Leymerie. Other fossils in this unit include *Nummulites globula* Leymerie, *N.* aff. *lucasana* Defrance in d'Archiac, and *Coskinolina balsillei* Davies. (1.0 m).

Brown, earthy, clay shale and soft impure limestone, generally with rare small black ironstone concretions and shark teeth. *Nummulites globula* Leymerie and *Ostrea turkestanensis* Romanovski. (1.2 m).

Light-gray marls and limestones with rare molluscs. (1.5 m).

Limits.—The base is at the contact of the gray marl and limestone of the basal Dammam with underlying calcarenite or limestone of the upper Rus formation.

The top is at the unconformable contact of any part of Dammam formation with the overlying, slightly sandy, compact limestone of the basal Hadrukh formation.

FOSSILS AND AGE

The upper Dammam formation includes a number of characteristic Lutetian (middle Eocene) species listed above under members in which they occur. In the type sequence, the lower 8.9 meters is characterized by an *Alveolina*-bearing fauna considered by Sander (1952) to be lower Eocene, probably Ypresian. Larger Foraminifera from the Dammam formation have been described by Sander (1952) and Smout (1954).

Hadrukh Formation (Lower Miocene?)

TYPE SECTION

Location.—In the south face of Jebel Hadrukh (27°05.1'N., 49°12.0'E.), and in a shallow well about 1,500 meters to the southeast.

Thickness.—51.5 meters in surface exposures below which a nearby well adds 32.5 meters down to the top of the Eocene, to give a total thickness of 84 meters.

Lithology.—Green and grayish-green, generally finely sandy clays, and green to gray calcareous sandstones. The latter commonly weathers to fine concretionary pellets. Minor amounts of cream to gray marl and one bed of gypsum are also present. Much chert occurs at a number of levels. The basal 5-10 meters is commonly cream sandy limestone.

Limits.—Generally the basal bed of the Hadrukh is a few meters of cream, dense, slightly sandy limestone. This rests unconformably on underlying marine Eocene rocks.

The top is at the base of the *Echinocyamus*-bearing limestone and marl of the basal Dam.

FOSSILS AND AGE

Rare nonmarine fossils including *Chara* are present. In some parts of central coastal Hasa a few thin layers near the top contain poorly preserved marine molluscs. Because of its apparent continuity with the overlying Dam formation, the Hadrukh is tentatively considered to be early Miocene.

Dam Formation (Miocene)

TYPE SECTION

Location.—In the eastern face of Jebel Dam (Lidam) (26°21.6'N., 49°27.7'E.) and in Jebel Lemaigher (26°17.7'N., 49°30.8'E.).

Thickness.—89.8 meters.

Lithology.—Pink, white, and gray marl and red, green, and olive, partly finely sandy clay predominating; minor sandstone in the upper part; white to cream marly and chalky limestone generally forms the lower part. Marine fossils are present at many levels but ordinarily are poorly preserved.

Limits.—The base is at the bottom of a marine limestone and marl unit generally characterized by *Echinocyamus* sp.

The top is at the contact of the marl and limestone with marine fossils of the uppermost Dam with overlying sandstone and gravel of the basal Hofuf. At the type locality of the Dam this is a more easily recognized boundary than at the type locality of the Hofuf.

FOSSILS AND AGE

Ostrea latimarginata Vredenburg, *Echinocyamus* sp., "*Archaias*" sp., and others are found. The fauna appears to indicate an approximate correlation with the lower Fars of Iraq, and the Dam is presumably about middle Miocene.

Hofuf Formation (Miocene or Pliocene)

TYPE SECTION

Location.—The southern end of Barq er Ruqban, 20 kilometers north-northwest of Hofuf (25°31.5'N., 49°31.0'E.).

Thickness.—78 meters in Barq er Ruqban but exposures to the west and northwest of the type locality add 17 meters additional upper beds to give a total of 95 meters. The total thickness of the unit varies substantially laterally.

Lithology.—Cream to white sandy marl and cream rubbly weathering, compact sandy limestone, with minor amounts of calcareous sandstone and shales. The basal beds in some areas are marly sandstone which include many quartz pebbles.

Limits.—The base is at the contact between the generally quartz pebble-bearing basal clastic phase of the Hofuf and underlying calcareous rocks of the Dam formation.

The top is at the upper limit of the exposures of the Hofuf area, commonly an old surface showing strong calcium carbonate enrichment of duricrust-caliche type.

FOSSILS AND AGE

Only occasional freshwater fossils, including *Lymnaea* and *Chara,* have been found and these are of no use in dating the Hofuf. As the closing unit of Arabian Tertiary deposits it may be either late Miocene or Pliocene.

STRUCTURE

The central and northeastern parts of the Arabian Peninsula are areas of unusual tectonic stability. Early in Saudi Arabian geological work, Max Steineke recognized two main structural elements in them. Central Arabia is primarily a homoclinal area of unusually uniform, generally low, dip to the east which he called the "interior homocline." Northeastern Arabia, structurally nearly flat with several long folds superimposed on it, was named by him the "coastal structural terrace." For the present, this purely structural subdivision is used rather than to try to distinguish the conventional tectonic elements of "stable" and "unstable shelf." Undoubtedly, the sedimentary area of central Arabia is typical "stable shelf." However, in northeastern Arabia the grand scale and minor if any direct tectonic control of the shifting patterns of sedimentation during various stages of the Mesozoic and Cenozoic make difficult recognition of just what part should be classed as "unstable shelf." Difficulty in the use of these tectonic terms elsewhere in the Middle East has already been recognized by Henson (1951b, p. 120–22).

The "interior homocline" is a great curved belt of outcropping older rocks in central Arabia. It is characterized by marked uniformity of dip, generally between 5 and 18 meters per kilometer, ordinarily, with the rate of dip related to the stratigraphic unit involved. Dip is commonly so uniform that unconformities are apparent by differences in average attitudes of beds over large areas. In part, the low angle of dip may be initial, but most of it seems to be the result of slight tilting at various times. A far better understanding of lateral changes in sedimentation and the environmental inferences to be drawn from these will be required before the amount of initial dip can be resolved. In one part of central Arabia some tensional faulting is present.

The term "coastal structural terrace" is applied to the northeastern Arabian Peninsula east of the "interior homocline." It is a slightly warped, nearly flat structural plane whose irregularities are limited to low gentle structural undulations of irregular form. On this is superimposed a pattern of several major north-south anticlinal axes, of which the most notable is the En Nala anticline, part of which is responsible for the Ghawar field. Dips on the flanks on some of these anticlines may be as steep as 7° to 10° at depth in older rocks but basically the axes are broad, low folds. Commonly, the axes show more or less coinciding positive gravity anomalies, in some places as strong reversals in gravity and in others as changes in gradient. Ghawar, Abqaiq, Qatif, Bahrain, and Dukhan (Qatar) are typical examples of the anticlinal axes. The origin of the axes is still obscure, although the best hypothesis seems to be that movements of blocks of the underlying basement are responsible for their uplift, rather than compressional forces acting within the sedimentary blanket.

Dammam Dome is exceptional in its oval shape, complex pre-Upper Cretaceous keystone faulting and strong negative gravity anomaly. It is regarded as the shallow expression of a deep-seated salt plug.

Further discussion of local structure is outside of the scope of this paper except that its development as related to the migration and accumulation of oil will be mentioned in a later section. Some of the features of individual fields are given by Eicher and Yackel (1951).

UNCONFORMITIES

Tilting of large parts of the central and northeastern peninsula at various times has left a clear record in the sedimentary sequence, generally as unconformities at at which truncation is primarily on a regional scale. Elimination of older rocks is observed, commonly over wide distances, at rates as low as a few tenths of a meter per kilometer. Important unconformities of this regional type are—pre-Khuff (Permian or older), pre-Marrat (Lower Jurassic), pre-Tuwaiq Mountain (pre-Upper Jurassic), pre-Buwaib (middle Lower Cretaceous), pre-Wasi'a (Middle Cretaceous), and in Central Arabia pre-Aruma (Upper Cretaceous). The extent of elimination of older rocks at these various unconformities shows moderate variation. Some of them are important only in certain moderately large areas, whereas others show gradual systematic truncation over much of the area of their outcrop. Less truncation, in general, is found in the area of the oil fields which much of the time must have been farther offshore and more basinal than central Arabia.

Two important unconformities manifest themselves mainly by local effects, primarily by truncation of beds on the crests of individual folds. These are the pre-Aruma (Upper Cretaceous) and pre-Miocene unconformities.

In spite of the intensity of late Tertiary diastrophism across the Persian Gulf in Iran, little trace of simultaneous tectonic activity has been found in the central and northeastern parts of the peninsula. Minor tensional fault movements in central Arabia seem to be as young as late Tertiary, but even most of this type of movement was apparently older. Miocene-Pliocene rocks are mainly terrestrial and so far it has been almost impossible to distinguish how much of their present very low average dip of 1-2 meters per kilometer toward the present Persian Gulf is the result of tilting and how much is to be regarded as initial dip. These younger rocks, however, have been widely disturbed by solution-collapse phenomena in parts of the Eastern Province.

JURASSIC ROCKS AND OIL[10]

JURASSIC LITHOLOGY AND PALEOGEOGRAPHY

The Jurassic outcrops of the "interior homocline" furnish an excellent cross-section of the western mainly shallow-water facies of the various units, but in the Eastern Province of Saudi Arabia no more than a dozen wells have been drilled a substantial depth into the pre-Arab Jurassic and only half of these were drilled completely through it. Nevertheless, the overall picture as now glimpsed seems to be a simple one, and the following remarks are based on this presumption.

LOWER JURASSIC

Marrat formation.—Lower Jurassic is represented in the outcrop area of central Arabia by the Marrat formation which in its typical development consists of upper and lower limestone-dolomite units separated by a red shale member locally with interbedded sandstone. The upper and lower carbonate units are dated by ammonite faunas as Toarcian (upper Liassic). In central Arabia an unconformity at the base of the Marrat separates it from the underlying Minjur sandstone. North of the general latitude of Riyadh, the middle red shale unit thins and the Marrat becomes almost entirely carbonate rocks. To the south, the middle red shale unit is progressively replaced by sandstone, and still farther southward the upper and lower carbonate units are also replaced by sandstone and shale. At about Latitude 22° North, the Marrat has become indistinguishable from the lower Dhruma and cannot be traced farther south.

In the Eastern Province the Marrat cannot be distinguished at present because of the lack of paleontological evidence. Its equivalent no doubt is found in a dark, dense, argillaceous limestone unit which separates dated Dhruma rocks and underlying presumed Triassic.

MIDDLE JURASSIC

Dhruma formation.—No attempt will be made to discuss the Dhruma formation in full because little oil has so far been located in it and the paleogeographic relations of the rocks are poorly understood. A shallow-water facies with abundant fossils crops out at the surface in central Arabia where the rocks are relatively well dated by ammonites. In the area along the western shore of the Persian Gulf a muddier, deeper-water facies prevails and the details of correlation and ages of the various portions of the unit are little understood.

In the Riyadh-Dhruma area, the Dhruma formation consists primarily of limestone with much shale in the upper and lower parts. In some places a thin gypsum layer occurs at the immediate base. Clean, current-washed calcarenites occur as thin members at various levels, commonly as units of high lateral persistance useful in setting up a framework of correlation. Five good ammonite zones have been found in the area so far (from bottom to top, *Dorsetensia, Ermoceras, Thambites, Micromphalites,* and *Dhrumaites*), although unfortunately the highest 100 meters of the unit has not yet furnished determinable ammonites. The ammonites are in general provincial, unusual types.

[10] R. A. Bramkamp and N. J. Sander.

Lateral changes in the Dhruma formation to the south along the outcrop as seen in the western face of Jebel Tuwaiq consist primarily of a gradual replacement of the carbonate rocks and shale by sandstones, so that in the vicinity of Wadi Dawasir only sandstones, apparently of continental type with several ironstone-bearing horizons and minor shale, make up the whole unit. The transition from limestones to sandstones takes place laterally in individual beds with very little if any passage through intervening shales, although the limestones become silty and impure as the transition is approached. A feature of considerable interest, still incompletely understood, is that bed tracing, both on the ground and by means of aerial photographs, shows that clean calcarenite units of normal marine facies are within a few meters the lateral equivalents of ironstone-bearing horizons presumably representing times of essentially nondeposition in the continental clastic facies.

North along the outcrop from the Riyadh-Dhruma area, the Dhruma shows a gradual increase in the proportion of shales until these dominate the unit about Latitude 25°20′ North. A short distance north of this it becomes a sequence of interbedded sandstones and shales. Although the trend toward a continental environment is obvious near Zilfi, the most northerly area in which the Dhruma is known with any adequacy, it is likely that considerable proportion of the unit is still marine in spite of the virtually complete dominance of fine and coarse clastics; and most of the time the actual strand line was still farther north.

Two facies of the Dhruma are recognizable in the Persian Gulf coastal area. In the central Eastern Province (Ghawar, Abqaiq, Dammam, Fadhili, etc.) fine-grained limestones commonly dense, and locally dark and argillaceous, dominate, but variable proportions of fine calcarenite generally with a fine muddy matrix, are interbedded. The oil of the Fadhili Zone of the Fadhili field is found in a persistent fine calcarenite a short distance below the top of the Dhruma. Elsewhere the same unit where penetrated even on the crests of structures, has proved water bearing, although stains and minor shows have been found in several wells.

In the northeastern part of the Eastern Province, the Dhruma is dark limestone, to varying extent argillaceous with interbedded gray and black shale, with subordinate pure limestone. In this area calcarenite layers are few, thin, and mainly with a mud-sized original matrix.

UPPER JURASSIC

Tuwaiq Mountain limestone and Hanifa formation.—These two units which are topographically important in the Jurassic outcrop belt of central Arabia are known so far only in that portion of Saudi Arabia. They cover the general time span from Callovian (on good ammonites) to Oxfordian? (presumed from the occurrence of *Somalirhynchia africana* [Weir]). Unfortunately only the Tuwaiq Mountain limestone has furnished adequate ammonites.

The Tuwaiq Mountain limestone is, in general, a great plate of coral-bearing, dense, pure limestone, at the base of which is a thin noncoraliferous well-bedded chalky unit. Most of the ammonites were found in this lower member. Small thicknesses of calcarenite are present near the middle in the latitude of Riyadh and increase in relative amount to the north. Thin calcarenite layers are also present near the top. In typical Tuwaiq Mountain limestone, coral heads are scattered, commonly in position of growth, but ordinarily they make up only a small proportion of the rock as a whole. Several small reefs up to 15 meters high and about 50 meters in diameter are present just west of Riyadh, but aside from these the unit is thickly, but well, bedded. To the north as far as its most northerly known outcrops near Aqibba (about Latitude 27°10′ North), the Tuwaiq Mountain limestone shows increased inclusion of soft limestone. To the south of the latitude of Riyadh, the unit progressively thins, the proportion of calcarenite increases, and corals become more and more concentrated into rubbly beds in which coral remains make up a substantial proportion of the rock. Branching stromatoporoids, few in the central area, increase in abundance and, in places in the south, become rock forming.

The Hanifa formation is another nearly pure carbonate unit, differing from the Tuwaiq Mountain in the presence of a number of layers of brown clean calcarenite, which are particularly prominent in the upper part, as well as by the presence of argillaceous limestone units.

A few coral-bearing horizons are present in the Hanifa of central Nejd, and in these the corals generally occur as isolated heads in position of growth in a fine-grained matrix. In some places coral heads are also present as rolled boulders in the calcarenite units. Much of the calcarenite of the Hanifa, particularly in the upper part is oölite, using this term in the narrow sense for more or less authigenic spherical carbonate bodies with well-developed concentric structure.

Both to the north and the south the Hanifa shows gradual progressive change primarily by an increasing development of red and gray dolomite units in the upper part, generally replacing calcarenite, and the presence of increasing amounts of rubbly coral rock. The general character of the Hanifa in southern Jebel Tuwaiq is shown in Figure 3.

In wells of the coastal Eastern Province, the Tuwaiq Mountain and Hanifa have not been recognized, and the presence of lower Kimmeridgian ammonites identified by W. J. Arkell,[11] within a hundred feet of a Dhruma fauna with *Posidonia ornati* (Quenstedt) leaves little doubt that any equivalent of the two units must either be highly condensed or entirely lacking. *Posidonia ornati* (Quenstedt) below and *P. somaliensis* Cox above, likely to be indicative respectively of Middle and of Upper Jurassic ages, bracket the contact to within three feet in another well. The evidence seems to be strong that over much of the Eastern Province Jubaila (Kimmeridgian) rocks rest directly on the Dhruma (Bajocian-Bathonian). The type of unconformity is difficult to understand inasmuch as the one cored example of the contact shows a visible lithologic contact, but clastic limestones, evidence of erosion, and other suggestions even of shallow-water deposition are lacking. The absence of wells penetrating the contact between the coastal Eastern Province and central Arabia prevents constructive suggestions as to how the apparent elimination of the Tuwaiq Mountain and Hanifa formations takes place from west to east.

Jubaila limestone.—Along its belt of outcrop in central Arabia, the Jubaila is dominantly of shallow-water facies. Figure 3 shows the general rock pattern in

[11] Letter of Sept. 25, 1951

southern Jebel Tuwaiq. Fine-grained limestone, with generally thin layers of cleanly washed calcarenite of generally high persistence, dominate the northern and central part of this section. In the upper portion of the Jubaila varying amounts of dolomite are interbedded. Corals and stromatoporoids, few in the Riyadh area, gradually become numerous at several levels both to the north and the south.

In the southern part of the Jubaila outcrop as shown in the section relatively marked changes occur, including the appearance of a major lens of marine sandstone in the lower part. This sand unit increases in thickness by grading laterally from limestone to sandstone from east to west up the dip of the beds. The lens also grades laterally into carbonate rocks north of about Latitude 22° North, and south of about 20°30′ North. In this same area the upper Jubaila is almost entirely replaced by thickly bedded dolomite which, particularly in its upper portion, includes a number of coral- and stromatoporoid-bearing biostromal layers.

Near Majma'a, the northernmost part of the outcrop of the Jubaila, replacement of the lower Jubaila by sandstone and dolomite takes place in a fashion similar to that found in southern Jebel Tuwaiq.

In the coastal part of the Eastern Province, the Jubaila likewise shows moderate lateral changes. The basal part is mainly dark, generally black, argillaceous limestone with varying amounts of black calcareous shale. In a few places these black shales have furnished fossils, including *Posidonia* and ammonites. *Idoceras* cf. *balderum* (Oppel), *Glochiceras* cf. *fialar* (Oppel), *Lithacoceras* cf. *stenocyclus* (Fontannes?) Schneid., *Ochetoceras* sp., *Ataxioceras semistriatum* Schneid., *A.* cf. *litorale* Schneid., *Streblites* sp. indet., and others have been determined in material from Fadhili No. 1 by W. J. Arkell,[12] who considers the fauna lower Kimmeridgian and roughly contemporaneous with the few ammonites recorded from the lower Jubaila of central Arabia. Presumably the black limestones and shales represent an euxinic environment, or at least an approach to this condition.

The remaining upper five sixths of the Jubaila is primarily fine-grained limestones interbedded with varying amounts of fine-grained, generally muddy calcarenite. Clean calcarenite, commonly coquinal, is limited to thin layers in the uppermost part, and such layers are present only in moderate areas, for example, the Ghawar oil field. Most calcarenite of the Jubaila has matrix which originally must have been carbonate mud of silt size. The fine-grained limestones must represent original lime mud, and therefore the Jubaila was almost entirely muddy sediment, probably deposits mainly laid down outside of the mud line of the period of time represented. The clean calcarenites near the top may be indicative of areas in which shallow-water conditions appeared in the closing phases of Jubaila deposition. Possibly deposition in that part of the basin had reached a sufficient level to bring the bottom up within periodic reach of wave base or currents.

The distribution of the muddy fine carcarenite bodies in the Jubaila varies from place to place. In the northeastern part of Eastern Province, for example at Abu Hadriya and Safaniya, the muddy calcarenite bodies are best developed low in the Jubaila, just about the basal black limestone-shale unit. In the Abqaiq-

[12] Letter of Sept. 25, 1951

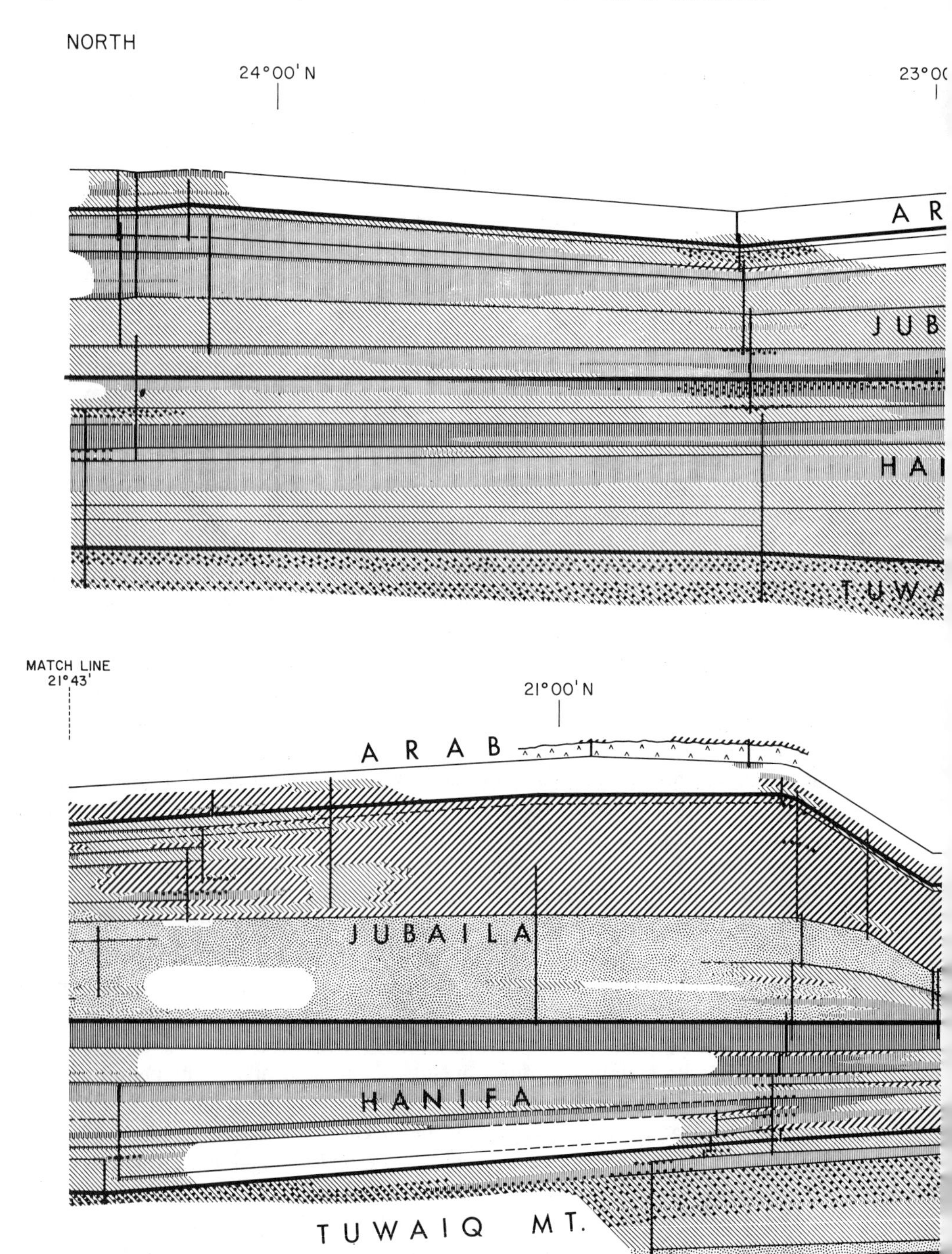

FIG. 3.—Reconstructed stratigraphic section, Upper Jurassic, southern Jebel Tuwaiq, Saudi Arabia.

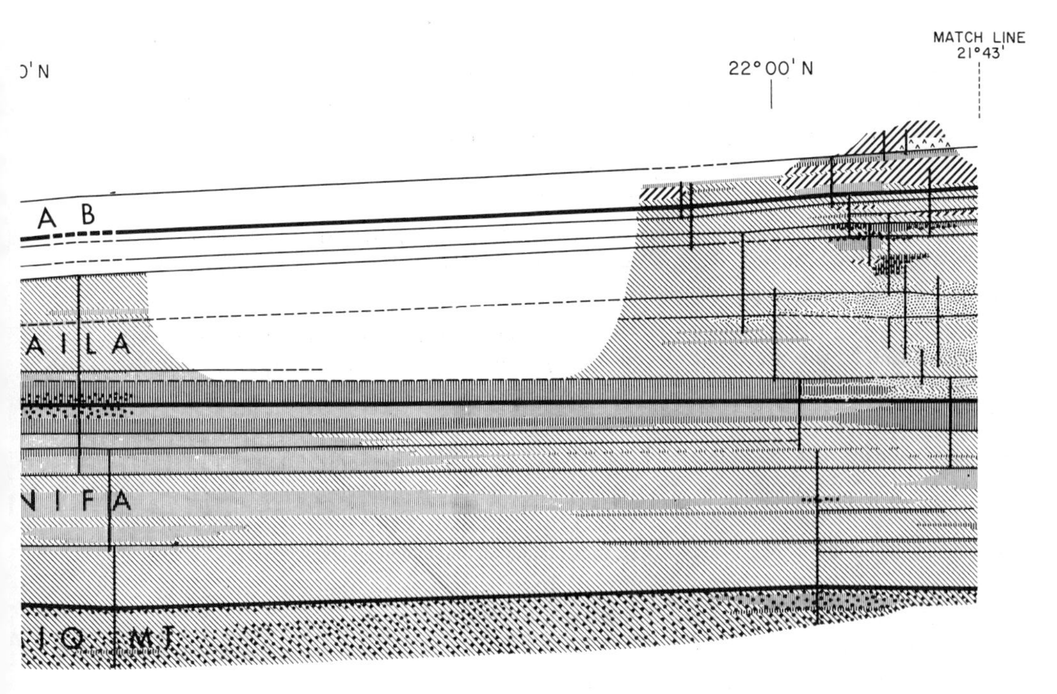
MATCH LINE
21°43'
22°00' N
A B
A I L A
N I F A
T U W A I Q MT
SOUTH
20°00' N

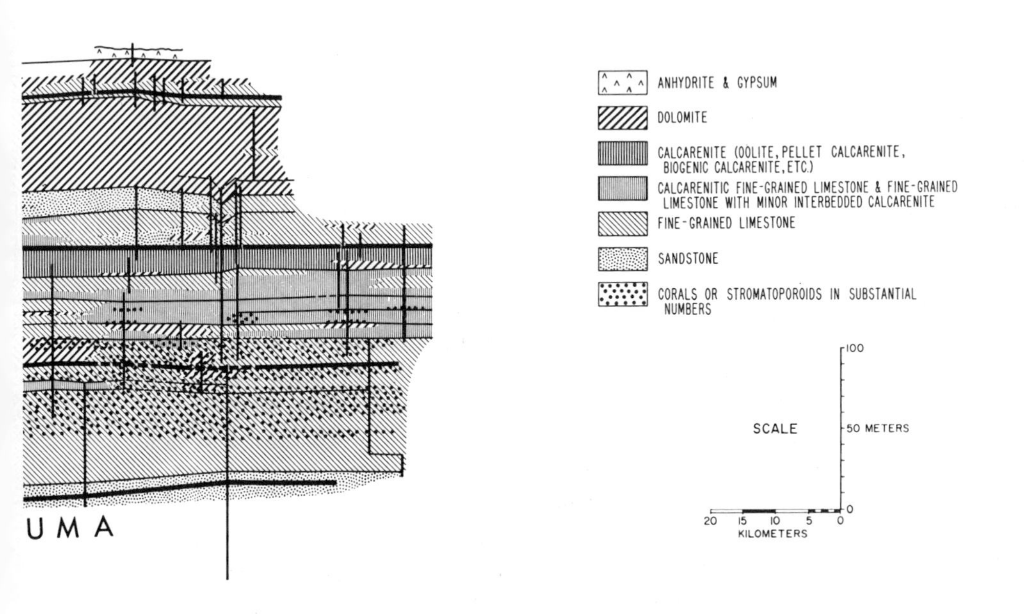
U M A
ANHYDRITE & GYPSUM
DOLOMITE
CALCARENITE (OOLITE, PELLET CALCARENITE, BIOGENIC CALCARENITE, ETC.)
CALCARENITIC FINE-GRAINED LIMESTONE & FINE-GRAINED LIMESTONE WITH MINOR INTERBEDDED CALCARENITE
FINE-GRAINED LIMESTONE
SANDSTONE
CORALS OR STROMATOPOROIDS IN SUBSTANTIAL NUMBERS
SCALE
100
50 METERS
0
20 15 10 5 0
KILOMETERS

Ghawar area calcarenite units are best developed near the middle. In the latter area coarser coquinal calcarenite layers become common in the uppermost Jubaila. To the east in the Dammam-Bahrain-Qatar area apparently calcarenite tends to be replaced to the east by fine-grained limestone, much of it of nearly lithographic texture. Some calcarenite still remains in the most easterly wells on which information is available.

The uppermost 10-20 meters of the Jubaila is a small unit whose lithologic pattern maintains unusual constancy over large areas. Immediate recognition of the position of the contact is possible along the whole length of nearly 700 kilometers of its outcrop in central Arabia. The contact also can be traced with reliability through the wells of much of the coastal Eastern Province although signature characteristics are less obvious in subsurface data. The persistence of the contact in wells escaped recognition until recently, because of the complexities introduced into the rock sequence and electric logs by irregular dolomitization, but its significance became obvious as diamond cores became available.

Thicknesses of the Jubaila in central Arabia ordinarily fall between 100 and 150 meters. In wells of the Eastern Province this ordinarily increases to about 300 meters. Possibly the increased thickness of Jubaila partly compensates for the lack of Tuwaiq Mountain and Hanifa in the eastern area.

Arab formation.—Rocks of the Arab formation show increased paleogeographic differentiation over those of the Jubaila, particularly in the Eastern Province, although the general pattern of lateral change likewise seems to be simple.

In central Arabia much of the Arab has been replaced by evaporites, presumably mainly or entirely anhydrite. Only a basal unit of carbonate rocks 15–25 meters thick remains, and even within this, along nearly the entire length of its outcrop, a persistent brecciated zone suggests that at least one thin, highly persistent evaporite layer was included. The rocks of the lower carbonate unit are dolomite and limestone, the latter including finely crystalline and fine-grained types, oölite, and pellet calcerenite. Aside from generally increased cementation, the basic rock types are closely similar to those of "D member" of the Arab of the wells of the Eastern Province. In a few local areas up to a meter of brown quartz sandstone occurs at the Jubaila-Arab contact. Determinable fossils other than *"Valvulinella" jurassica* Henson and a few other Foraminifera have not been found in central Arabia in spite of rather intensive search.

Above the basal carbonate unit, little is known of the true rock sequence as good exposures are few and fragmentary because extensive solution-collapse phenomena have almost entirely eliminated outcrops of the anhydrite and possibly other soluble evaporites equivalent to the Arab of the Eastern Province, as well as the overlying Hith anhydrite. It is for this reason that a subsurface sequence is taken as type for the Arab instead of one on the surface like those which have been used for the other main stratigraphic units. Only in the southern part of the solution-collapse zone are anhydrite outcrops common. Mainly the area in which the Arab and Hith would be expected to crop out is a nearly hopeless jumble of low hills representing complexly settled rocks. Dropped masses of the younger Cretaceous rocks occupy the eastern part of the solution-collapse zone, and dolomite and other carbonate rocks representing thin carbonate members, possibly equivalent to "A," "B," or "C members" of the Arab of the Eastern Province, show at the surface in the western part. The amounts of carbonate rocks residual from the solution effects are not large and presumably the interval must have been mainly evaporites. Apparently more residual carbonate rocks are present in the solution-collapse zone in the south than in the north.

The solution-collapse zone, representing the belt in which the Arab and Hith should crop out, can be traced from Latitude 20°45′ North in the south, north to about Latitude 26° North, a distance of approximately 550 kilometers. Rough estimates of the total thickness of Arab and Hith are 200–300 meters through much of this distance.

In the Eastern Province, field and exploratory wells have allowed a reasonable picture of the Arab and its lateral changes in certain areas, although great gaps in the over-all picture remain. Structurally low wells are few, but it seems doubtful that the deposition of Arab rocks was seriously influenced by individual structures as they are now known. Areas of recent drilling, particularly the Ghawar field, have furnished especially useful information on the lower Arab because of extensive diamond coring. The Qatar sequence has been recently described by Daniel (1954).

The Arab of the Eastern Province seems to represent four main cycles of deposition, each starting with more or less normal marine carbonate rocks and closing with anhydrite. The Hith anhydrite over much of the drilled area is regarded as the closing evaporite unit of the fourth and upper cycle, although possibly in areas to the east of present drilling other similar cycles may be present in beds equivalent to the Hith, in which the carbonate rock members were sufficiently less extensive that they failed to reach the drilled area.

The carbonate portions of the four Arab cycles have been denominated informally "A," "B," "C," and "D members," in order from top to bottom. Lateral changes in the various members are roughly the same in character, but the geographic locations at which these changes took place differ from member to member.

"D MEMBER"

Over fifty suites of diamond cores of "D member" are available. Figure 4, showing a typical Ghawar well, gives an example of the pattern in which the various carbonate rock types occur.

Lateral changes shown by "D member" of the Arab in the Hasa oil fields are roughly as follows.

1. The proportion of anhydrite between "C" and "D members" progressively increases at the expense of carbonate rocks below, from northeast to southwest through the Abqaiq-Ghawar area; nevertheless the overall thickness of "D member" plus that of the anhydrite between "C" and "D members" is nearly constant over this same area.

2. "D member" of the Abqaiq-Ghawar area is primarily calcarenite. Pellets, as this term has been used by Illing (1954), make up a high proportion of the sand-sized sedimentary particles. True oölite is nearly absent in central and southern Ghawar, but gradually increases to become well developed, although not to dominate, in northeastern Ghawar and Abqaiq. In obvious contrast to this, "D member" of Dammam, Qatif, Bahrain, and Qatar shows extensive replacement of these clastic carbonates by fine-grained limestones, and in these fields cleanly washed calcarenites are present only in small amount. This trend is most apparent in Qatif where "D member" contains much lithographic limestone. "D member" in its finer-grained facies in these fields must have been originally mainly lime mud. Part of the mud contained abundant

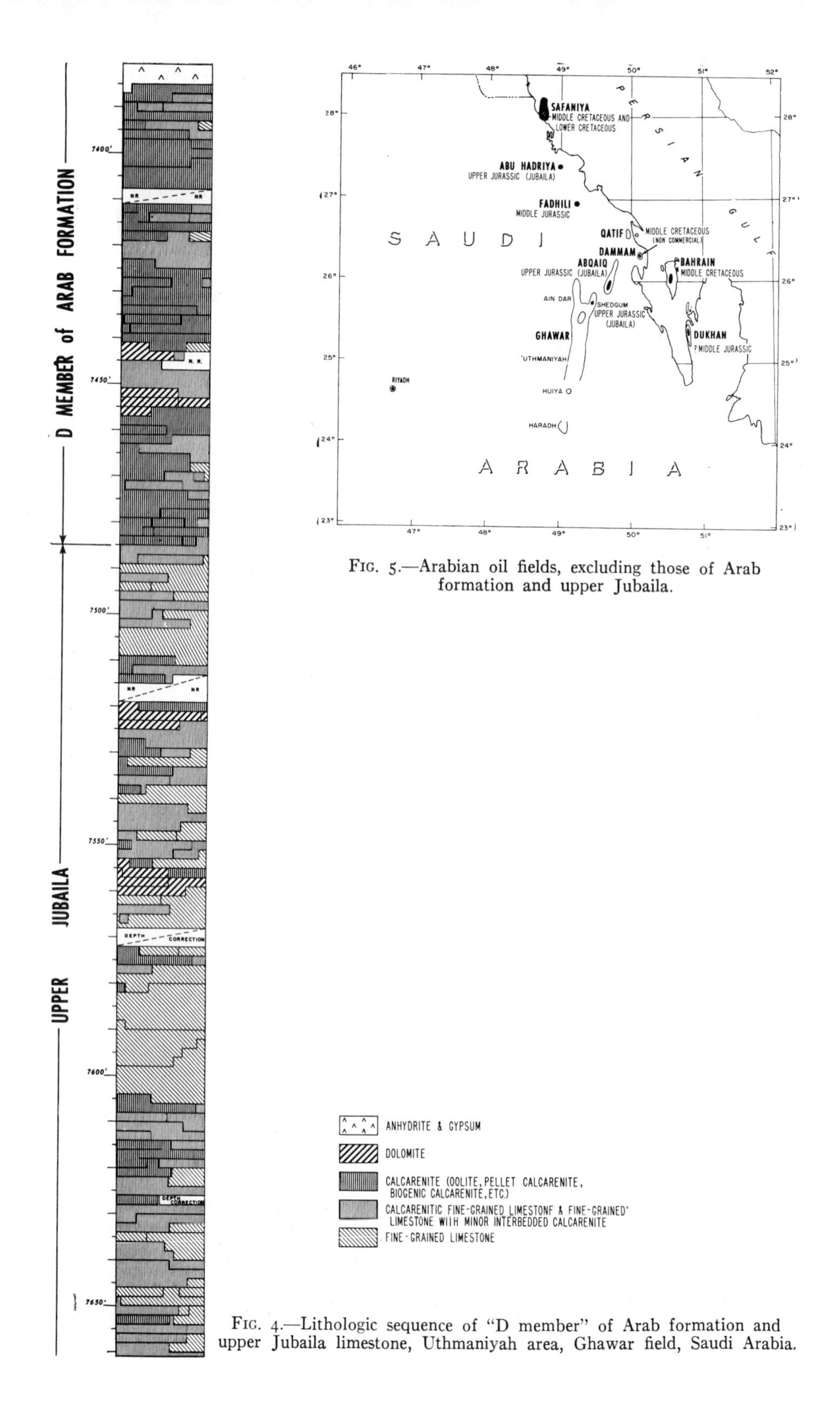

Fig. 5.—Arabian oil fields, excluding those of Arab formation and upper Jubaila.

Fig. 4.—Lithologic sequence of "D member" of Arab formation and upper Jubaila limestone, Uthmaniyah area, Ghawar field, Saudi Arabia.

included calcarenite particles, in places sufficient to make it muddy calcarenite. Much of these rocks have sufficient remaining original porosity to be moderately productive.

3. Evidence from various sources suggests that a minimum thickness of "D member" carbonate rocks, and maximum replacement of the upper part of the member by anhydrite, are reached just west of the Ghawar field and that this situation persists from there west to the outcrop of the beds. This facies seems to have been lagoonal with the beds equivalent to the upper part of "D member" in this western area replaced by evaporites.

These basic relations are indicated schematically in Figure 6*D*.

Information now available on "D member" suggests that the area of thick development of clean, current-washed calcarenite was a bar or the margin of a shelf. Apparently to the west, rocks of lagoonal facies replace the calcarenite, whereas a deeper water facies, composed mainly of muddy deposits, seems to have been present to the east. It is tempting to assume that the change from dominantly sand-sized particles of the calcarenite-rich area to mainly mud sized toward the east represented the mud line of the shallow sea of that time. Sedimentation in the modern Persian Gulf, however, suggests such an assumption is dangerous because of the demonstrable importance of tidal and/or other currents well below present wave base.

"A", "B", AND "C MEMBERS"

These upper, thinner limestone units of the Arab show far less extensive oil accumulation than "D member." They also appear to show more rapid replacement of limestone by anhydrite to the south and west. In the eastern part of the oil-field area, "C member" whose total thickness is approximately 100 feet, includes a lower unit primarily composed of calcarenite and coquina, and an upper portion of fine-grained limestone. Farther to the west there is an increase in the proportion of fine-grained limestones, together with gradual replacement of the upper part of "C member" by anhydrite, with a consequent reduction in its over-all thickness.

The pattern of "A" and "B members" resembles that of "C." Figures 6*A* and 6*B* show schematically the trends of lithologic variation in the three members.

Arabian Jurassic Oil Accumulations and Their Possible Origin

The locations of Arabian oil fields are shown on Figures 5 and 6.

LOWER JURASSIC

Marrat.—No oil has been found in the Marrat of central Arabia. Oil stains and minor traces of hydrocarbons are reported in the coastal part of the Eastern Province in beds which may be equivalent.

MIDDLE JURASSIC

Dhruma.—A number of the wells in the Eastern Province have shown stains and minor shows of oil in the middle and upper part of the Dhruma but only one field, Fadhili, has so far been discovered. The oil of the Fadhili field occurs mainly in fine calcarenite in the upper part of the Dhruma. Equivalent beds with moderate porosity in other closed structures have been found to be water-bearing. The general sedimentary situation controlling Dhruma oil accumulation is not yet apparent. The relatively low permeability of much of this part of the rock sequence offers some support to the idea that Dhruma oil probably originated in or near the reservoir rocks in which it is now found.

UPPER JURASSIC

Tuwaiq Mountain and Hanifa.—As indicated elsewhere, these two units occur in central Arabia but apparently are lacking in the coastal Eastern Province. No oil has been found in either of the units.

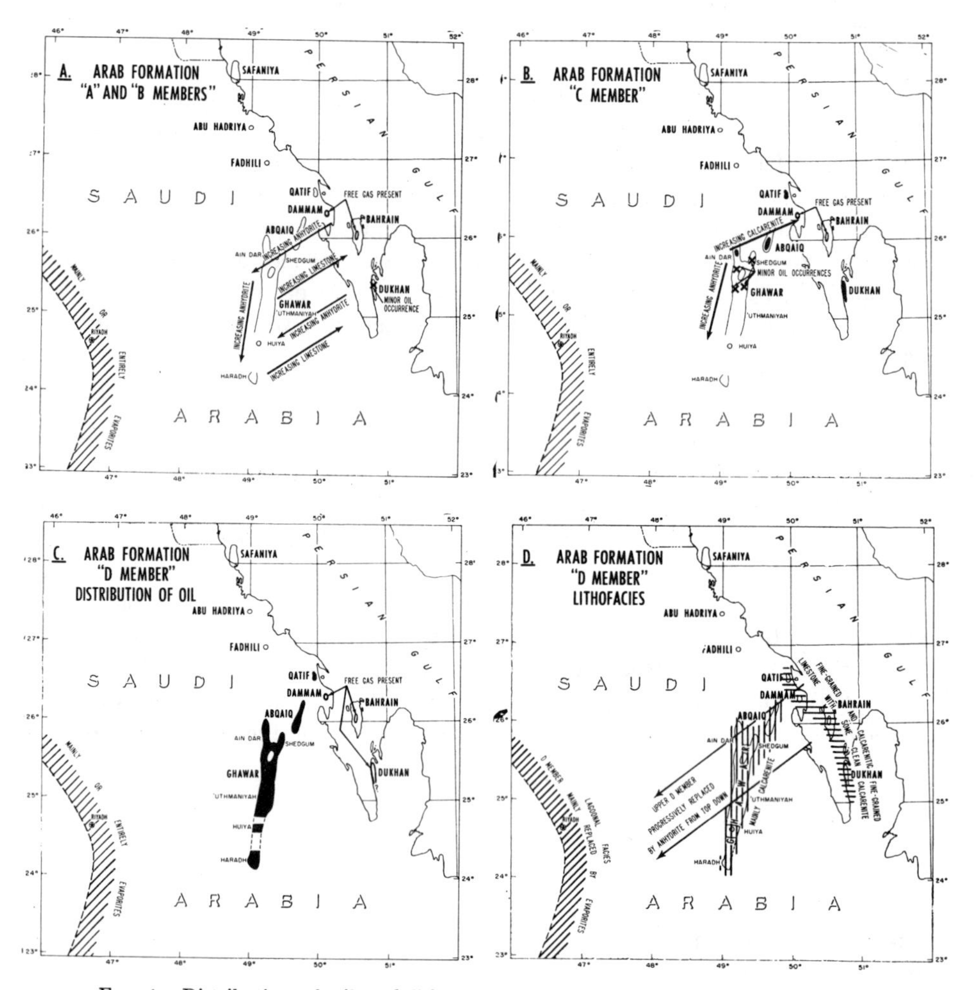

FIG. 6.—Distribution of oil and lithofacies of the Arab formation, Saudi Arabia.

Jubaila.—Oil accumulations are present in the Jubaila under the following circumstances.

1. Lower Jubaila in the Abu Hadriya field ("Hadriya Zone"), a calcarenitic limestone member composed of clastic carbonate particles in a fine-grained matrix.
2. Middle Jubaila of the Abqaiq field and the Shedgum area of Ghawar, mainly in calcarenitic limestone.
3. Uppermost Jubaila, mainly in the Ghawar field, in various types of porous limestone including calcarenite (some of it coquinal, and possibly a little of it biostromal), calcarenitic limestone and fine-grained limestone which are interbedded with dense limestone.

Although the detailed lateral variation in Jubaila lithology is well known only in the upper part of the unit in the developed oil fields, porous rocks in the Jubaila in general are apparently characterized by much lenticularity. The lateral equivalents of some beds productive in one structure have been found to be water bearing in other closed structures. Jubaila sediments as a whole must have been originally muddy and compaction of its sediments prior to lithification probably was large, if similar modern sediments in the Persian Gulf can be used as a standard. This makes it reasonable to assume that the Jubaila oil already found was derived from the calcarenite and calcarenitic rocks, as well as the adjacent fine-grained limestone and concentrated into its present reservoirs during compaction and lithification. Evidence that jointing plays any role in Jubaila oil accumulations has not been detected.

Arab.—The distribution of oil and gas in the Arab formation is shown in Figure 6. Much of the Arab reserves are in "D member," although substantial amounts also occur in "C." The area of maximum calcarenite development in "D member," Ghawar and Abqaiq, is likewise that of the largest oil accumulations, with much of the oil contained in pore space probably remaining from the time of original deposition of the sand-sized carbonate particles. Recrystallization, dolomitization, and cementation have all modified the original pattern of the openings but not to an extent to fill or greatly decrease their volume.

The anhydrite above "D member" and the apparently capricious distribution of oil and water in porous beds of the Jubaila favor a hypothesis which confines potential source beds of "D member" oil to the Jubaila and "D member" itself. The oil now found in the uppermost Jubaila is apparently in hydrostatic continuity with that of "D member" and hence this oil is likely to have had the same general source as that of "D." Although there is an increase in sand sized and coarser carbonate particles in the Jubaila from northeast to southwest over the oil field area, the Jubaila must have been laid down primarily as muddy sediments. Modern lime muds in the Persian Gulf, comparable to some Jubaila rocks, show compaction considerably in excess of 40 per cent when they are dried without load. However, the modern Gulf offers only an incomplete parallel to Jubaila deposition as it is doubtful if any bodies of fine-grained limestones of the type so prominent in the Jubaila are now accumulating there. No doubt the Jubaila had an appreciable original organic content and it must have compacted sufficiently to displace much or all of any generated hydrocarbons into adjacent potential reservoirs.

The calcarenite and fine-grained limestone of "D member" itself are also potential source rocks. Circulation of fluids in the calcarenites immediately after deposition could have caused oxidation and dissipation of any original organic matter present in these rocks, but in the modern Persian Gulf an almost negligible amount of silt-sized material has been found to allow anaerobic bacterial action below a few millimeters of oxidized surface sediment. Hence the rocks of "D member" cannot be eliminated as a potential source even where they are in coarse-grained facies. Evidence has not yet been uncovered to favor the derivation of "D member" oil from Jubaila or Arab rocks alone, and perhaps both contributed significantly.

The time of origin of the individual structures in which "D member" oil is now found has not yet been accurately determined. The first unambiguous evidence of the growth of the structures seems to be in early Upper Cretaceous time, although there probably were insignificant local structural disturbances during the Lower Cretaceous. If oil was generated soon after the time of deposition of Jubaila and Arab rocks, considerable time must have elapsed prior to localization of oil into the structures in which it is now trapped. It now seems possible that the great calcarenite lens or lenses of "D member" of the Arab may have acted as a stratigraphic trap or traps to hold the oil in the general vicinity until the appearance of the structural traps as they are now known. Thus, the preferred hypothesis at present is that the major "D member" fields are the final result of reconcentration by later folding of oil temporarily contained in one or more earlier large stratigraphic traps.

In "C member" of the Arab, oil and locally gas are present in significant amounts in Dammam Dome, Bahrain, Qatar, Qatif, and Abqaiq. Small amounts of oil are present in the low crestal closures of middle and northern Ghawar where porosity is adequate. Brine with shows of oil is found in southern Ghawar. The occurrence of oil in "C member" seems to show a rude correlation with the extent to which carbonate clastics are present, although this may be unrelated coincidence. Because of the presence of anhydrite members confining "C member," it seems necessary to assume that oil found in it is indigenous. Reconcentration of the oil by later folding comparable to that suspected for "D member" is the best hypothesis yet available to explain its present distribution.

Oil and gas in "A" and "B" members are even more restricted geographically than in "C member." Here also, it seems necessary to assume that this oil is indigenous to the members in which it is found.

REFERENCES

ARKELL, W. J., (1952), "Jurassic Ammonites from Jebel Tuwaiq, Central Arabia, with Stratigraphical Introduction by R. A. Bramkamp and Max Steineke," *Philos. Trans. Roy. Soc. London,* Ser. B., Vol. 236, pp. 241–313.
BAKER, N. E., AND HENSON, F. R. S., (1952), "Geological Conditions of Oil Occurrence in Middle East Fields," *Bull. Amer. Assoc. Petrol. Geol.,* Vol. 36, No. 10, 1885–1901.
BLANCKENHORN, MAX, (1914), "Syrien, Arabien und Mesopotamien," *Handbuch der Regionalen Geologie,* Bd. V. Abt. 4, Heft 17.
COX, L. R. (1933) in *The Empty Quarter,* by H. St. J. B. Philby, pp. 383–89.
DANIEL, E. J., (1954), "Fractured Reservoirs of Middle East," *Bull. Amer. Assoc. Petrol. Geol.,* Vol. 38, No. 5, pp. 774-815.
EICHER, D. B., AND YACKEL, M. P., (1951), "Exploration Gets Results in Saudi Arabia," *Oil and Gas Jour.,* Vol. 50, No. 33 (December 20), pp. 208, 212, 293.
HENSON, F. R. S., (1951a), "Oil Occurrences in Relation to Regional Geology of the Middle East," *Tulsa Geol. Soc. Digest,* Vol. 19, pp. 72–81.
———, (1951b), "Observations on the Geology and Petroleum Occurrences of the Middle East," *3d World Petrol. Cong.* Section I, pp. 118–40.
ILLING, LESLIE *V.,* (1954), "Bahaman Calcareous Sands," *Bull. Amer. Assoc. Petrol. Geol.,* Vol. 38, No. 1, pp. 1–95.
KERR, R. C., (1953), "The Arabian Peninsula," *Sci. of Petrol.* Vol. 6, Pt. 1, pp. 93–98.
LAMARE, PIERRE, (1936), *"Structure Geologique de l'Arabie,"* Ch. Beranger. Paris.
PHILBY, H. ST. J. B., (1922), *The Heart of Arabia,* 2 vols., Constable & Co. Ltd., London.
———, (1928), *Arabia of the Wahhabis,* Constable & Co. Ltd., London.
———, (1933), *The Empty Quarter,* Constable & Co. Ltd., London.
———, (1939), *Sheba's Daughters,* Methuen & Co. Ltd., London.

QUENNELL, A. M., 1951). "The Geology and Mineral Resources of (Former) Trans-Jordan," *Colonial Geol. and Min. Res.,* Vol. 2, No. 2, pp. 85–115, London.
SANDER, N. J., (1952), "La Stratigraphie de l'Eocene le long du Rivage Occidental du Golfe Persique," Unpublished Doctoral Thesis, Faculté des Sciences de l'Université de Paris.
SMOUT, A. H., (1954), *Lower Tertiary Foraminifera of the Qatar Peninsula,* Brit. Mus. Nat. Hist., London.
STANISLAS-MEUNIER, M., (1891), in *Journal d'un Voyage en Arabie* (1883-84), by S. Huber, pp. 759–70, Paris.
STEINEKE, MAX, AND YACKEL, M. P., (1950), "Saudi Arabia and Bahrein," in "World Geography of Petroleum," by Wallace E. Pratt and Dorothy Good, *Amer. Geogr. Soc. Spec. Publ. No. 31,* pp. 203–29.
VON WISSMANN, H., RATHJENS, C., AND KOSSMAT, F., (1943), "Beiträge zur Tektonik Arabiens," *Geol. Rundschau,* Bd. 33, pp. 221–353.
WELLINGS, F. E., (1954), "Middle East Oil Sources and Reserves," *The Post-war Expansion of the U.K. Petroleum Industry,* Inst. of Petrol., London, pp. 1–24.
WRIGHT, C. W., (1952), "A Classification of the Cretaceous Ammonites," *Jour. Paleon.,* Vol. 26, No. 2, pp. 213–22.

BULLETIN OF THE AMERICAN ASSOCIATION OF PETROLEUM GEOLOGISTS
VOL. 43, NO. 2 (FEBRUARY, 1959), PP. 434-454, 8 FIGS.

GHAWAR OIL FIELD, SAUDI ARABIA[1]

ARABIAN AMERICAN OIL COMPANY STAFF[2]
New York, New York

ABSTRACT

Ghawar field is located about 50 miles inland from the western shore of the Persian Gulf, in Hasa Province, Kingdom of Saudi Arabia.

The field was discovered in 1948 by a wildcat well at Ain Dar. Other wildcats, Haradh (1949), 'Uthmaniyah (1951), and Shedgum (1952) proved to be on subsidiary closures of the same great anticline, and parts of the same oil field. Oil-finding has been primarily through structure drilling, although surface mapping furnished initial clues and gravity mapping proved helpful.

The oil field is a structural accumulation along at least 140 miles of the north-south-trending En Nala anticlinal axis. The fold is simple in the south, develops two low marginal crestal closures in the center, and is subdivided into two adjacent anticlines in the north. Seven crestal closures have been found, but oil appears to be continuous from Ain Dar to Haradh.

Ghawar oil is Upper Jurassic, occurring in the shallow-water carbonate sediments of the upper Jubaila and lower Arab formations. It is confined above by the lowest of the three anhydrite members of the Arab formation. Only minor shows of oil have been found in higher parts of the Arab. The most prolific production is derived from calcarenites in which much original pore space survives. Other rocks making up the reservoir are fine-grained limestone, calcarenitic limestone, dolomitic limestone, and dolomite. Partial dolomitization, usually porphyroblastic, is extensive.

The oil accumulation, conservatively estimated at 875 square miles in area, has a maximum vertical oil column of about 1,300 feet. Gravities range from 36° API in the north to 33° API in the south. The crude is undersaturated, with saturation pressures decreasing to the south.

The oil-water contact rises to the south. The reservoir water ranges from 240,000 ppm total solids in the north to 38,000 ppm in the south. On the west side of the structure elevations of the contact are higher and salinities of the water are lower than on the east.

Incipient growth of the Ghawar structure apparently started in the Lower Cretaceous but the structure in its present form first became apparent in truncation at an unconformity between the early and late Upper Cretaceous (roughly post-Cenomanian and pre-Maastrichtian). Growth continued into the middle Eocene with minor disturbances possibly continuing to the Miocene. No fundamental structural disturbance, except perhaps regional tilting, seems to have taken place during and after Miocene time.

The source of Ghawar oil apparently is in the Arab and Jubaila formations. The oil volume forces an assumption of substantial lateral migration. Temporary accumulation of the oil in a stratigraphic trap prior to its localization in the present oil field is the preferred hypothesis.

INTRODUCTION

The fundamental structure of Ghawar field is relatively simple. Widely spaced structure holes in addition to field wells have been used to outline the long north-plunging, in part compound, anticlinal axis which has been proved productive over a distance in excess of 140 miles.

For better understanding of the great area involved, intensive study of lateral lithologic changes has been required. This is particularly true of the Arab-*D* member and the upper part of the Jubaila formation, the main producing interval of the field. Although these studies are still in their early stages, some of the preliminary findings are presented in this paper. The following discussion is basically descriptive: a factual summary of available data from which a few obvious conclusions have been drawn.

[1] Read before the Association at Los Angeles, March 13, 1958. Manuscript received, July 21, 1958. Published by permission of Arabian American Oil Company.

[2] Numerous individuals have been involved in the collection and processing of the data and formulation of the ideas expressed here. It is not possible to cite their contributions individually, but it should be noted that each has added to our knowledge of Arabian geology.

HISTORY

The first clue to the presence of the En Nala anticlinal (Fig. 1) axis was detected by Steineke and Kock[3] in 1935 during surface mapping of Miocene-Pliocene rocks. Later detailed surface mapping showed closure in the Haradh area, including one very small area where Eocene rocks were exposed, but along the remainder of the axis only the same minor irregularities noticed by Steineke and Koch were found. In 1941 widely spaced structure holes were drilled which confirmed the presence of a major anticlinal axis. Gravity mapping of northern Ghawar was done prior to World War II.

Immediately following World War II, structure drilling supplemented by gravity-magnetic mapping was used to define wildcat locations in the Ain Dar and Haradh areas. Oil was found in the Arab-*D* member and in the uppermost Jubaila, at Ain Dar in June, 1948, and at Haradh in February, 1949. These wildcats were followed by development wells near the first two discoveries and two more wildcats, 'Uthmaniyah in April, 1951, and Shedgum in August, 1952, both of which found oil. Development drilling, continuing to the present, blocked out the structure as it is now known. The Ain Dar, Shedgum, and 'Uthmaniyah areas are adequately mapped, but the southern half of the field is only roughly defined by widely spaced wells and shallow structure drilling. The limit of the field north of Ain Dar has not been reached.

Seismic work has been generally unsuccessful on crestal Ghawar although flank information has been obtained.

Production, started in 1951, averaged nearly 600,000 barrels per day in 1957.

TOPOGRAPHY AND GEOMORPHOLOGY

The structure of the 'Uthmaniyah, Ain Dar, and Shedgum areas of Ghawar is approximately reflected at the surface by a dissected plateau capped by a resistant limestone layer, the plateau surface extending a considerable distance out over the west flank. The preservation of this plateau seems to be the result of carbonate enrichment of its surface over the structure, giving it a greater resistance to erosion there than in the synclinal areas. The present plateau surface appears almost completely resistant to wind erosion, although perhaps there is very slow destruction of it by carbonate solution. In southern Ghawar a generally high, shallowly dissected area covers the structure. Here the resistant capping layer is not developed to the extent of that farther north and only patches of it remain.

The whole area of the Ghawar field lacks integrated drainage. Small flash floods, occurring periodically, have carved minor wadian (valleys or dry-water courses) which carry the water, a few kilometers at most, to scattered closed basins. Over large areas even wadian are lacking, and the little rain water which falls usually soaks in only to be lost by evaporation during the following summer.

[3] Geologists, Arabian American Oil Company.

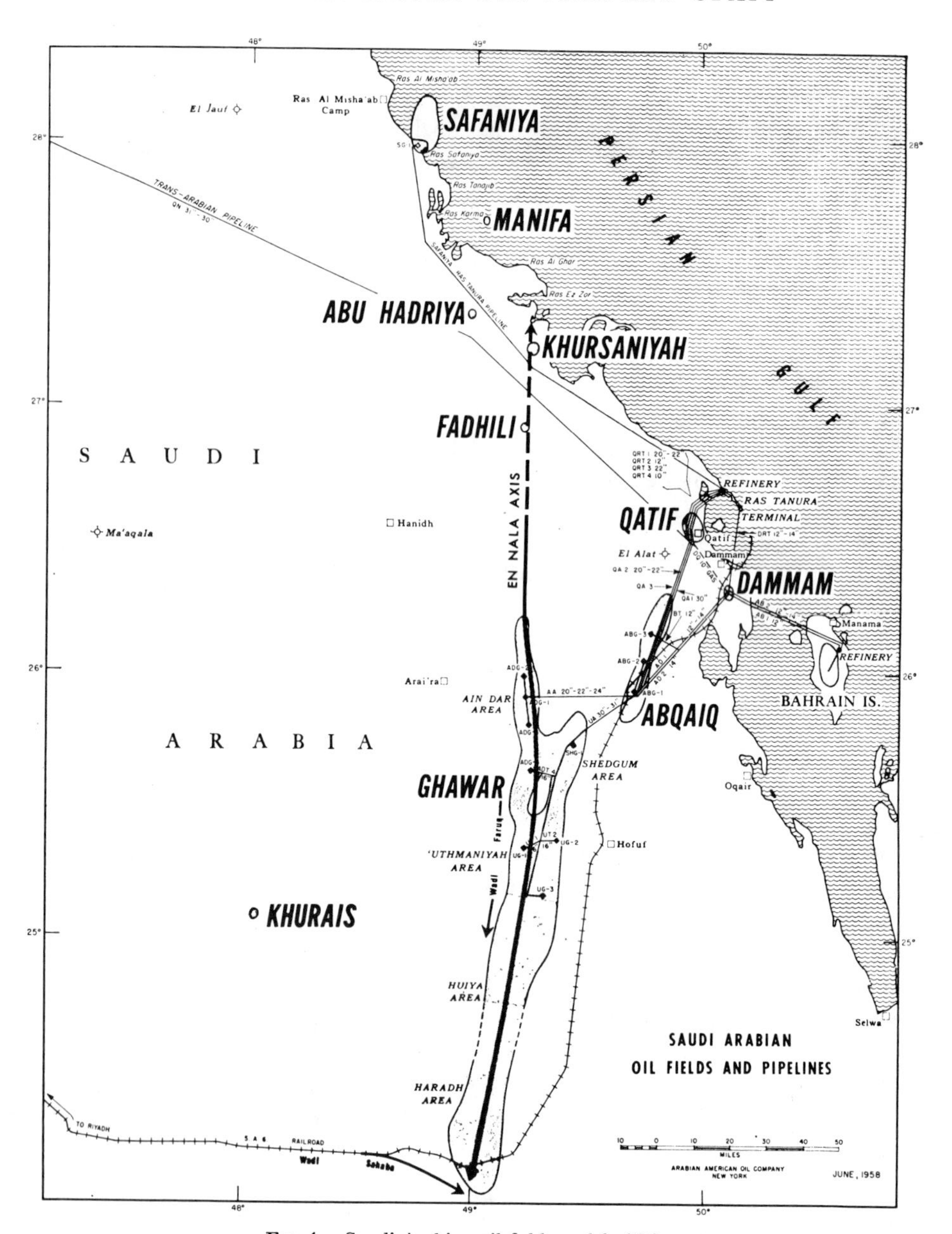

FIG. 1.—Saudi Arabian oil fields and facilities.

However, small amounts may be led down to static ground water by the rare sink holes. The main agency for the removal of material from the area is wind, but this takes away only material which has been previously loosened and prepared by rain wash. A few small areas of sand dunes occur in northern Ghawar, and considerable additional eolian sand is scattered over the area as thin sheets and hummocks which are usually in topographic shadows or are held by small shrubs. A trip to Ghawar on a windy day leaves no doubt that, in spite of the absence of major surface accumulation, considerable amounts of sand are moving through the area.

This brief account does not do justice to the obviously complicated Quaternary erosional history, part of which is reflected in the topography of the area. Geomorphological evidence of recent tilting is the main interest, but so far it has proved elusive. The only suggestions of disturbance, and these are virtually unsupported, are found in the Wadi Sahaba area where drainages, originally primarily toward the north and northeast, may have been diverted toward the east and perhaps southeast. The time of this tilting should be early Pleistocene and perhaps even late Pliocene. Regional slopes are so gentle, however, that effects on the oil field must have been very slight.

SURFACE GEOLOGY

Except for two very small Eocene outcrops in the center of Haradh, the Ghawar field is blanketed by Miocene-Pliocene continental deposits, mainly sandy "marl" and sandy limestone with subordinate shale and sandstone. The top of the group of Miocene-Pliocene sediments is a hard resistant limestone, in many places vertically fluted, which appears to be an old carbonate-enriched surface. This hard layer is responsible for the very stable plateau surfaces which cover much of middle and northern Ghawar.

Correlation of the continental Miocene-Pliocene is difficult and uncertain. No satisfactory map of the area on a single datum has been made, but it is known that general regional dips toward the east are interrupted by an irregular terrace over the Ghawar field and that closure is demonstrated only in the Haradh area. There is some tendency for an increase in amount of carbonate rocks in the Miocene-Pliocene over the En Nala axis.

The best information at present suggests that Ghawar's minor reflection in Miocene-Pliocene rocks is due primarily to its influence on initial depositional dips rather than to true deformation of the Miocene-Pliocene rocks themselves. Structure drilling shows that the Eocene rocks of the En Nala axis were a prominent topographic high at the start of Miocene-Pliocene deposition, but this high had progressively less influence as the Miocene-Pliocene sediments accumulated. The final result is the very minor topographic reversal of only a few meters in the resistant capping limestone layer which originally led Steineke to suspect the presence of the structure.

Topographically low areas on each side of the En Nala axis, Wadi Faruq, etc., on the west and a gravel plains area on the east have resulted from more rapid erosion of the softer rocks in the synclinal areas. The gravels must have been derived from the interior west of En Nala. One general drainage course spread north along a topographic low west of En Nala, while a second crossed the

En Nala axis in the vicinity of Wadi Sahaba, from which point it spread as a great fan toward the north, east, and south. During this time the Ghawar axis stood as a topographically high area above the gravel-covered flood plains. These gravels, although including some rocks of local origin, contain a high proportion of quartz and other minerals and rocks which must have been derived originally from the pre-Cambrian basement rocks of the Arabian shield.

Prior to the spread of the gravel sheets and channels the Miocene-Pliocene rocks were let down in various types of solution-collapse structures, usually oval or circular sinks. These sinks are particularly common on the west flank of the En Nala axis. They result mainly from the solution of underlying soluble rocks, although it is possible that plastic flow of wet clayey rocks may also have been involved.

The effects of the En Nala anticline on Miocene-Pliocene and later sedimentation and erosion are now reasonably well known. Much of this is obvious only in retrospect, however, as many of the observed facts had alternate explanations during early exploration work. In general, the late Tertiary and Quaternary geology has served to confuse, rather than clarify, the search for oil.

STRATIGRAPHY

Figure 2 outlines the stratigraphic sequence of rocks down to the base of the productive units. The significance of changes in thickness of these units taken as a possible clue to the history of the growth of the Ghawar structure is discussed later.

The informal terms in the Lower Cretaceous sequence are used advisedly. The formational contacts were defined from type sections in outcrops 60 and more miles toward the west, so that doubt remains about the exact position of some of the contacts in Ghawar wells. Future wells in the intervening gap will undoubtedly settle these problems; in the meantime, temporary unit names are used.

Miocene-Pliocene.—The first indurated sediments beneath the desert sand mantle are of Miocene-Pliocene age. The rocks show a wide lateral variation range, lithologically, from unconsolidated to poorly cemented sandstone and sandy marl in the upper part to hard sandy limestone and well cemented sandstone in the lower part. Their arenaceous nature distinguishes them from the underlying Eocene.

Eocene-Paleocene.—The arenaceous Miocene-Pliocene sediments rest unconformably on middle Eocene non-sandy limestones and marls. In the crestal areas of Ghawar the unconformity has truncated beds of the Dammam formation and in some places the Rus and Umm er Radhuma formations.

The Dammam formation is composed of five recognizable sedimentary units through most of Ghawar field. From the top down the units are: (1) the Alat limestone, typically white granular porous limestone, but in places compact and siliceous; (2) the "Orange marl," a yellow to gray fine-grained in part dolomit-

GEOLOGIC SECTION - Ghawar Field				
AGE		**ROCK UNIT**	**LITHOLOGIC CHARACTER**	**THICKNESS (ft.)**
MIOCENE-PLIOCENE		MIOCENE-PLIOCENE (UNDIFFERENTIATED)	SANDY LIMESTONE AND SANDY MARL, LOCALLY WITH SANDSTONE AND CLAY	550
		UNCONFORMITY		
MIDDLE EOCENE		DAMMAM FORMATION	LIMESTONE, DOLOMITE AND SUBORDINATE CLAY	140
LOWER EOCENE		RUS FORMATION	MARL, LIMESTONE AND CLAY, LOCALLY INCLUDING ANHYDRITE AND GYPSUM	75
PALEOCENE		UMM ER RADHUMA FORMATION	LIMESTONE AND DOLOMITE, IN LOWER PART WITH SUBORDINATE SHALE	800
UPPER CRETACEOUS	MAASTRICHTIAN & CAMPANIAN (?)	ARUMA FORMATION	LIMESTONE, DOLOMITE AND SUBORDINATE SHALE	550
		UNCONFORMITY		
	CENOMANIAN & ALBIAN (?)	WASI'A FORMATION	SHALE, LIMESTONE AND SANDSTONE WITH SUBORDINATE DOLOMITE AND DOLOMITE-SIDERITE ROCK	1230
--?--		**UNCONFORMITY**		
LOWER CRETACEOUS		UNNAMED UNIT	DOLOMITE	250
		BIYADH FORMATION	SANDSTONE AND SHALE, WITH SUBORDINATE LIMESTONE, DOLOMITE AND DOLOMITE-SIDERITE ROCK	700
		BUWAIB FORMATION	LIMESTONE AND SUBORDINATE SHALE	450*
		UNCONFORMITY**		
	VALANGINIAN	YAMAMA FORMATION	CALCARENITE AND FINE-GRAINED LIMESTONE	260
LOWER CRETACEOUS (?)		SULAIY LIMESTONE	FINE-GRAINED LIMESTONE, USUALLY WITH A BASAL CALCARENITE UNIT	310
		?DISCONFORMITY		
UPPER JURASSIC	KIMMERIDGIAN AND/OR YOUNGER	HITH ANHYDRITE	ANHYDRITE WITH SUBORDINATE-FINE-GRAINED LIMESTONE, DOLOMITE AND CALCARENITE	500
		ARAB FORMATION	CALCARENITE, CALCARENITIC LIMESTONE, FINE-GRAINED LIMESTONE AND DOLOMITE, WITH THREE ANHYDRITE MEMBERS	560
	KIMMERIDGIAN	JUBAILA FORMATION	FINE-GRAINED LIMESTONE, WITH SUBORDINATE CALCARENITE, CALCARENITIC LIMESTONE AND DOLOMITE	1200

* Lower part includes beds, possibly an unnamed unit, missing at the Yamama - Buwaib unconformity in Nejd.

** An unconformity is present between the Yamama and Buwaib in central and southern Saudi Arabia but has not been detected in the Eastern Province.

Reference of Lower Cretaceous rock units to European stages, except for the Yamama are under study.

Thickness are those in a typical crestal well in northeastern 'Uthmaniyah.

FIG. 2.—Sequence of rocks in Ghawar field.

ic marl; (3) the Khobar dolomite, a variable dolomite and limestone; (4) the "Alveolina zone," a limestone and marl; and (5) the Midra shale, a thin earth-brown shale which in some places contains associated impure pyritic limestone.

The Rus formation is characterized by upper and lower granular chalky limestone units separated by a middle marly shale section which in some flank wells contains anhydrite. The anhydrite in the Rus formation influences higher horizons to the extent that they do not accurately reflect deep structural configuration. The top of the Umm er Radhuma formation ("base of the Rus formation" in company terminology) is thus a critical stratigraphic horizon, as it is the first marker encountered which can be used to interpret deep structure.

The Umm er Radhuma formation is composed of two units: an upper dolomitic limestone of lower Eocene age and a lower limestone-shale unit of Paleocene age. The Eocene-Paleocene contact can be identified by foraminifera when cuttings are available, but because this section is penetrated without circulation in most Ghawar wells the Umm er Radhuma is usually considered as a single rock unit, with no attempt made to determine the Eeocene-Paleocene contact in routine wells. The basal Umm er Radhuma is limestone except in Haradh, where it is shale and dolomite.

Cretaceous.—The Aruma formation is composed of an upper limestone-dolomite unit, predominantly gray to grayish tan in color, a middle gray shale unit (called lower Aruma shale), and a basal crystalline limestone or dolomite. The top of the Aruma formation is determined at the first appearance of gray to black, lignitic, pyritic shale. An unconformity is present within the lower Aruma shale.

The contact between the Aruma and the underlying Wasi'a formation is an angular unconformity with as much as 300 feet of section truncated in crestal areas of the field. The contact is usually characterized by the occurrence of numerous limonite nodules, glauconite and free pyrite. The Wasi'a is divided into several units for local correlation. Lithologically, however, it is composed of two major units, an upper limestone and shale and a lower massive sandstone. The upper unit ranges in thickness from more than 300 feet in northern Ghawar to nearly zero in the Haradh-Huiya area where the Wasi'a-Aruma unconformity truncation is greatest.

The next unnamed unit (informally called the "Dolomitic limestone") is predominantly a massive crystalline dolomite, though in widely separated areas, and usually in flank wells, it is in a sand-shale-lignite facies. When present, the sand-shale-lignite facies can be distinguished from the overlying Wasi'a by its resistivity pattern on the electric log.

The Biyadh formation in Ghawar field is predominantly clastic and can be divided roughly into two units; the upper is composed of greenish gray, rather calcareous shales with interbedded sandstone, limestone and dolomite; the lower is greenish gray calcareous shale and chalky, pyritic, glauconitic limestone with subordinate thin sandstones. Regionally, there is an increase in clastics toward Haradh.

The Buwaib formation is a persistent massive limestone, mostly fine-grained, compact and pyritic, with a few streaks of calcarenite in the lower part.

In its typical development the Yamama formation is a soft fine-grained poorly indurated calcarenite. A facies change takes place in the upper part of the formation; the typical porous calcarenite is 40 feet lower stratigraphically in 'Uthmaniyah than in north Ain Dar.

The Sulaiy limestone is a hard fine-grained erratically calcarenitic limestone.

STRUCTURE

Ghawar oil field occupies the southern and highest part of a major anticline, the En Nala axis (Fig. 1). The Fadhili and Khursaniyah oil fields are located on the northern part of the structure. Total length of the axis is about 250 miles, with Ghawar representing somewhat more than half of this.

The structure of the Ain Dar, Shedgum, and 'Uthmaniyah areas of the field are fairly well understood from the results of oil-well drilling (Fig. 3b). Well spacing in Huiya and Haradh in the south and in the area north of Ain Dar is still too wide for detailed mapping.

The results of oil-well drilling are supplemented by a considerable number of structure wells completed at the top of the Umm er Radhuma formation (lower Eocene). These shallower data are helpful, and, in fact, were responsible for the location of the wildcats discovering Ghawar and its subdivisions. Nevertheless, they give an inadequate, and sometimes distorted, version of the deeper structure. Dips in the lower Eocene are much gentler than those of the Arab formation.

Flanks.—Long stretches of the flanks of Ghawar are straight or slightly curved, with flank dips in the Arab formation a generally uniform 5°–8°, but in places reaching 10°. A substantial offset of flank trends takes place in the southeastern 'Uthmaniyah area, with a shift of the nearly straight east flank of Haradh and Huiya 10–15 kilometers east, where it resumes its linear characteristics along the east flank of 'Uthmaniyah and Shedgum. A similar shift is present in the west flank, but incomplete information beyond the limits of the oil field prevents a full analysis of its significance.

Crestal area.—The crest of Ghawar differs somewhat in its northern and southern parts. Data from the widely spaced wells of the Huiya-Haradh area indicate that here the field is a simple anticlinal axis. The height of the crest varies somewhat, but so far there is no suggestion that it is compound. In contrast, in the Ain Dar, Shedgum, and 'Uthmaniyah areas there is a crestal depression with closures along each rim. 'Uthmaniyah shows a low anticline having about 450 feet of relief along the eastern edge and a very low swell of less than 50 feet of relief along the western edge; in Ain Dar and Shedgum both crests rise about 400 feet above the central sag. There is a large oval structurally low area between Ain Dar, Shedgum, and 'Uthmaniyah which must be a fundamental feature of the axis, as it is shown by paleostructure maps of rocks as old as Aptian-Albian.

In the 'Uthmaniyah area the crestal closures are parallel with the flanks and oriented very nearly north-south; Shedgum and south Ain Dar trend N. 15° E.; north Ain Dar resumes the north-south orientation seen on the south. The oblique trend on Shedgum and south Ain Dar approximates the orientation of Abqaiq, which lies a short distance northeast of Shedgum. No doubt there is a common causal relation.

Faulting.—In four north Ain Dar wells there are normal faults of small displacement which, as they do not alter the fundamental characteristics of the structure, are considered to be insignificant shears.

Gravity-magnetic mapping.—The part of the En Nala axis surmounted by Ghawar shows a strong positive gravity anomaly (Fig. 3a.). In general, the coincidence between gravity mapping and structure is remarkably close. In some areas, south Ain Dar in particular, gravity control is better than shallow struc-

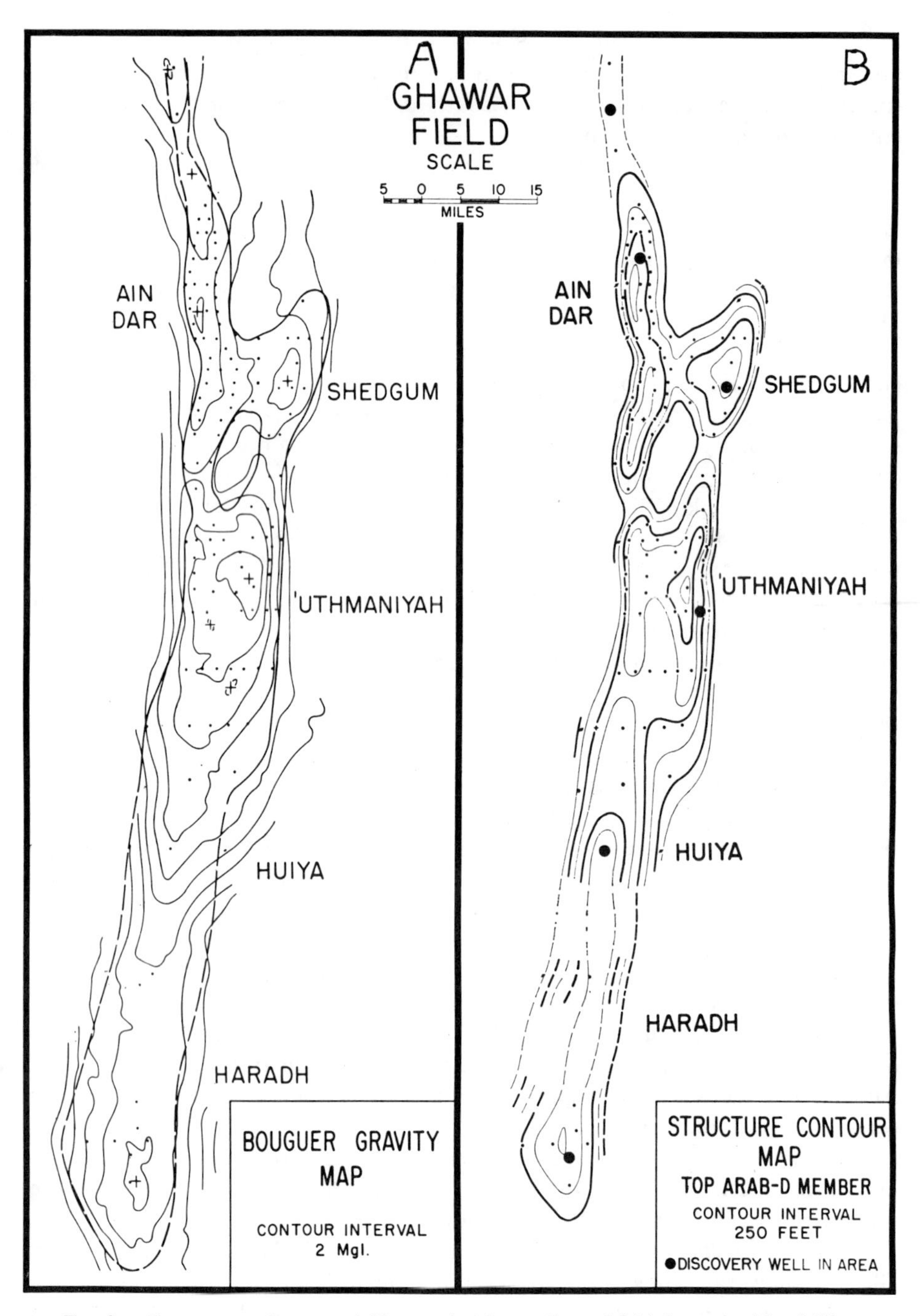

FIG. 3a.—Bouguer gravity map of Ghawar showing outlines of field determined by drilling.
FIG. 3b.—Structure contour map of top of Arab-*D* member.

ture drill data. The raw positive anomaly is of the order of magnitude of 16 milligals.

The southern half of Ghawar has been covered by magnetics, but maps of the results are essentially featureless.

The strong gravity anomaly requires the assumption of actual basement displacements beneath the En Nala axis. The form of the anomaly suggests a horst. So far, attempts to compute the form and magnitude of the basement feature have not given a unique solution.

Synclinal areas.—As deep well data are lacking, the nature of the synclinal areas on each side of Ghawar must be inferred from partial structure-drill and seismic coverage. The area east of Ghawar is interpreted to be structurally nearly flat. The flank dips from Abqaiq, as well as Ghawar, flatten and merge into this area. The syncline on the west is narrower with its axis close to Ghawar.

Possible origin of Ghawar structure.—Interpretations of the origin of the Ghawar structure must be largely speculative. The facts now available are best fitted by the hypothesis that Ghawar represents the upward reflection, through a pile of sediments, of basement shears which have formed a horst-like mass. In southern Ghawar this displacement may have been rather simple, but in central and northern Ghawar and Abqaiq it is more complex. The most interesting feature is that the east flank of 'Uthmaniyah and Shedgum is nearly in line with the west flank of Abqaiq, suggesting the possibility of reverse in displacement along this particular line of shear. Actual fault displacements must be far below the depth of any of the present wells, as the magnitude of the gravity anomaly is difficult to reconcile with the structure as mapped in the Upper Jurassic, and as paleostructure maps have failed to indicate anything approximating these structures prior to Aptian time. The solution of this puzzle remains for the future.

GHAWAR RESERVOIR

Upper Jubaila-Arab formation reservoir rocks.—The rocks of the Arab formation represent the transition from continuous carbonate deposition in the Jubaila formation to the nearly pure evaporite facies in the Hith formation. Four obvious cycles, each of which starts with more or less normal marine sediments and closes with evaporities, can be recognized in the Arab formation throughout much of east central Saudi Arabia. The Hith anhydrite, which actually may include several cycles in more basinal areas, is the closing phase of the evaporite deposition.

The carbonate members of the Arab formation have been denominated, from the top down, by the letters *A* to *D*, and the hyphenated form (for example *C-D* anhydrite) has been used for the separating anhydrite units.

The relation of the top of the Jubaila formation to these cycles is dubious. In many areas the uppermost Jubaila is fine-grained limestone, representing an episode of muddy deposition over a wide area, and thus it may be indirectly related to the Arab formation cycles. Except as the evaporitic facies is approached the range of environments in the Arab-*D* member differs little from that of the upper Jubaila.

Arab-A, Arab-B and Arab-C members.—Each of these carbonate members, separated from the others by anhydrite, shows considerable lateral change, but, in general, they represent units consisting of varying proportions of calcarenite and fine-grained limestone. Dolomite and calcarenitic limestone are present as minor constituents. Recrystallization and dolomitization are variable. All three members show progressive facies change from carbonate to anhydrite from northeast to southwest.

The Arab-*A* and Arab-*B* members lack oil in the Ghawar field, except for minor staining. The Arab-*C* member does not contain oil in southern Ghawar, but significant shows have been encountered in Shedgum, south Ain Dar, and east and west 'Uthmaniyah, and a minor accumulation is present in north Ain Dar. The distribution of oil roughly parallels that of substantial amounts of calcarenite.

Upper Jubaila-Arab-D member.—In general, the productive beds of Ghawar are found in the first 220–240 feet of the upper Jubaila-Arab-*D* below the *C-D* anhydrite. This has been parly or completely diamond-cored in 60 wells. The upper part, the Arab-*D* member, decreases in thickness from northeast to southwest by gradual facies change from carbonate to anhydrite in its upper part (Fig. 4). In this same general direction there is a gradual change from pre-

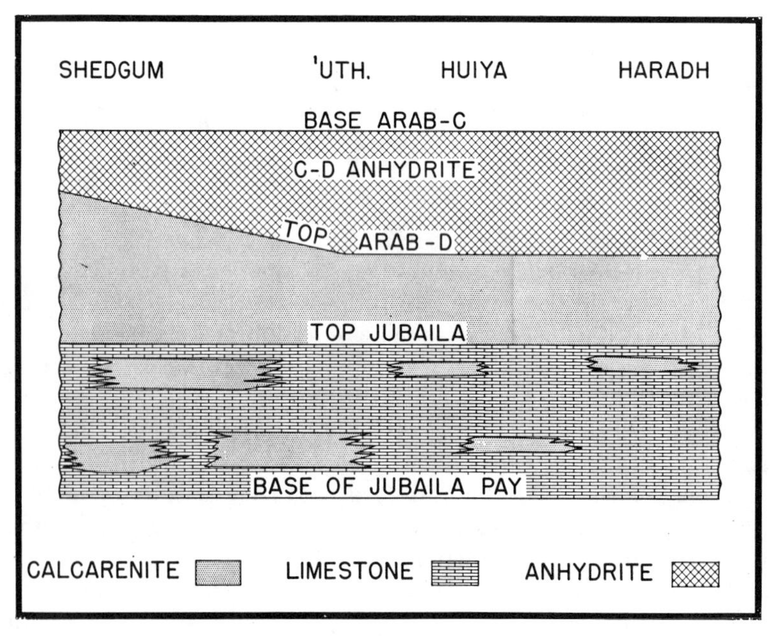

FIG. 4.—Diagrammatic section showing major variations in upper Jubaila and Arab-*D* producing zones.

dominantly calcarenite in the northeast to mixed calcarenite and fine-grained limestone in the southwest. The interval from the base of the Arab-*C* member to the top of the Jubaila is surprisingly constant throughout the Ghawar field; the wedging-down of the Arab-*D* member toward the southwest is compensated by increase in thickness of the *C-D* anhydrite.

The upper Jubaila is usually interbedded permeable and impermeable rocks, and has much less vertical continuity of reservoir than the Arab-*D* member. The permeable layers are calcarenite, calcarenitic limestone, and dolomite; the tighter rocks are fine-grained limestone, carcarenitic limestone, and dense dolomite. The upper Jubaila shows a less obvious but recognizable trend of increasing number of porous beds, distributed over an increasing thickness of the upper Jubaila, from northeast to southwest.

Lithology of Arab formation carbonate rocks.—The carbonate rocks making up the upper Jubaila and Arab are all shallow-water types. They include sediments whose original grain size ranged from silt, and possibly even finer particles, through sands to conglomerates.

Figure 5 shows a typical, somwhat generalized sample of the carbonate sedimentary pattern of the upper Jubaila-Arab-*D* member of Ghawar. High oil productivity goes in general with the calcarenites. Dolomite locally shows high permeability, but much of it is tight. Calcarenitic limestone (the flow properties of which are determined by the matrix) and fine-grained limestone range from low to high porosities, but their average permeabilities are low.

Primary subdivision of Arabian carbonate rocks has been based on the nature of the original sediment. These groups are further subdivided on type and degree of recrystallization and dolomitization, and on visually estimated porosity.

Nature of original sediments.—Fine-grained sediments have been placed in two main categories: those with, and those without included sand-size carbonate particles. These fine-grained limestones were deposisted as calcium carbonates silts and muds in a relatively quiet environment, either below wave base or where the bottom configuration afforded protection from currents and wave agitation. Part of the included sand-size and coarser carbonate particles may have been washed in from nearby current-dominated areas, but a certain proportion must have developed in place by organic and inorganic processes.

Sand-size carbonate sediments, the calcarenities, are important constituents of the upper Jubaila and Arab-*D*. Where the preservation is adequate for determination, these are commonly found to be made up of pellets (using this term in the sense of Illing, 1950). In certain areas outside of Ghawar—North Abqaiq and Qatif for example—true oöliths are present and in some places even dominate. Particles resulting from the attrition of organic remains and pre-existing carbonate rocks usually are minor in amount. The critical feature of the calcarenites is their lack of original fine-grained matrix, signifying that they were laid down under conditions of current and wave agitation which removed the finer sediments for deposition elsewhere. The possibility that part of the carbonate

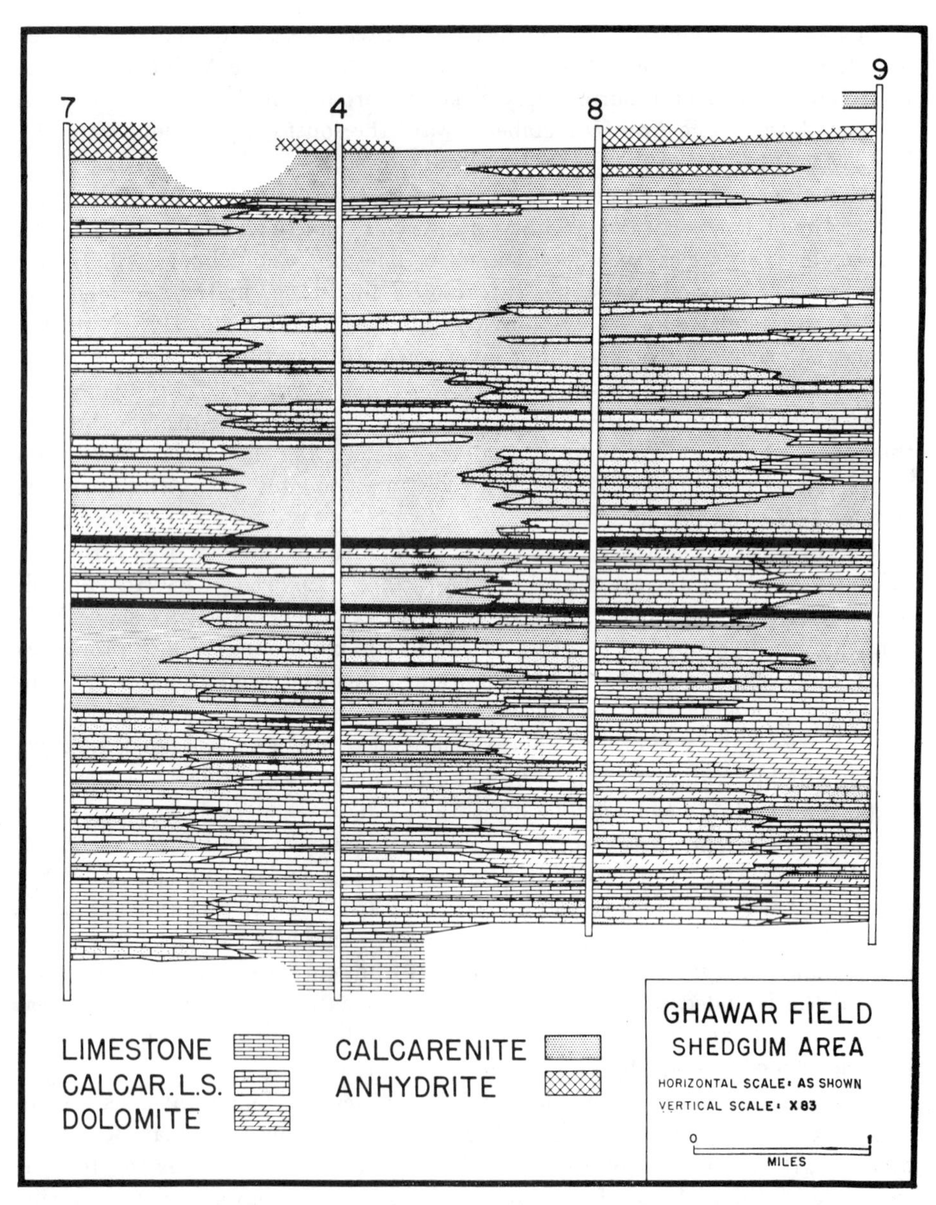

FIG. 5.—Schematic stratigraphic reconstruction of upper Jubaila and Arab-*D* producing zones in profile across northern part of Shedgum area. Upper heavy black line is top of Jubaila formation.

particles may have formed in or near the place in which they are now found does not affect this conclusion.

The conglomeratic sediments of the Arab are still poorly understood. The coarser particles often are pieces of rock similar to types found above and below. These are assumed to have resulted from wave or current destruction of pre-existing, partly lithified bottom. Other large carbonate fragments may be concretionary; still others are of obvious organic origin. In the studies made so far only a small proportion of the coarse carbonate clastics have been classified as to origin.

A little of the Arab can be considered biostromal. However, actual reef structures have not been distinguished, and it is supposed that the fragments of reef organisms represent colonies which were distributed over a relatively level bottom similar to some of the large coralliferous areas on the western shoals of the Persian Gulf today. As more detailed studies are made it is reasonable to expect that small bioherms will be found. The common reef-building organisms in the Arab are stromatoporoids.

The calcarenites and coarse carbonate clastics are usually closely associated, and usually the coarser sediments have a calcarenite matrix. The general depositional environment, like that of the modern Persian Gulf, must have included rubbly and rocky bottom areas on which were scattered clumps of various types of reef-builders.

Cementation.—Detailed petrographic studies to bring out the exact sequence of cementation and recrystallization have not been made; however, many thin sections show that one of the early events in the lithification of the sediments was the deposition of finely crystalline calcite cement, usually as a drusy coating of the carbonate particles. Cementation of this type ranges in degree from a very thin "skin" of each carbonate particle to complete filling of the original pore space.

Recrystallization and dolomitization.—The original sedimentary textures of most of the carbonate sediments of the upper Jubaila and Arab have undergone some alteration, either by recrystallization or dolomitization or both. Recrystallization often produced a more or less holocrystalline texture by solution and redeposition of the original material. The original texture, according to the degree of alteration, may have been nearly or entirely obliterated. Commonly, but not invariably, dolomitization accompanied this type of recrystallization.

The dominant type of dolomitization was by development of porphyroblastic rhombs, which increased in number as the dolomitization progressed. The indiscriminate growth of rhombs in orginal sediment particles, in matrix and in some places in cement, leaves little doubt that the dolomitization was "secondary." The ultimate end members, both by recrystallization of holocrystalline type and dolomitization by increasing numbers of porphyroblasts, are granular crystalline dolomites. In some samples traces of the original texture can be seen in the final product, but in others, there is no clue to the nature of the original sediments from which they were derived.

Less commonly, anhydrite cementation and replacement, found mainly in transitions from carbonates to evaporites, were involved in the recrystallization. Relatively large porphyroblasts of anhydrite were occasionally developed.

Silicification is virtually absent from upper Jubaila-Arab reservoir rocks.

Working carbonate rock classification.—Figure 6 (Bramkamp and Powers,

ARABIAN CARBONATE RESERVOIR ROCK CLASSIFICATION							
	ORIG. TEXTURE NOT VISIBLY ALTERED (except by cementation)	ORIGINAL TEXTURE ALTERED					ORIGINAL TEXTURE OBLITERATED
		MODERATELY		STRONGLY			
	LESS THAN 10% DOLOMITE	LESS THAN 10% DOLOMITE	MORE THAN 10% PORPHYROBLASTIC DOLOMITE	LESS THAN 10% DOLOMITE	**10-75% DOLOMITE	MORE THAN 75% DOLOMITE WITH RELIC TEXTURE	MORE THAN 75% DOLOMITE
GROUP 1 *FINE-GRAINED LIMESTONE* Fine-grained limestone with less than 10% sand size or coarser than sand size clastic carbonate fragments *	1A, 1B, 1C FINE-GRAINED LIMESTONE	11AR, 11BR, 11CR PARTIALLY RECRYSTALLIZED, FINE-GRAINED LIMESTONE	11AD, 11BD, 11CD PARTIALLY DOLOMITIZED, FINE-GRAINED LIMESTONE	12AR, 12BR, 12CR CRYSTALLINE LIMESTONE	12AD, 12BD, 12CD DOLOMITIC LIMESTONE	13A, 13B, 13C CALCITIC DOLOMITE	
GROUP 2 *CALCARENITIC LIMESTONE* More than 10% sand size or coarser than sand size clastic carbonate fragments * in more than 10% original fine-grained matrix	2A, 2B, 2C CALCARENITIC LIMESTONE	21AR, 21BR, 21CR PARTIALLY RECRYSTALLIZED, CALCARENITIC LIMESTONE	21AD, 21BD, 21CD PARTIALLY DOLOMITIZED, CALCARENITIC LIMESTONE	22AR, 22BR, 22CR CRYSTALLINE, CALCARENITIC LIMESTONE	22AD, 22BD, 22CD DOLOMITIC, CALCARENITIC LIMESTONE	23A, 23B, 23C CALCARENITIC DOLOMITE	**GROUP 4** 4A, 4B, 4C CRYSTALLINE DOLOMITE
GROUP 3 *CALCARENITE* Sand size clastic carbonate fragments * in less than 10% original fine-grained matrix	3A, 3B, 3C CALCARENITE	31AR, 31BR, 31CR PARTIALLY RECRYSTALLIZED CALCARENITE	31AD, 31BD, 31CD PARTIALLY DOLOMITIZED CALCARENITE	32AR, 32BR, 32CR STRONGLY RECRYSTALLIZED CALCARENITE	32AD, 32BD, 32CD DOLOMITIC CALCARENITE	33A, 33B, 33C CALCARENITE DOLOMITE	
GROUP 5 *COARSE CLASTIC CARBONATE* Coarser than sand size clastic carbonate fragments * in less than 10% original fine-grained matrix	5A, 5B, 5C COARSE CLASTIC CARBONATE	51AR, 51BR, 51CR PARTIALLY RECRYSTALLIZED, COARSE CLASTIC CARBONATE	51AD, 51BD, 51CD PARTIALLY DOLOMITIZED, COARSE CLASTIC CARBONATE	52AR, 52BR, 52CR STRONGLY RECRYSTALLIZED, COARSE CLASTIC CARBONATE	52AD, 52BD, 52CD DOLOMITIC, COARSE CLASTIC CARBONATE	53A, 53B, 53C COARSE CLASTIC CARBONATE DOLOMITE	

* Clastic carbonate fragments include sand size and coarser than sand size grains irrespective of origin, for example, particles of older carbonate rock, ooliths, pellets and biogenic trash.

** Dolomitic and calcitic portions of rock so intimately intergrown as to be generally visually indistinguishable.

KEY TO RESERVOIR ROCK CLASSIFICATION SYMBOLS

1st Numeral - Designates rock groups. Groups 1,2,3, & 5 distinguished on basis of original textural differences; Group 4 on complete alteration of any of these to crystalline dolomite.

2nd Numeral - Indicates degree of alteration in original texture, excluding compaction and cementation. **1** = Moderately altered original texture. **2** = Strongly altered original texture. **3** = Relic original texture.

1st Letter - - - Shows amount of visual porosity. **A** = Porous and/or poorly cemented. **B** = Compact and/or moderately cemented. **C** = Dense and/or strongly cemented.

2nd Letter - - - Type of alteration. **R** = Recrystallization involving rearrangement of materials already present. **D** = Recrystallization brought about by introduction of dolomite.

FIG. 6.—Carbonate rock classification used in geological and reservoir engineering studies. Key symbols facilitate machine data processing. Modified from Bramkamp and Powers (1958).

1958) shows the carbonate rock classification used for the study of the upper Jubaila and Arab-*D* member rocks. Its purpose is to bring out information needed for both geological and reservoir studies. The classification is applied to cuttings and cores by the use of a hand lens and low-power binocular microscope. Thin sections are made for more detailed geological studies; and various physical properties, such as porosity and permeability, are measured for reservoir study purposes. The key symbols are for machine data processing.

RESERVOIR CHARACTERISTICS

Ghawar oil.—The gravity of Ghawar crude increases from south to north. Average API gravities are 32°–33° in Haradh, 33°–34° in Huiya, 33°–35° in 'Uthmaniyah, 34°–36° in Ain Dar, and 35°–36° in Shedgum. Ghawar crude is strongly undersaturated. PVT studies show a gradation in bubble points from about 1,920 psia (215°F.) in Ain Dar to 1,720 psia (205°F.) in Haradh. No vertical change in bubble point has been found, but enough determinations have not been made to rule out the possibility. An initial gas cap was not present.

Oil-water contact.—The oil-water contact has been cut in a few structurally low wells and calculated in others which encountered water in the producing zone (Fig. 7a). The elevation of the contact rises nearly 450 feet from the area north of Ain Dar to northern Haradh, and it is consistently higher on the west flank than on the east. The general slope of the contact is thus northeast.

Oil-field waters.—Total solids in the reservoir water range from more than 240,000 parts per million in Shedgum Well 11 to about 38,000 parts per million in Haradh Well 9 (Fig. 7b and Table I). No wells structurally low enough to reach water have been drilled in the southern part of the Haradh area. The little data available show a gradient northeast similar to the slope of the oil-water contact.

Some type of static equilibrium is indicated by the high content of dissolved solids in the water, but its nature is not yet understood. As high brine concentrations have been found only in structurally low areas, there may be a possibility of large-scale vertical salinity stratification.

Reservoir temperatures.—Reservoir temperature ranges, in each case measured at the midpoint of the productive zone, are summarized in Figure 7b. Temperatures higher than 220°F. are confined to the eastern closure of 'Uthmaniyah; some of the temperatures over 230°F. are in crestal wells. An explanation for this area of elevated temperature has not been found.

Well potentials.—Well potentials are shown in Figure 8b. Rates are for oil delivered by the well under its own pressure to the gas-oil separator plant. Productivity of wells decreases progressively south from Ain Dar. The reduction toward the south is a result of (1) a gradual reduction in net reservoir thickness, (2) a corresponding reduction of porosity and permeability of the reservoir rocks, and (3) an increase in the viscosity of the oil. Average reservoir pressures are now approximately 2,500 in Ain Dar, 2,650 in Shedgum, and 2,915 in 'Uthmani-

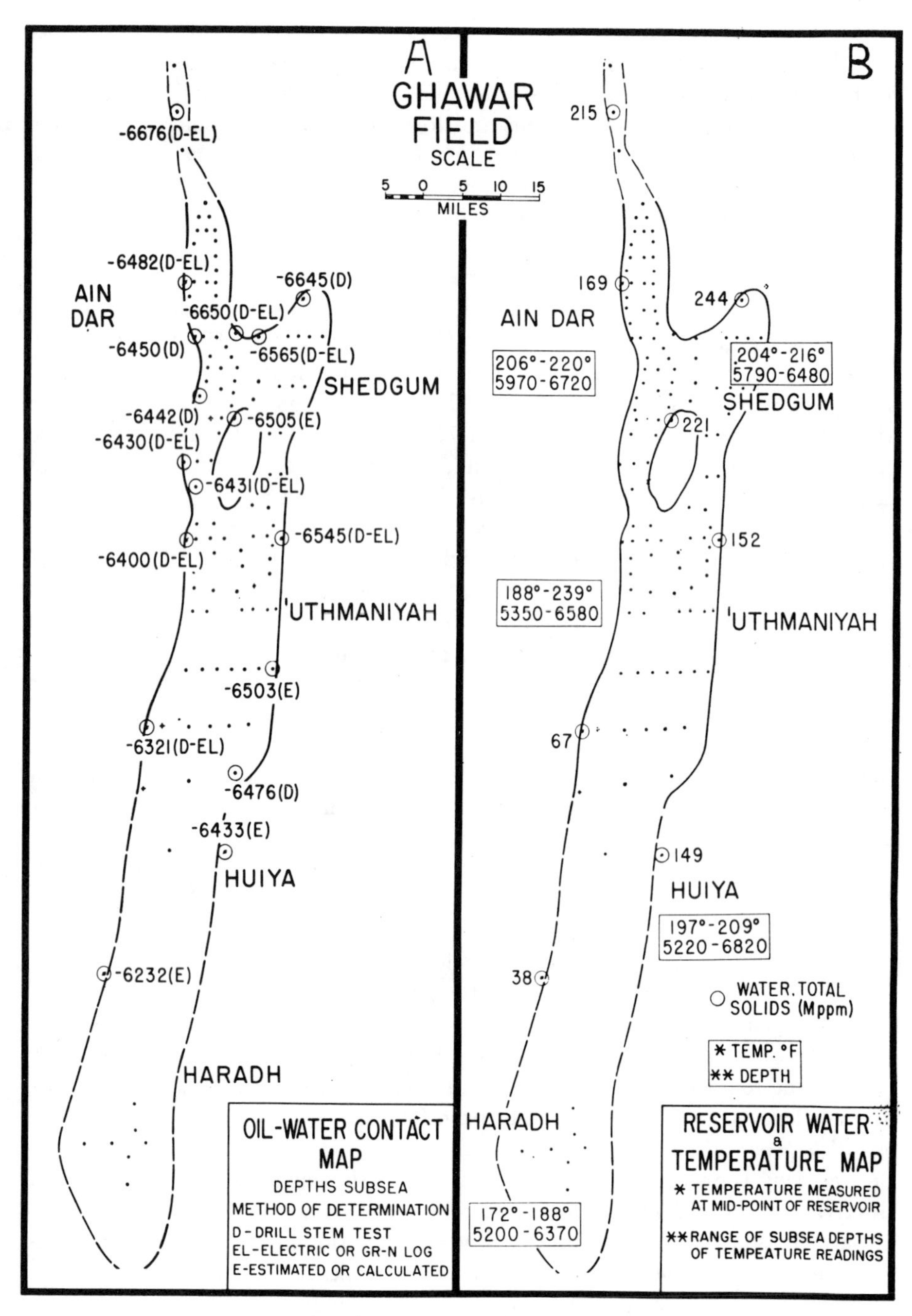

FIG. 7a.—Oil-water contact map.
FIG. 7b.—Reservoir water and temperature map.

TABLE I. TYPICAL ARAB-*D* WATER ANALYSES, GHAWAR FIELD

Well Depth	*Haradh 9 DST 7,416–7,477 Feet*		*'Uthmaniyah 37 DST 7,386–7,445 Feet*		*Ain Dar 8 Bottom-Hole Sample at 7,260 With 7,259–7,261 Ft. Perforated*		*Ain Dar 56 DST 7,194–7,325 Feet*	
	West Flank		*West Flank*		*West Flank*		*N. Plunge of Ain Dar*	
	ppm.	*%R.V.*	*ppm.*	*%R.V.*	*ppm.*	*%R.V.*	*ppm.*	*%R.V.*
Na*	9,616	32.00	17,298	32.00	51,992	38.37	61,988	35.98
Ca	3,586	13.69	6,486	13.77	11,032	9.35	17,600	11.72
Mg	685	4.31	1,209	4.23	1,634	2.28	2,098	2.30
SO_4	1,296	2.06	810	0.72	57	0.02	73	0.02
Cl	21,868	47.19	40,470	48.45	104,370	49.96	32,415	49.84
CO_3	0		0		0		0	
HCO_3†	592	0.75	1,037	0.72	55	0.02	622	0.14
Total solids*	37,643		67,310		169,140		214,796	
pH	6.9		6.5		5.3		6.1	
Sp. gr. (60°F.)	1.0290		1.0468		1.122		1.1510	

Well Depth	*Huiya 2 DST 7,611–7,643 Feet*		*'Uthmaniyah 9 Bottom-Hole Sample Perfora-ations at 7,338 Feet*		*Shedgum 11 Bottom-Hole Sample at 7,200 Feet with Unknown Interval Perforated*		*Ain Dar 42 DST 7,558–7,601 Feet*	
	East Flank		*East Flank*		*N. Plunge of Shedgum*		*W. Side of Sink*	
	ppm.	*%R.V.*	*ppm.*	*%R.V.*	*ppm.*	*%R.V.*	*ppm.*	*%R.V.*
Na*	34,629	28.72	43,929	35.09	65,157	33.16	61,094	34.33
Ca	18,166	17.29	12,214	11.52	23,840	13.93	19,764	12.73
Mg	2,546	3.99	1,538	2.39	3,024	2.91	2,767	2.34
SO_4	2,430	0.96	16	0.01	57	0.01	29	0.01
Cl	90,880	48.88	93,720	49.94	151,000	49.93	137,030	49.94
CO_3	0		0		0		0	
HCO_3†	488	0.16	159	0.05	317	0.06	256	0.05
Total solids*	149,139		151,576		243,625		220,940	
pH	6.3		5.3		5.6		6.1	
Sp. gr. (60°F.)	1.0965		1.111		1.1726		±1.154	

* Computed values.
† HCO_3 shows great variation in Arab waters, probably as result of sampling procedures.

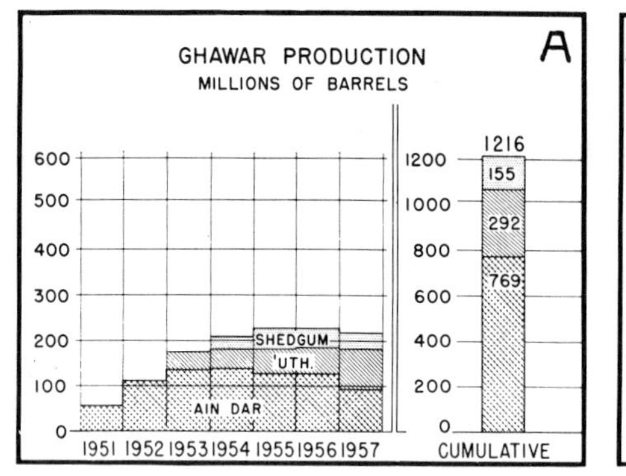

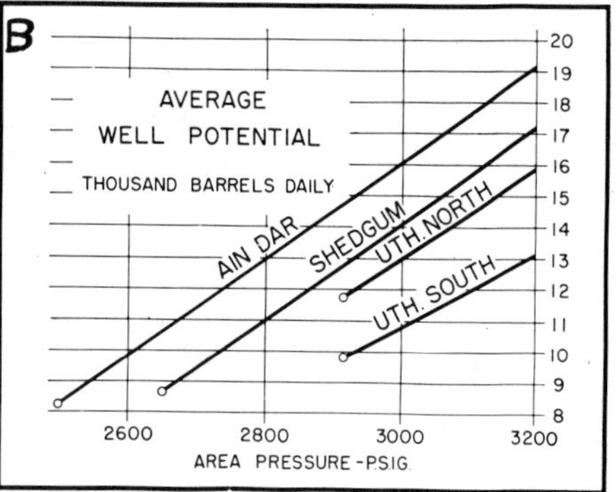

FIG. 8a.—Ghawar production.
FIG. 8b.—Potentials of Ghawar wells. All pressures are psig at 6,100 feet subsea.

yah (all pressures are psig at 6,100 feet subsea). The pressure in Ain Dar was 3,183 psig when production began, but estimated initial reservoir pressure was 3,212 psig. Withdrawal effects of production of Abqaiq field had already spread into the area prior to first pressure measurements in Ain Dar.

ORIGIN OF GHAWAR OIL ACCUMULATION

Whatever its history, it is obvious that Ghawar is a structural trap and that the overlying anhydrite is the seal. Information on the oil-water contact is sparse, but the little that is available shows the contact sloping northeast. The great size of the reservoir, the varying properties of both the water and the oil, the lithologic changes and the associated changes in capillary properties, are a few of the factors which obscure the significance of the sloping contact.

Possible source beds.—A presumption that upper Jubaila and Arab-*D* oil originated in the rocks in which it is now found has already been expressed. This opinion was based on the apparently capricious distribution of oil and water in porous units of the middle and lower Jubaila. Further evidence of this type has been accumulated, but a conclusive case still can not be made. The amount of this suggestive evidence, however, is sufficient to give strong support to the hypothesis.

Distribution of porous rocks.—Porous rocks of the Ghawar field are roughly proportional to the amount of calcarenite in the upper Jubaila and Arab-*D*. In the Arab-*D* member the field is thought to be a part of a north-south porosity trend, which includes Abqaiq, limited by a tighter lagoonal facies toward the west and a fine-grained somewhat deeper-water facies toward the east. In the upper Jubaila there is a gradual increase in the proportion of calcarenitic rocks and calcarenite from northeast to southwest, but the strong differentiation shown by the Arab-*D* does not seem to exist. Direct structural control of sedimentation has not been substantiated for either the Arab-*D* or upper Jubaila but studies now in progress may do so. It is thought that porosity and permeability are controlled, to some extent, by structure.

Effective reservoir porosity decreases, as does the calcarenite, from northeast to southwest through the field. Porosity variation in the upper Jubaila is similar to that of the Arab-*D* member but less pronounced. The effects of this porosity distribution on oil migration and accumulation is problematical, but certainly in Arab rocks the Ghawar field lies in a north-south-trending area of enhanced porosity and permeability.

History of Ghawar structure.—Lithologic changes indicative of the times of growth of the Ghawar structure have not been detected. The evidence dating growth is found almost entirely in thickness variations.

Cumulative isopach maps are the best index of structural growth. The first structurally controlled irregularities in thickness occur in Buwaib-Biyadh rocks, but these variations are insignificant. A paleostructure map of the Arab-*D* member on a late Wasi'a datum shows, however, a low but reasonably clear beginning of the Ghawar structure, except in the Sedgum area, as it is now known. By Wasi'a-Aruma unconformity time the structure is well developed, though its relief is still low. A considerable part of the now-existing relief is shown on a map using a lower Eocene datum. Some further growth is indicated in middle Eocene time and during some part of the time interval represented by the Eocene-Miocene unconformity. Suitable horizons for gaging Miocene and later growth have not been found, but so far available evidence suggests that the growth of Ghawar was completed by Miocene time, and that the very minor dips found in the Miocene-Pliocene continental rocks are either initial or are the result of solution-collapse.

Time of oil accumulation.—Dips sufficient to localize oil accumulation into anything even approximating the Ghawar structure do not appear to have been significant until mid-Cretaceous time, although the first indications of the Ghawar structure are in the Lower Cretaceous. The hypothesis previously expressed still appears to be the best explanation of the observed facts; that is, that oil was localized and held in the area by a stratigraphic trap controlled by the distribution of porous rocks in the upper Jubaila and the Arab, and then, during Cretaceous time, concentration into the field as it is now known started. So far nothing has been found to suggest when accumulation was completed.

DRILLING AND PRODUCTION

Wells are between 6,500 and 7,500 feet in depth. A typical casing program has $18\frac{5}{8}$-inch in the Dammam formation, $13\frac{3}{8}$-inch in the lower part of the Aruma formation, $9\frac{5}{8}$-inch in the lower part of the Biyadh formation and 7-inch in the anhydrite immediately above the porous portion of the Arab-*D* member. Most wells are open-hole completions. Water and water-oil observation wells are completed through perforations in a 5-inch liner set through the producing zone.

TABLE II. STATUS OF WELLS, DECEMBER 31, 1957

Area	*Producing*	*Shut-In or Standing*	*Observation*	*Suspended*	*Abandoned*	*Total*
Ain Dar	44	5	2	0	1	52
Shedgum	14	1	1	0	1	17
'Uthmaniyah	30	13	3	1	2	49
Huiya	0	1	1	0	0	2
Haradh	0	7	1	1	0	9
						129

Numerous lost circulation zones are the determining factor in the number of casing strings required. Table II shows the status of wells as of December 31, 1957.

Production facilities.—Figure 1 shows production facilities now in operation. Each gas-oil separator plant (GOSP) has a nominal capacity of 100,000 BPD. Normally, 10–13 wells flow to each separator through 6- and 8-inch flowlines. Average flowline length is 15,000 feet; however, flowline lengths up to 56,000 feet are in use. Data for fluid flow problems were obtained from the 56,000-foot line. In all installations flow to the GOSP is by wellhead pressure. Pipeline outlets from Ghawar consist of a 20–22–24-inch line from the three north Ain Dar GOSP's and a 30–31-inch line serving one Ain Dar, one Shedgum, and three 'Uthmaniyah GOSP's.

Production.—Ghawar production is shown in Figure 8a. Average daily production during 1957 was as follows: 255,734 BPD in Ain Dar, 96,217 BPD in Shedgum, and 232,037 BPD in 'Uthmaniyah. The total is 583,988 BPD for Ghawar.

REFERENCES

BRAMKAMP, R. A., AND POWERS, R. W., 1958, "Classification of Arabian Carbonate Rocks," *Bull. Geol. Soc. America*, Vol. 69, No. 10 (October), pp. 1305–17. Table 2.

STEINEKE, MAX, BRAMKAMP, R. A., AND SANDER, N. J., 1958, "Stratigraphic Relations of Arabian Jurassic Oil," *Habitat of Oil*, Amer. Assoc. Petrol. Geol., pp. 1294–1329.

BULLETIN OF THE AMERICAN ASSOCIATION OF PETROLEUM GEOLOGISTS
VOL. 48, NO. 2 (FEBRUARY, 1964), PP. 191-206, 3 FIGS.

LOWER CRETACEOUS-UPPER JURASSIC STRATIGRAPHY OF UMM SHAIF FIELD, ABU DHABI MARINE AREAS, TRUCIAL COAST, ARABIA[1]

F. T. BANNER[2] AND G. V. WOOD[2]
Sunbury-on-Thames, Middlesex, England

The Umm Shaif field, located in the Abu Dhabi Marine Areas Concession, offshore of the Sheikdom of Abu Dhabi on the Trucial Coast of Arabia, was recently brought into production.

Independent studies of the petrography and paleontology arrived at similar conclusions as to the geological history of the Lower Cretaceous-Upper Jurassic rocks, and these different lines of approach are co-ordinated in this joint paper.

In the Umm Shaif area, the rocks assigned to the Upper Jurassic and Lower Cretaceous represent four major and separate sequences of deposition. The basal Cretaceous (Lowest Neocomian) beds are thought to be represented by a basal conglomerate which is transgressive across and lies disconformably on the marine Upper Jurassic (Callovian-Oxfordian). The Neocomian marine sedimentation was interrupted by a period of regression which resulted in the deposition of a dolomite-anhydrite formation at Umm Shaif.

The Lower Cretaceous-Jurassic succession in Umm Shaif is compared with that in Qatar. In the light of the recently observed subaerial formation of anhydrite, the stratigraphical and paleogeographical relationships of the anhydritic rocks of the Persian Gulf are discussed and compared with similar successions in nearby areas. The need for similar reconsideration of the rocks of this age in the western Persian Gulf is emphasized.

INTRODUCTION

Location and sampling.—The Umm Shaif field is situated 22 miles east-northeast of Das Island, 95 miles northwest of Abu Dhabi, in the Abu Dhabi Marine Areas Concession ($\frac{2}{3}$ British Petroleum Company Ltd., $\frac{1}{3}$ Compagnie Française des Pétroles), off the Trucial Coast of Arabia (Fig. 1). At the time of writing ten wells have been completed and the first tanker left the loading terminal on Das Island bound for the BP Refinery at Aden during July, 1962.

This paper is based on a study of thin sections prepared from both core and cuttings samples from five wells at Umm Shaif. The average sampling interval was 5 feet, but in the critical cored sections samples were examined every 2 feet. Whenever foraminiferal identifications were in doubt the rocks were exhaustively examined, and oriented thin sections were prepared of the fossil in question.

PREVIOUS WORK AND STATEMENT OF PROBLEM

Very little published information on the Mesozoic stratigraphy of the western Persian Gulf exists in the literature although the oil companies interested in this area have considerable evidence in their confidential files. The purpose of this paper is to point out the anomalies in the available publications and to offer faunal and petrographic evidence together with our interpretation of the Upper Jurassic-Lower Cretaceous stratigraphy of Umm Shaif and Qatar areas.

Early in the studies of these well sections, it became apparent that the provisional dating of the beds encountered, based on broad correlation of the succession with formations described in publications from adjacent areas, was at some variance with the internal paleontological evidence. The most obvious discrepancy was the discovery that fossils, believed to be restricted to beds of Cretaceous age (or younger), occurred *in situ* below a dolomite-anhydrite formation which, on general lithological similarities, had been correlated with the Hith anhydrite formation of eastern Saudi Arabia. This Hith Formation was originally defined by Steineke and Bramkamp (1952) and had subsequently been considered, by correlation with similar beds over much of the Arabian-Iraqi area, to be an isochronous index formation either separating the Cretaceous from the Jurassic (Owen and Nasr, 1958; Hudson and Chatton, 1959; Arabian American Oil Company Staff, 1959) or even being intercalated within the Jurassic (Dunnington et al., 1959).

Steineke, Bramkamp, and Sander (1958, p. 1297, 1306 1308) redescribed the Wadi Hanifa

[1] Manuscript received, May 4, 1963, and published by permission of the chairman and directors of the British Petroleum Co. Ltd.

[2] British Petroleum Co. Ltd., BP Research Centre, Exploration Division, Sunbury-on-Thames, Middlesex, England.

Many colleagues on the staff of British Petroleum, both at Sunbury and Umm Shaif, have been of great assistance, especially F. E. Eames and W. J. Clarke. The writers are indebted to staff of the Iraq Petroleum Co. Ltd. for discussions.

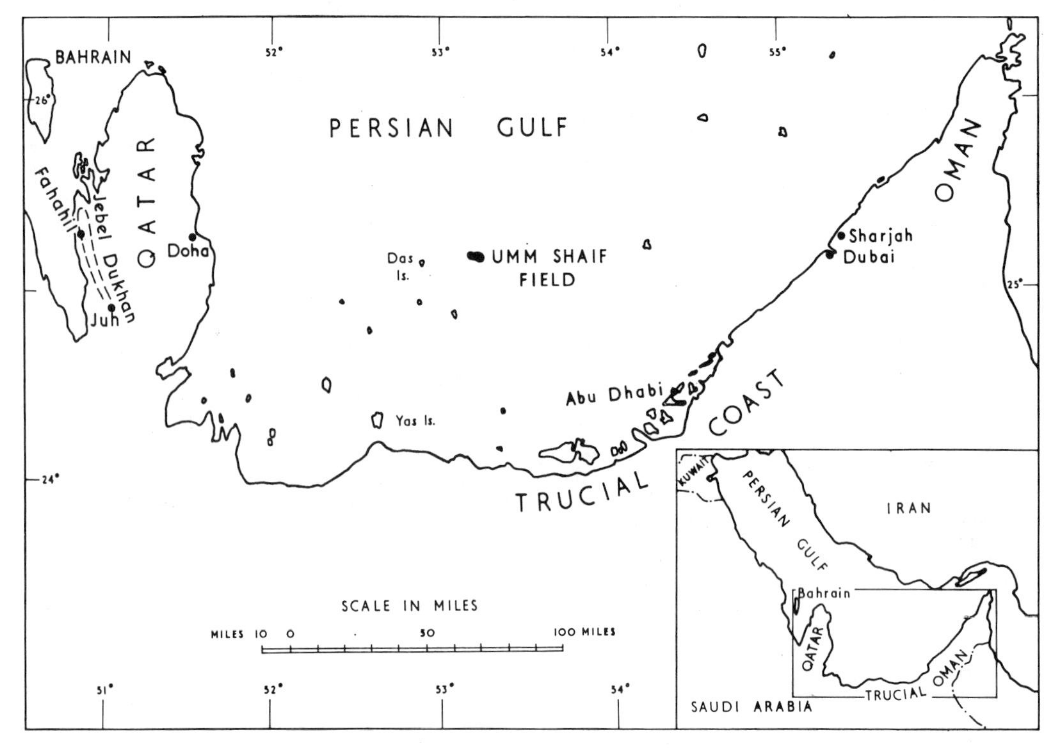

FIG. 1.—Map of Persian Gulf to show location of Umm Shaif field.

sections of central Arabia; there, the type Jubaila Formation (of Early Kimmeridgian age) is followed by the Riyadh "formation" (i.e., lower part of the Riyadh Group) and then by the Hith anhydrite, neither of which, apparently, was found to contain diagnostic fossils. The Hith anhydrite was also considered by these authors to be referable to the upper part of the Riyadh Group, the lower part of the group being believed to be equivalent to the Arab Formation, which occurs between beds in the Dammam wells which were correlated with the Jubaila and the Hith. The Arab Formation again provided no direct evidence of age; however, Steineke et al. (1958), stated that it has (un-named) fossils common to the underlying Jubaila equivalent. This, together with the belief that sedimentation had been continuous from the Jubaila equivalent into the Arab (even though the division between the formations was considered to be represented by "the contact of the basal calcarenitic rocks of the Arab with underlying dense limestone of the upper Jubaila"), seemed sufficient evidence to include the Arab Formation within the Upper Jurassic. The apparent grading upward between the Arab and the Hith Formations was then taken as evidence for the inclusion of the Hith Formation in the uppermost Jurassic.

Current opinion about the age of the Hith anhydrite formation has been greatly influenced by the assumed correlation of this formation with the Gotnia anhydrite formation of North Iraq. The Gotnia, which has its type locality at Awasil, approximately 1,000 miles away from Umm Shaif, has been dated as being of Late Jurassic age. This problem, and its effect upon the stratigraphy of Umm Shaif, is discussed later in this paper, together with a summary of recently published results on a closely comparable problem—the conflict of opinion on the age of the anhydritic-gypsiferous Gaurdak Formation, approximately 1,200 miles away in Uzbekistan. The Gaurdak, once thought to have been formed during a Late Jurassic marine regression, is now believed to be a deposit of Early Neocomian age transgressive across rocks ranging from Late Jurassic to Paleozoic in age.

In the Mesozoic section penetrated in the Umm Shaif field, an anhydrite-dolomite formation exists which has been tentatively identified with the Hith (but not necessarily with the Gotnia of Iraq), and below this, a considerable thickness of limestones has been found, containing a fossil assemblage believed to be of Early Neocomian age. This is described in the principal part of this paper, and this appears to be the first public suggestion that anhydrite developments, resulting in rocks comparable with the Hith, petrographically at least, may be present within a Neocomian series in the Persian Gulf area, in spite of the fact that supporting evidence for such a phenomenon has long been available from Iran.

Ammonites, discovered by geologists of the Anglo-Iranian Oil Company Ltd. (now the British Petroleum Company Ltd.) in outcrop sections at Khaneh Kat and Tang-i-Darbat at levels below a dolomite which has been assumed to be the correlative of the Hith anhydrite (unpublished records), were identified by Professor J. A. Douglas of Oxford University, and include *Berriasella* (which ranges from Tithonian to Berriasian) and "*Steueroceras*" auctt. *non* Cossman, a restricted Berriasian form now referred to the restricted genus *Cuyaniceras* Leanza, 1945 (Berriasian only, *teste* Arkell et al., 1957).

The data appear to allow only one of the following conclusions to be reached:

(1) that the dolomite-anhydrite formation of Umm Shaif is not the equivalent of the Hith Formation, in which case the development of anhydrite in this approximate part of the stratigraphical column can no longer be relied on as an isochron, even within the Arabian-Persian Gulf province, and the paleogeographic concepts of this part of the Mesozoic in the Persian Gulf area must be considerably revised; or

(2) that the Umm Shaif dolomite-anhydrite formation is equivalent to the Hith Formation of Arabia, in which case the age of the Hith Formation must be revised. From this, either (a) the age of the Riyadh Group and Arab formations also must be revised, or (b) the isochronous Hith anhydrite formation was deposited upon strata of widely different ages—representing a middle Neocomian marine regression at Umm Shaif, but possibly the same earth movements resulting in contemporaneous flooding of other areas over older Jurassic rocks. In each case paleogeographic concepts must be emended, or

(3) that the whole suite of fossils obtained from the Umm Shaif wells, from between acknowledged Jurassic and the dolomite-anhydrite formation, must have their geological ages (and stratigraphical ranges) revised to include their presence in the Late Jurassic—this in spite of the fact that many are of well known and widespread Cretaceous occurrence, but unknown in the Jurassic, whereas not a single proved restricted Jurassic index fossil has yet been found to occur indigenously in the assemblage.

Consequently, the generally accepted picture of the stratigraphy of the Persian Gulf is anomalous. Because of the apparently contradictory nature of the evidence, it was proposed to undertake detailed but independent petrographic and paleontological studies of the Jurassic and Lower Cretaceous sediments in Umm Shaif. G. V. Wood carried out the petrography, and F. T. Banner the micropaleontology. The studies were undertaken with a minimum of preconceptions and it was found that our interpretations of the sedimentary history of the sequences of rocks studied coincided very closely. The joint appraisal is given. The results are offered in the hope that they may contribute to an objective assessment of the Mesozoic pre-Cenomanian stratigraphy of the western Persian Gulf.

Nomenclature

Formation Taxa

In an *ab initio* study of this nature, in which doubts are raised as to the correlation of the rocks of Umm Shaif with neighboring areas, the use of the current stratigraphic nomenclature is undesirable. The original idea was to establish a petrographic framework on which to base the paleontological detail. Four sequences of deposition were established in the Lower Cretaceous-Upper Jurassic rocks of Umm Shaif; each is essentially homogeneous in rock-type, any variations being small and gradual, but is delimited above and below by abrupt lithological changes. During the study these sequences were referred to by capital letters until it became apparent that they corresponded with geological time units (Fig. 2). The lithological subdivisions of each sequence are termed "members" and are referred to by the dominant rock type. These "members" may well be of widespread lateral persistence but a standard conventional nomenclature must await further work, including discussion and evaluation of the results given in this present paper.

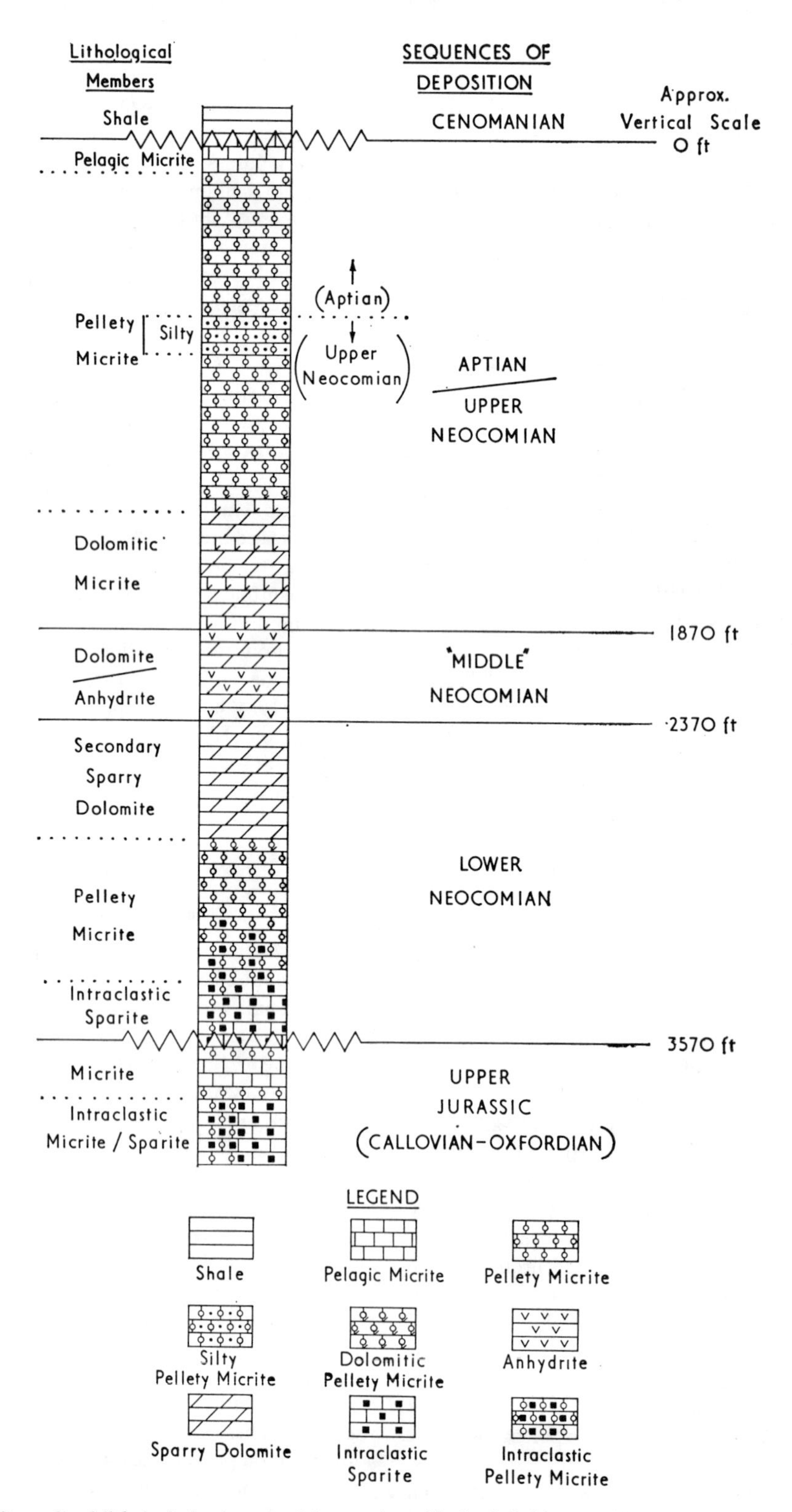

FIG. 2.—Generalized lithological column and its stratigraphical subdivision, within Upper Jurassic and Lower Cretaceous of Umm Shaif fields. Thicknesses, not depths, are given.

ROCK NOMENCLATURE

Until recently, carbonate nomenclature has involved the use of a confusing variety of unrelated descriptive terms which are frequently interpreted in different ways. This confusion has been such that adjacent sections containing identical rock-types, when described by different geologists, have proved difficult to equate.

We feel that any practical classification of rocks should be as nearly objective as possible and for sedimentary rocks should clearly differentiate the grains, matrix, and cement (as in Krynine, 1948). Deductions as to origin are usually dependent on the bias of the individual author and are frequently a matter of controversy; in our opinion these should come after the objective naming of the rock.

Following this line of argument, the only limestone classification that, so far, meets our requirements is that of Folk (1959). We have found it necessary to modify Folk's definitions by distinguishing between the allochemical modifiers "intraclast" and "pellet" on the basis of the internal morphology of the grains. For the purposes of this paper an "intraclast" is defined as a grain showing heterogeneity of texture, whereas a "pellet" is a grain which is homogeneous in texture. The genetic definition of "intraclast," suggested by Folk, in this way usually, but not necessarily, follows.

The only other minor modification of Folk's classification which was found to be necessary was the dropping, on philological grounds, of the scheme of using abbreviated forms of the allochemical terms as a prefix to either "micrite" or "sparite." For example, instead of the term "oopelsparite" of Folk, we prefer "oolitic pellety sparite."

The inferences made in the following discussion as to the relative depth of water during the accumulation of the various types of limestone follow Illing (1959). Work on recent carbonate sedimentation in the Persian Gulf by Houbolt (1957) clearly illustrates the control of the depth of accumulation on the rock-type, although, as Plumley *et al.* (1962) point out, water agitation is not related to depth of water in restricted environments. For the sediments other than the dolomite and anhydrite, we have no reason to suppose the depositional environment to have been so restricted.

STRATIGRAPHICAL GEOLOGY

A generalized lithologic column of the Lower Cretaceous-Upper Jurassic rocks encountered in Umm Shaif is given in Figure 2, and each sequence of deposition is described under the relevant geological time units. The distributions of the important foraminifera and algae, throughout these sequences and their lithologic members, are recorded in Figure 3.

JURASSIC

UPPER JURASSIC

The lowest Jurassic limestones yet penetrated in the Umm Shaif wells belong to a lithological member consistently recognizable in each well of sufficient depth. This consists of intraclastic and pellety sparites and micrites, interbedded and intergrading. In both the sparites and the micrites, the intraclasts consist of bioclastic and foraminiferal micrites, which contain a fauna indistinguishable from that apparently *in situ* in the matrix of the rock. The whole fossil assemblage appears to belong to one natural fauna, and it is believed that repeated penecontemporaneous "cannibalism" of the sediments, in an environment often well within the limits of active wave erosion, was responsible for the partial reworking, transportation, and redeposition of many of the fossils. We believe that this member represents a transgressive depositional phase, the sediments being laid down on an unstable sea-floor, the depth of which fluctuated with time.

Although there is no doubt that the fossil fauna is of Jurassic age, evidence is as yet insufficient to enable its precise dating in terms of the European standard stages. L. R. Cox has identified a molluscan fauna obtained from cores of this member; *Mytilus* (*Falcimytilus*), which appears to belong to the widespread Kimmeridgian-Oxfordian species *M.*(*F.*) *jurensis* (Roemer), and *Liostrea*, apparently indistinguishable from the Callovian-Portlandian species *L. dubiensis* (Contejean), both indicate a Late Jurassic age. Some of the foraminifera, however, suggest a greater, perhaps Middle Jurassic, age; these include an undescribed species of *Orbitopsella* and rare forms close to *Lituosepta* and *Haurania deserta.* They contrast with the evidence of other foraminifera restricted to this member: *Pseudocyclammina*, probably identical with *P. ukrainica* Dain, described from the Kimmeridgian, and *Trocholina* close to *T.*

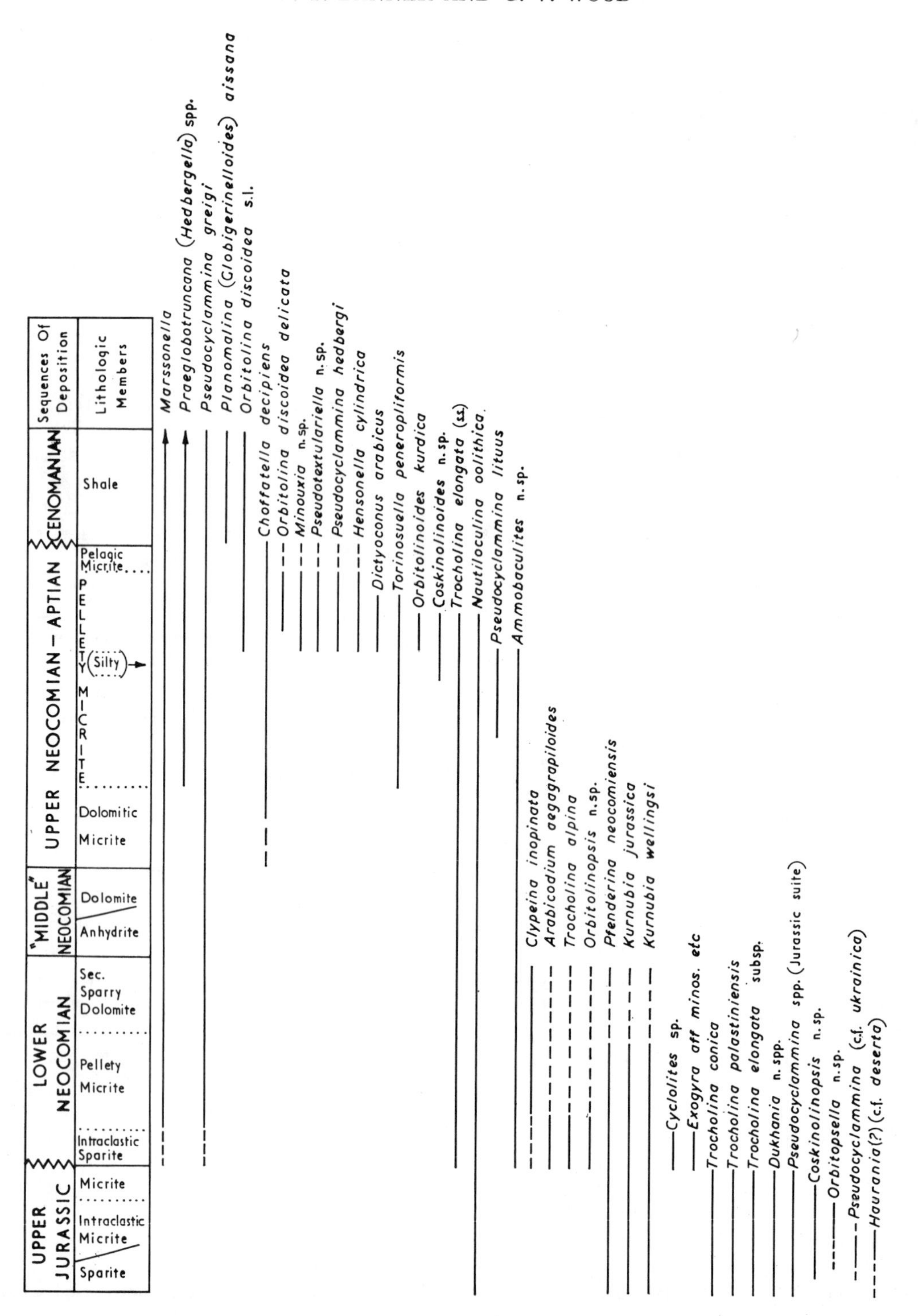

FIG. 3.—Distribution of significant microfossils in Upper Jurassic and Lower Cretaceous of Umm Shaif field, in relation to lithological column. Broken lines indicate possible extensions of range, suggested by poorly preserved specimens or specimens doubtfully *in situ.*

feifeli of the Late Jurassic, indicate a Late Jurassic age. This is supported by the fact that most of the foraminiferal species (*Nautiloculina oolithica*, *Pfenderina neocomiensis*, *P. trochoidea*, *Kurnubia jurassica*, *Trocholina conica*, *T. palastiniensis*, a restricted Jurassic subspecies of *T. elongata*, *Coskinolinopsis* sp., two undescribed species of *Dukhania*, and two new species of *Pseudocyclammina*, etc.) persist *in situ* into the overlying, more certainly Upper Jurassic member.

The specimens referred here to *Trocholina palastininesis* have been compared directly with topotype specimens from the Callovian-Oxfordian of the Kurnub anticline, Israel, and it is noteworthy that the *Trocholina* assemblage from the beds at Kurnub seems very similar to that discussed here from the Upper Jurassic of Umm Shaif.

The intraclastic and pellety sparite and micrite member grades up into the overlying micrite member with progressive decrease in the frequency of sparite and intraclasts, and, in the highest levels, a similar decrease even of pellets. This is interpreted as representing a progressive deepening of the Jurassic sea, until only micritic limestones were being deposited at, or even below, the lower limits of wave action. This micrite member, which, when developed in its typical form, consists of a dense aggregate of pellets in a micrite matrix contains a foraminiferal fauna dominated by valvulinids and lituolids, together with scattered mollusca, echinoid debris, and encrusting algae. Among the lamellibranchs, Cox was able to identify (from a core) *Chlamys* cf. *curvivarians; C. curvivarians* (Dietrich) is known to range from Callovian to Kimmeridgian. The foraminifera include the species listed within parentheses in the foregoing paragraph, *Kurnubia wellingsi*, and a few additional forms, as yet undescribed, of *Dukhania* and *Pseudocyclammina*. The paleontological affinity with the underlying member is great, and we see no reason to suppose that their ages are significantly different. This confirms the belief that both members are merely contrasting expressions of one sequence of deposition, probably of early Upper Jurassic (Callovian-Oxfordian) age. This age determination is supported by the strong faunal similarity of these members to the "*Trocholina* limestones" and "Beni Zaid limestone formation," described from Oman by Hudson and Chatton (1959), which were dated by these authors as Callovian-Oxfordian. They also show a very strong relation, both faunally and lithologically, to the Najmah limestone formation of Iraq, which was also dated, by Dunnington (1959, p. 207–211), as Upper Jurassic and believed by him to be transgressive at its base.

The top of the micrite member in the Umm Shaif field is marked by a thin development of pellety sparite; this rock is paleontologically indistinguishable from the underlying beds, except for a decrease in the numbers of valvulinids and an increase in the numbers of *Trocholina palastiniensis* and *T. conica*—a change which is probably due to the same change in environment which caused the development of the sparite. The rock consists of pellets, similar to those in the underlying micrite, cemented by sparry calcite; no intraclasts occur, and we interpret this sediment as merely representing a shallowing of the Jurassic sea prior to emergence (probably in post-Oxfordian, possibly in Kimmeridgian time), the evidence suggesting that the post-Oxfordian stages of the Jurassic are missing at Umm Shaif.

CRETACEOUS

LOWER NEOCOMIAN

The sparry last phase of the Jurassic micrite member is followed by a sharp lithological and paleontological break. The surface of this pellety *T. palastiniensis* sparite is eroded and fissured, and is succeeded by a series of alternating intraclastic micrites and sparites, commonly of calcirudite grade, containing much material clearly reworked from the underlying formations. It appears to be an excellent example of a basal conglomerate within a limestone sequence. Both (1) direct reworking and redeposition of geologically much older bioclastic rocks and (2) "cannibalism" of penecontemporaneously deposited bioclastic sediment, appear to have occurred; intraclasts of Jurassic *Trocholina palastiniensis*-rich micrites and sparites, strongly pyritized, are commonly found contained inside other micrite intraclasts, both in micritic and sparitic beds, in some places indicating up to three sequences of erosion and redeposition. The third group of fossils (foraminifera, algae, and corals), which are those believed to be indigenous to this formation, are commonly (but not invariably) abraded and thus have every appearance of allochthoneity. We interpret this sediment as representing a shallow-water deposit, the whole being deposited well within the zone of active wave action. The alternation between sparite and micrite would result from relative oscillation of the sea-floor within this zone—older sediments being eroded, transported, and redeposited continuously but with varying intensity. The presence of allochthonous coral and codiacean algal debris itself indicates the presence of shallow-water reefal conditions in a current-swept nearby area.

In many places the microfacies of these basal beds is indistinguishable from that of the *Trocholina*-rich beds of the Berriasian Öhrlikalk of the Altenalp Türm, as illustrated and described by Grasmück-Pfluger (1962, p. 422–444, 430–431, 436, fig. 7).

Among these corals are numerous specimens of the genus *Cyclolites*, and examples occur even close to the base of the formation. This simple coral is of almost world-wide distribution (Eurasia, North Africa, West Indies) within the Cretaceous and Eocene, but has never been authentically recorded from the Jurassic, even though species are well known to occur in the Neocomian (e.g., of France and the Crimea; Kuzmicheva, 1960); in fact, the whole family to which *Cyclolites* belongs is unknown before the Cretaceous (Wells, 1956, p. F 386; Vaughan and Wells, 1943, p. 134–135). The specimens have been examined by H. Dighton Thomas (British Museum) who agrees with us that their biocharacters are those of the genus *Cyclolites* and not of the Jurassic genus *Chomatoseris*. Fragments of the alga *Arabicodium aegagrapiloides* also occur; this monotypic genus has been recorded by Elliott (1957) only from the Lower Cretaceous of Oman. He has kindly identified some of the specimens of this genus for us, and all the others found have been carefully compared with his ideotype.

In these basal sparites, no indigenous forms believed to be restricted to the Jurassic have yet been found. The only foraminifer which has previously been recorded exclusively from Jurassic beds, and which appears to occur indigenously at this horizon, is *Kurnubia wellingsi*. We believe that this form may well range up into the Lower Neocomian, previous suggestions that it does not do so being due to its apparent absence in European sediments of this age (the whole of the Pfenderinidae shows marked geographical restriction—see Smout and Sugden, 1962). So far, this species has been found to occur only very rarely even in these beds at Umm Shaif. *Kurnubia jurassica* is known to occur in the Valanginian of southern France (Marie and Mongin, 1957) and the Neocomian of the Middle East (Hudson and Chatton, 1959); *K. wellingsi* is a morphologically more advanced species and it should not be surprising that it would range equally high.

The indigenous foraminiferal fauna of these basal intraclastic beds strongly contrasts with that of the formation which disconformably underlies it. The Jurassic *Coskinolinopsis*, *Pfenderina trochoidea*, *Orbitopsella*, *Trocholina* spp. (*T. elongata* subsp., *T. conica*, *T. palastiniensis*, etc.), *Dukhania* spp., *Pseudocyclammina* spp. (*P.* cf. *ukrainica*, etc.), previously mentioned, are missing and are replaced by an assemblage dominated by *Trocholina alpina* and *T. elongata* (sensu stricto) (= *T. altispira* (*pars*) of Henson), accompanied by rarer *Neotrocholina* sp. nov., *Orbitolinopsis*, and *Pseudocyclammina* cf. *greigi*. *Pfenderina neocomiensis* and *Nautiloculina oolithica* persist from the underlying formation (only the latter ranging up into the Upper Neocomian).

The *Trocholina* assemblage of the Lower Neocomian of Umm Shaif is very similar, in both the species which it contains and in the sequence in which these species occur, to that described by Guillaume (1963) from the Lower Neocomian of the Jura; in both cases *Trocholina alpina*, the true *T. elongata* and the form referred by Guillaume to *T. molesta* Gorbatchik occur abundantly at lower levels (noticeably without *T. conica* or *T. palastiniensis*), and at higher levels in the Valanginian *Neotrocholina valdensis* appears.

With progressive decrease in the frequency of sparite and intraclasts, the formation grades up into the succeeding micritic limestone member. This appears to reflect the increasing establishment of deeper-water conditions, at, or near, wave base—a virtual repetition of the sequence of physical sedimentary conditions seen in the preceding Jurassic deposits. Thin beds of intraclastic and pellety micrites still occur sporadically throughout this deeper-water phase, and in many places they contain debris apparently transported from nearby reefal areas sporadically established during temporary phases of relative shallowing. Among the bioclasts of these thin beds are fragments of both *Clypeina inopinata* and rare *Milleporidium crassum;* the former was originally described from the Valanginian of the central Jura by Favre (1932), who stated that the species *C. jurassica* ranged from Rauracian to basal Valanginian (in both North Africa and Europe) and, at the top of its range, evolved into and was replaced by *C. inopinata*, which then persisted throughout the Valanginian. *M. crassum* is believed to be characteristic of the Valanginian in southern Europe (Schnorf, 1961).

The micrite which is the dominant rock type of this member is pellety and pyritic; many of the individual pellets are mainly composed of pyrite. In this sediment, the fauna appears wholly autochthonous, and consists only of small miliolids, textulariids, valvulinids, polymorphinids, *Glomospira*, etc. A significant form is *Marssonella*, a genus which has never been proved to occur below the Cretaceous, and which is generally agreed to appear first in the Berriasian (Moullade, 1960).

The fauna and flora of the near-type Berriasian and Valanginian of the south of France have been described recently by Marie and Mongin (1957) and by Cotillon (1960). They showed that *Pfenderina neocomiensis* (which was first described, as *Eorupertia neocomiensis*, from the southern French Valanginian, and has, to now, been considered to have an anomalously short range in the Persian Gulf area—see Smout and Sugden, 1962) is there associated with *Kurnubia jurassica*, *Trocholina alpina*, *T. elongata* (*s.s.*), *Nautiloculina oolithica*, and *Clypeina jurassica* (*vel C. inopinata*). This assemblage is exactly parallel with that of the formation here described from Umm Shaif. Marie and Mongin (1957) showed that this microfossil assemblage is associated, in Les Bouches-du-Rhône, with characteristic, stratigraphically restricted, Valànginian mollusca (e.g., *Natica leviathan* = *Leviathania leviathan* (Pictet and Campiche) = *L. sautieri* Coquand; *Leviathania*, as a genus, is restricted to the Neocomian). L. R. Cox has identified mollusca obtained from cores of the formation here described from Umm Shaif; although the material was not adequate for precise specific determination, it was clear that the whole assemblage was quite different from that found in the underlying Jurassic. It included *Exogyra*, probably belonging to the widespread Neocomian-Aptian/Albian species *E. minos* (Coquand) ("no comparable species is known from the Jurassic"—Cox, private report), *Chlamys* possibly belonging to the Neocomian species *C. robinaldina* (d'Orbigny), and *Camptonectes* spp. which could well be referred to the Neocomian species *C. cottaldinus* (d'Orbigny) and *C. striatopunctatus* (Roemer).

All the evidence available to us indicates that the Jurassic rocks of the Umm Shaif area are disconformably succeeded by transgressive deposits of Berriasian and(or) Valanginian age, representing the basal Cretaceous transgression, and that, in the latter part of Valanginian time, marine sedimentation was well established, depositing a series of micritic limestones in a relatively deep-water environment.

The higher part of these Lower Neocomian limestones is increasingly dolomitized, and is now, in consequence, only patchily fossiliferous. Rare *Neotrocholina* cf. *valdensis* have been found in association with small miliolids, textulariids, and valvulinids (*N. valdensis* has been reported by Reichel (1956) only from Valanginian sediments). The dolomite is increasingly abundant and coarse-grained at progressively higher horizons, and each rhomb has a distinct dark core. Interdigitation with intensely dolomitic micrite and some pellety sparites also occurs. We believe that this development of dolomite is due to secondary dolomitization of limestones which were originally closely similar to those of the underlying micrite member. It is probable that the magnesium had its origin in the overlying primary dolomites (ascribed to the middle part of the Neocomian) which probably formed due to a marine regression at the end of the Lower Neocomian sequence of sedimentation.

"MIDDLE NEOCOMIAN"

This consists of an essentially evaporitic sequence of primary dolomite and anhydrite, with accessory gypsum. Four dominant rock-types have been recorded in this unit.

(1) Massive felted anhydrite.

(2) Laminae of micritic dolomite, showing structures attributed to primary deposition, with intercalated fibrous anhydrite. Dark pellets are commonly recorded in the dolomite but they show no evidence of replacement after calcite.

(3) Large euhedral anhydrite poikilitically enclosing closely spaced, coarse, clear rhombs of sparry dolomite.

(4) Coarse, clear dolomite mosaic with intersertal anhydrite.

A possible mode of origin for this unit is a restriction of the Neocomian sea (a "proto-Persian Gulf"), perhaps by the elevation and emergence of barriers, in the vicinity of the known structurally positive and tectonically active Oman Mountains (Morton, 1959). Little or no evidence exists for the presence of normal marine organisms; the few small arenaceous foraminifera which have been found and which are believed to be *in situ* in this unit may well have been able to exist under conditions of abnormal salinity. The age of this dolomite-anhydrite sequence in Umm Shaif can only be inferred from the fact that it lies below Upper Neocomian beds and above beds dated as Early Neocomian.

UPPER NEOCOMIAN-APTIAN

The primary dolomite and anhydrite sequence is succeeded by beds of sparry secondary dolomite, which have intercalations of dolomitic micrite. The dolomite rhombs of both rock-types possess cloudy internal cores, and this feature has been found by us to be a reliable indication of a diagenetic origin for the dolomite. The dolomitic micrites contain an apparently impoverished microfauna of small arenaceous foraminifera (*Glomospira*, etc.), small miliolids and polymorphinids and rare, specifically indeterminable, more complex lituolids. This fauna is essentially marine, and, in part at least, could exist under conditions of abnormal salinity; its paucity could also be due to the deposition below wave base of these rocks, a suggestion fully compatible with the non-pellety nature of the micrite.

The frequency and intensity of dolomitization decrease upward, and the micrite becomes pellety. At the top of the beds ascribed here to the Neocomian, silt-size detrital quartz grains appear and become common, although nowhere exceeding 5 per cent of the rock-volume. The appearance of the pellets and allochems probably corresponds with a shallowing of the sea, which is reflected in an increase in the numbers of microfossils. The microfauna of these beds has some affinity with the underlying Lower Neocomian (e.g., *Trocholina elongata*, *Pseudocyclammina greigi*, *Nautiloculina oolithica*, and a distinctive and undescribed species of *Ammobaculites*) but lacks the distinctive Valanginian forms described in preceding paragraphs. It also shows relationships with the overlying Aptian, in its possession of the oldest found *Praeglobotruncana* (*Hedbergella*), *Choffatella decipiens*, and small distinctive species of valvulinids (e.g., *Coskinolinoides* sp.). Microfossils characteristic of the succeeding Aptian (e.g., *Hensonella*, *Pseudocyclammina hedbergi*, *Orbitolina*, *Dictyoconus arabicus*, etc.) are absent. *Pseudocyclammina lituus* occurs sporadically; this species, which has often been mentioned in the Middle East to characterize the Valanginian, is now known to be of long geological range, occurring from Kimmeridgian at oldest to Aptian (e.g., Maync, 1950, 1959) in many localities throughout the Tethys and in Central America, and it appears to be prone to considerable ecological restriction. *P. lituus* has been found by A. N. Thomas (BP private report) in beds with *Choffatella decipiens*, *Heteraster obliqua*, and *Hinnites* cf. *renevieri* which occur in the Kuh-i-Mund hinterland of southwestern Iran at approximately the same stratigraphical horizon in a sequence comparable with that described here. It is probable that both are of uppermost Neocomian age and that the occurrence of quartz grains in these beds at Umm Shaif reflects an uplift of neighboring emergent areas during earth movements at the close of Neocomian time.

Apart from the absence of quartz grains, the limestones of the lower part of the Aptian are petrographically indistinguishable from those of the top of the Neocomian; no gross lithological break occurs, both being pellety micrites usually of calcarenite grade. So, as no environmental change can be deduced from the petrography, and as this is confirmed by the presence in both limestones of foraminifera of the same closely related groups (i.e., *Pseudocyclammina*, valvulinids, *Praeglobotruncana*, etc.), the paleontological change which occurs at the top of the limestone with quartz grains is believed to be of stratigraphical value, and to represent the Aptian-Neocomian boundary.

The pellety limestone which, with an intermediate deeper-water phase of non-pellety deposition, comprises the Lower and Middle Aptian at Umm Shaif, contains throughout *Hensonella cylindrica*, *Orbitolina discoidea delicata*, *Choffatella decipiens*, *Pseudocyclammina hedbergi*, and distinctive valvulinids which include the oldest known representatives of the genera *Minouxia* and *Pseudotextulariella*. Occurring in the lower part only of these beds, which probably represent the Early Aptian, arc *Orbitolina kurdica*, *Dictyoconus arabicus*, and the lamellibranch *Venericardia* (*Xenocardita*) *lacunaris* (known from the Aptian of the Lebanon), together with the highest observed rare specimens of *Nautiloculina oolithica*.

This sequence of benthonic faunas closely corresponds with that described by Reiss (1961) from surface exposures of northern Galilee. There, too, it was found that *Hensonella* occurred only in Aptian beds, and that the highest *Nautiloculina oolithica* was associated with *Pseudocyclammina hedbergi*, *Choffatella decipiens*, and "*Praecuneolina*" (apparently the same form as that referred to here as *Pseudotextulariella*) in the lower part only of the Aptian.

The highest beds of the Aptian show a rapid transition into a dense, non-pellety micrite, with a microfauna clearly dominated by specimens of *Praeglobotruncana* (*Hedbergella*). The benthonic foraminifera common in the older Aptian beds occur only rarely, probably due to their ecological exclusion by subsidence again below the limits of active wave action. The Late Aptian age of this pelagic phase of deposition is confirmed by the presence of the restricted Late Aptian ammonites *Cheloniceras* and *Colombiceras*.

These pelagic micrites constitute the last phase of the Lower Cretaceous limestone depositional sequence at Umm Shaif. They are followed abruptly and apparently non-sequentially, by shales with a deep-water globigerinid-lituolid microfauna, which are dated as Early Cenomanian on the occurrence of the typically Early Cenomanian *Nucula sorianoi*, together with the association of abundant *Hemicyclammina sigali*, *Planomalina* (*Globigerinelloides*) *aissana*, *Praeglobotruncana* (*Hedbergella*) *washitensis*, rarer *Orbitolina concava*, *O. discoidea*, and the ammonites *Spathiceras* and *Ficheuria*. The paleontological break thus indicated by the fossils is paralleled by the sharp lithological break between the Aptian limestones and the Cenomanian shales, as seen both in cores and on electric logs. Consequently, it is believed that, although the Late Aptian was represented by a short period of limestone deposition in water deeper than that of the preceding Middle Aptian, its termination was marked by uplift and non-deposition in Albian time, a completely new sedimentary cycle beginning with the first Early Cenomanian shales (which, in their turn, grade up into limestones with *Rotalipora* and an increasingly neritic Middle Cenomanian fauna).

COMPARISON WITH QATAR PENINSULA

We have been able to examine (through the courtesy of the Qatar Petroleum Company) the continuously cored sequence in the Qatar Petroleum Company's well, Juh No. 1 (=Dukhan No. 51), and the cuttings and core samples from the same company's well, Dukhan No. 65. These two wells clearly penetrated very similar lithological sequences, the beds containing similar microfossil assemblages. The following notes, however, refer directly to the Juh No. 1 well, which provided a more continuously cored sequence, and which is referred to by Sugden (ms.). In this well, the lowest Jurassic present (referred by Q.P.C. to the lower and middle parts of the Araej Formation, by correlation with the type section of that formation in Q.P.C. well Kharaib No. 1) agrees well in lithologic character and fauna with the Jurassic sequence in Umm Shaif.

Between the middle part of the Araej Formation (the Uwainat equivalent) and the anyhdrite formation (believed by Sugden, ms., to be the Hith equivalent), the following formations occur: Qatar, Fahahil, Darb (type), Diyab (type), and upper Araej. All these are being formally described by Sugden (ms.).

The upper part of the Araej Formation appears to grade from the highest beds ascribed to the Uwainat, and the foraminiferal faunas of the two formations are qualitatively very similar—the principal distinctions being the apparent extinction of *Pfenderina trochoidea* before the upper Araej and the appearance within these higher Araej beds of the oldest *Pseudocyclammina jaccardi* (*s.l.*). The fauna of the pellety, micritic upper Araej contrasts with that of the sparry, intraclastic basal Neocomian beds of Umm Shaif by its lack of all the Cretaceous elements, and by possession of a distinctly Jurassic foraminiferal assemblage. *Dukhania* spp., *Pseudocyclammina* cf. *ukrainica*, *Trocholina conica*, and *Trocholina palastiniensis*, for example, occur indigenously, and the *Trocholina elongata* (*s.l.*) assemblage contains both *elongata* (*s.s.*) and the subspecies found in Umm Shaif only in the lower Araej and Uwainat, appearing to represent a fully intermediate morphological assemblage, consistent with the view that the upper Araej assemblage of Qatar is distinctly younger than the lower Araej of Qatar (and its probable equivalent in Umm Shaif) but older than the beds which were deposited on top of the Uwainat equivalent at Umm Shaif.

The upper Araej formation grades up into the poorly fossiliferous micrites of the Diyab Formation; this is also conformably and apparently gradationally succeeded by the pellety micrites of the Darb Formation. At this horizon, *Pseudo-*

cyclammina jaccardi (*s.s.*) is abundant, a species which Maync (1960) considered to be restricted to the Late Oxfordian (*s.l.*) and Early Kimmeridgian (i.e., Lusitanian, approximately). *Pseudocyclammina* spp., believed to be conspecific with *P. kelleri* and *P. virguliana*, also occur commonly, and, again, the Cretaceous fossils known below the anhydrite in Umm Shaif have nowhere been observed.

The Darb Formation is succeeded in the Juh No. 1 well by secondary sparry dolomite of the basal Fahahil Formation. We have not been able to study either this or the succeeding Qatar Formation in this well, but Sugden (ms.) has shown that the Fahahil and its equivalent contain stromatoporoids, *Clypeina jurassica* and other algae, all indicating a Kimmeridgian age. Such age indications have not been found to occur beneath the anhydrite formation at Umm Shaif. The Qatar Formation of Qatar is also stated by Sugden (ms.) to contain Late Jurassic (probably Kimmeridgian) macrofossils.

From the foregoing, it appears that the lithological and paleontological succession between the Uwainat Formation and the sparry-dolomite-anhydrite formation at Umm Shaif is distinctly different from that of Qatar, although the lithologic character and fossils are remarkably similar both above and below this interval. In Qatar it seems that the Callovian-Lower Oxfordian (*s.l.*) middle Araej rocks are succeeded, without a break, by Upper Oxfordian (*s.l.*)-Kimmeridgian limestones, and that anhydrite-dolomite deposits rest directly on them (or, on their dolomitized highest members). In Umm Shaif, however, a similar Callovian-Lower Oxfordian sequence of beds is abruptly and disconformably followed by limestones which contain a flora and fauna of decidedly Lower Neocomian aspect, sediments which could well be considered to represent the basal Cretaceous transgression over an eroded Jurassic surface.

The stratigraphical and paleogeographical relationships of Qatar (as shown by the Juh No. 1 well) and Umm Shaif during Late Jurassic and Neocomian time must, therefore, be briefly considered, even though our knowledge is still far from complete. Much valuable paleontological information has probably been destroyed by the intense secondary dolomitization of the limestones within, and directly below and above, the anhydrite and primary dolomite deposits. Nevertheless, if the evidence of the fossils, as deduced by ourselves, our colleagues and the authorities consulted is to be followed, together with the strong supporting evidence of sedimentology in those horizons where secondary dolomitization has not affected the rocks, then it must be assumed that a pre-anhydrite marine succession of Lower Neocomian age is present at Umm Shaif, but absent at Qatar.

It is possible that movements in mid-Oxfordian (*s.l.*) time caused shallowing and formation of the upper Araej deposits in Qatar but emergence, followed by non-deposition, at Umm Shaif. Cessation of positive movement, followed by renewal of such movement, could have led to the deposition of the sediments which now form the Diyab and succeeding Darb formations, respectively, the latter perhaps being preliminary to final emergence at the end of Jurassic time, or at least to a marine regression which led to the formation of the evaporitic deposits.

It may be that the anhydrite-dolomite deposits of Qatar, referred by Q.P.C. to the Qatar Formation and the Hith Formation, represent a condensed sequence comprising uppermost Jurassic and lower to "middle" Neocomian rocks; this would imply that while normal marine sediments were deposited in Early Neocomian time at Umm Shaif, their equivalent in Qatar were evaporite and(or) relatively thin marine deposits (e.g., oölitic limestones) now heavily dolomitized. If the current Q.P.C. correlation with Saudi Arabia is to be followed, it would suggest that the lower part of the Riyadh Group (the Arab Formation) is of Late Jurassic age, and the succeeding upper Riyadh (Hith Formation) is of Early (to "middle"?) Neocomian age.

An alternative hypothesis may be that the increasingly shallow-water conditions of deposition in Qatar already seen in the contrasting lithologic features of the Darb and Fahahil Formations reached a maximum in the Qatar Formation with partial or entire marine cut-off resulting in alternations of anhydrites, dolomites, and limestones in that Kimmeridgian formation. Such conditions of extreme shallowness and fluctuation between marine and evaporitic conditions could easily lead to complete cut-off and non-deposition.

A resumption of deposition with shallow-water oölites (recorded by Sugden, ms., as being preserved in dolomite near the base of the Hith in

Qatar), dolomites and anhydrites could then occur without any noticeable break in the lithological sequence even though such a hiatus could in fact comprise the whole of Early Neocomian time. Thus, the basal Cretaceous sediments below the anhydrite of Umm Shaif and below its equivalent at Khaneh Kat and Tang-i-Darbat in Iran may not have been deposited in Qatar.

Comparison with Other Areas

The very complex question of the age and correlation of the Iraq anhydrites is clearly beyond the scope of this paper; however, Dunnington (1959, p. 133) has stated that the Hith anhydrite formation of Saudi Arabia, although not recognized in Iraq, "is certainly correlative" with the upper part of the Gotnia Formation. The Gotnia anhydrite formation has been dated (p. 117) as Late Jurassic (Upper Callovian? to Kimmeridgian to pre-Middle Tithonian); this dating was acknowledged to be based, not on internal evidence, but on correlation of overlying beds (after the intervening Makhul and Zangura Formations) with the Garagu Formation of the Amadia district (North Iraq, 250 miles north of the subsurface type locality of the Gotnia, p. 103), which may be dated as Valanginian and ?Early Hauterivian on its possession of *Neocomites*, etc.

We must point out, however, that we believe that the direct correlation of one anyhdrite with another, without supporting paleontological evidence from underlying and overlying beds, should be treated with caution, especially where the sections to be correlated may be subsurface only and separated by scores or hundreds of miles. The occurrences of the anhydrite formations, both believed to be one and the same Gotnia, at Awasil in Iraq and the Burgan deep test in Kuwait, are recorded as being separated not only by hundreds of miles but also by erosional and(or) depositional breaks (Dunnington, Pl. IV). Even in Iraq, the anhydrite formation is well known to be locally absent. The Barsarin Formation of Kurdistan, thought to be the Gotnia equivalent (Dunnington, p. 57–59), contains no anhydrite. At its type locality, the Najmah Formation (p. 207–211), a limestone of believed pre-Late Gotnia age (at least), directly underlies (probably unconformably) the Lower Neocomian Garagu Formation equivalent.

It is believed by Dunnington (p. 117) that the Gotnia anhydrite formation, in its lower part, is equivalent to the upper part of the Najmah Formation, and (p. 118) passes laterally into it; on Plate IV he indicates his belief that the lower part of the Najmah, at least, is equivalent to the Tuwaiq Mountain Formation of Saudi Arabia. Sugden (ms.) considers that the Tuwaiq Mountain Formation is a close correlative of the Diyab Formation of Qatar. The cored section of the type Diyab Formation (Juh No. 1 well, Qatar), which we have examined in detail, overlies the Uwainat Formation there but is separated from it by the upper Araej formation as stated by Sugden (ms.). The microfauna of the Uwainat Formation is identical, apparently, with that of the Najmah; this appears from our own observations on the Uwainat of Qatar, the microfaunas listed for the beds by Sugden (ms.), the microfauna published by Dunnington (1959, p. 208) for the Najmah Formation (and from what we have seen from that formation), the comparison drawn by Smout and Sugden (1962, p. 581–591) and from oral discussion with A. H. Smout. In the successions in Iraq, Qatar, and Oman, only one formation in each case appears to have this distinctive foraminiferal assemblage—the Najmah, Uwainat, and the Beni Zaid limestone formations, respectively. As each of these formations occurs in the same apparent place in the over-all faunal and lithological sequence in each area, there appears to be no *a priori* reason why they should not be correlated. A. H. Smout has assured us that there is no additional faunal evidence of age of the type Najmah or Uwainat Formations to be considered, but that their datings, as currently published, do deserve reconsideration.

In summary, it may be seen that

(a) the Hith anhydrite formation of Arabia (and its presumed equivalent in Qatar) is currently believed to be equivalent to the Gotnia anhydrite formation of Iraq, in part at least;

(b) the lower part of the Gotnia anhydrite is currently equated with the upper part of the Najmah Formation; this latter formation, in turn, is very probably equivalent, in part at least, to the Uwainat Formation of Qatar;

(c) the lower Najmah is considered to be the equivalent of the Tuwaiq limestone formation of Arabia, and the latter formation is, in turn, believed to be equivalent to the Diyab Formation of Qatar.

The preceding statements are, apparently, incompatible. The Hith equivalent is separated from the Uwainat by five formations in Qatar; the type Diyab Formation is separated from the underlying type Uwainat in their type sections by the interpolating upper Araej formation; also, the Diyab and Uwainat are paleontologically distinct, the former being dated by Sugden (ms.) as probably Middle Callovian-Early Oxfordian and the latter as possibly Middle Bathonian-Early Callovian. Hudson and Chatton (1959, p. 82–84) dated the Beni Zaid limestone formation as Oxfordian *s.s.* (=Early Oxfordian *s.l.*); as shown above, we believe the Uwainat Formation equivalent of Umm Shaif to be of Callovian-Early Oxfordian *s.l.* age on independent evidence, and the type Diyab Formation to be referable to part at least of the "Lusitanian" (=Late Oxfordian *s.l.*-Earliest Kimmeridgian).

Of the foregoing statements, (b) at least, is supported by lithological, paleontological, and stratigraphical evidence; (a) is based primarily on the assumption that only one major anhydrite formation is present in this part of the Mesozoic throughout the Iraq-Persian Gulf area, and that this formation is isochronous throughout its lateral extent.

Such an isochronous formation, of so great a lateral extent, would need uniform sedimentary conditions over this wide area—conditions which have often been assumed to be those of a regressive or shallowing sea. However, it appears, from the most recent literature available to us (e.g., Deer et al., 1962, p. 213–215, 223–224), that there is no conclusive evidence of anhydrite ever having been formed by direct precipitation during desiccation of sea-water. Its origin is usually attributed to the dehydration of primary gypsum, a process that must result in a decrease in volume. No evidence of such dehydration, or decrease of volume, has been observed by us, or has ever been recorded, from the Hith anhydrite or its equivalent in Umm Shaif, and to postulate its origin by the marine desiccation process would be without observational foundation.

Recent work on the coastal flat (the "Sabhka") of the Sheikhdom of Abu Dhabi (Curtis et al., 1963; Shearman, personal communication) has shown that anhydrite is being formed now, under subaerial conditions, adjacent to the shallow sea where aragonitic sediments are currently being deposited. The interpretation of this phenomenon is that the precipitation of aragonite provides a ground water relatively rich in Mg^{++} and SO_4^{--}. This ground water then percolates to the surface of the coastal flat, and provides a source for the dolomitization of the limestones found beneath the anhydrite, the latter presumably having been formed by efflorescence. This process provides a mechanism for the origin of primary anhydrite rocks which removes the difficulties inherent in considering anhydrite as part of a marine desiccation cycle, and would explain some of the anomalies emphasized for the Hith anhydrite formation during the course of this paper. Not only does it now appear that anhydrite deposition need not be a result of a marine regressive phase, but it indicates that it is, in some cases at least, a result of complete emergence. This resolves immediately theoretical difficulties in the mechanism of anhydrite (as opposed to gypsum) formation under subaqueous conditions. It also indicates that anhydrite deposits could well occur below transgressive marine deposits—in fact, that any marine deposit found to occur above, or within, bedded anhydrite should be suspected of being transgressive.

A case which is both closely analogous with our problem in the Persian Gulf, and which is believed to confirm the foregoing conclusion, is that of the Gaurdak Formation of Turkmeniya S.S.R. and the Uzbekistan S.S.R. The nature and current solution of the problem are here summarized.

Gypsum deposits occur in almost continuous outcrop along the foothills of the Gissar Range in Uzbekistan S.S.R., east of Kugitang-Tau. The gypsiferous beds overlie thick Oxfordian-Callovian limestone (dated on macrofossils), are themselves succeeded by Neocomian shales, and have generally been referred to the Kimmeridgian (Rudik, 1960), the top of the gypsiferous sediment being taken as the top of the Jurassic. The basic criterion for believing the Jurassic-Cretaceous boundary to occur at this horizon has been the idea that the formation of the gypsiferous deposits occurred during a general regression of the late Jurassic sea over a large area of central Asia. This concept was not rejected even though it was long known that the gypsiferous beds grade upward into the overlying Neocomian shales. In fact, some authors (e.g., Zhukovski et al., 1957) were forced by this latter consideration to place the Jurassic-Cretaceous boundary even higher in the succession, referring the lower part of the (Neocomian) shales to the Late Jurassic.

The gypsiferous beds continue south into the

Tadzhik Depression and the Kugitang-Gaurdak region, where they were studied (as the Gaurdak Formation, type), by Bratash et al. (1959), and were then shown to overlie the Upper Jurassic with angular unconformity; a limestone breccia was found to occur in places (but not everywhere) at the base of the Gaurdak Formation. The same Gaurdak Formation, still characteristically gypsiferous and anhydritic, is now known to occur subsurface as far northwest as Darganata, on the border of Turkmeniya, about 300 miles from Kugitang-Tau. Subsurface sections of this formation have been studied in detail by Rudik (1960) in the Bukhara area, where the formation is approximately 150–220 feet thick, has a clearly recognizable base, and is closely correlatable from field to field. Rudik showed that, in the Bukhara-Khivin area, the Gaurdak Formation lies unconformably on beds ranging in age from Late Jurassic to Paleozoic (probably Permo-Carboniferous). In each place, the Gaurdak Formation is conformably succeeded by Neocomian shales as in the Gaurdak district itself. Farther northwest, at Sultansandzhar, the Gaurdak Formation is missing, the Neocomian resting with a basal conglomerate directly on Middle Jurassic. This obviously pre-Cretaceous (and pre-gypsum-anhydrite!) unconformity is attributed by Rudik (1960) to a basal Cretaceous transgression subsequent to an uplift and folding phase of latest Jurassic age, such a phase of folding already being known in the Mangyshlak region.

Additional evidence for the Early Cretaceous age of the gypsum-anhydrite beds comes from the southwest foothills of the Gissar Range, where Rudik (1960) states that the Gaurdak Formation grades both vertically and laterally into the Neocomian shales of the Karabil Formation. Consequently, he concluded that a break in deposition occurred over a very large area of central Asia in Kimmeridgian time, and that this was followed by a marine transgression (initiating a new sedimentation cycle) which led to the formation of gypsiferous-anhydritic deposits of considerable lateral extent in the Bukhara-Gissar Range-Tadzhik Depression area, and that this transgression was contemporaneous with the Valanginian transgression in Mangyshlak and other areas. Consequently, contrary to previous generally held opinion, he considers that "there is sufficient basis for believing that the gypsum-salt and anhydrite-carbonate sediments of the Gaurdak formation are younger in age, that is, Lower Cretaceous, and are to be included in the Valanginian stage."

We believe that a similar explanation of the age and origin of the Hith is also valid.

Conclusions

If the ranges of several groups of fossils are not to be extended from Cretaceous into Jurassic, an extension which would be contrary to all evidence known to us elsewhere, and without confirmatory evidence of Jurassic fossils in the strata concerned, the stratigraphical position of a group of anhydrite-dolomite-limestone beds, used for many years as a correlative index in the Persian Gulf, must be reconsidered. Current correlations based on the belief of complete isochroneity of a single major mid-Mesozoic anhydrite in this area already leads to some stratigraphical difficulties; we think these are susceptible to solution on the grounds outlined in this paper.

References

Arabian American Oil Company Staff, 1959, Ghawar oil field, Saudi Arabia: Am. Assoc. Peteroleum Geologists Bull., v. 43, p. 434–454.

Arkell, W. J., Kummel, B., and Wright, C. W., 1957, Mesozoic Ammonoidea, *in* Treatise on invertebrate palaeontology, Pt. I, Mollusca 4: Geol. Soc. America, p. 80–465.

Bratash, V. I., Sokolov-Kochegarov, S. A., and Khasina, G. I., 1959, Kharakter izmeneniya strukturnikh planov Yuri, Mela i Paleogena Tadzhikskoi Depressii i metodika poiskov i razvedki Melovikh i Yurskikh struktur (Character and change in the structural plans of the Jurassic, Cretaceous, and Palaeogene of the Tadzhik Depression, etc.), VNIGNI pub.

Cotillon, P., 1960, Description d'une coupe de la partie supérieure des "calcaires blancs" de Provence au Nord du département du Var: Soc. géol. France, Compte Rendu Somm. des Séances, fasc. 3, p. 60–61.

Curtis, R., et al., 1963, Association of dolomite and anhydrite in the Recent sediments of the Persian Gulf: Nature, v. 197, p. 679–680.

Deer, W. A., Howie, R. A., and Zussman, J., 1962, Rock-forming minerals, v. 5 (non-silicates): London, Longmans.

Dunnington, H. V., Wetzel, R., and Morton, D. M., 1959, Lexique stratigraphique international: v. III. Asie fasc. 10a (Iraq), 333 p.

Elliott, G. F., 1957, New calcareous algal floras from the Arabian Peninsula: Micropaleontology, v. 3, p. 227–230.

Favre, J., 1932, Présence d'une nouvelle éspèce d'algue calcaire siphonée dans le Valanginien du Jura Central, *Clypeina inopinata* n. sp.: Eclog. géol. Helvet., v. 25, p. 11–16.

Folk, R. L., 1959, Practical petrographic classification of limestones: Am. Assoc. Petroleum Geologists Bull., v. 43, p. 1–38.

Grasmück-Pfluger, M., 1962, Mikrofazielle beobachtungen an den Öhrlischichten (Berriasian) der

Typuslokalitat: Eclog. géol. Helvet., v. 55, no. 2, p. 417–442.
Guillaume, S., 1963, Les trocholines du Crétacé Inférieur du Jura: Rev. de Micropal., v. 5, no. 4, p. 257–276.
Houbolt, J. J. H. C., 1957, Surface sediments of the Persian Gulf near the Qatar Peninsula: The Hague, Moulton & Co.
Hudson, R. G. S., and Chatton, M., 1959, The Musandam limestone (Jurassic to Lower Cretaceous) of Oman, Arabia: Notes et Mém. sur le Moyen Orient, Paris, v. 7, p. 69–93.
Illing, L. V., 1959, Deposition and diagenesis of some Upper Palaeozoic carbonate sediments in Western Canada: 5th World Petroleum Cong. Sec. 1, Paper 2, 28 p.
Krynine, P. D., 1948, The megascopic study and field classification of sedimentary rocks: Jour. Geology, v. 56, p. 130–165.
Kuzmicheva, E. I., 1960, On the morphology of the genus *Cyclolites* (in Russian): Paleont. Zhurnal, no. 3, p. 52–56.
Marie, M., and Mongin, D., 1957, Le Valanginien du Mont-Rose de la Madrague (Massif de Marseilleveyre, Bouches-du-Rhône): Soc. géol. France Bull., sér. 6, t. VII, p. 401–424.
Maync, W., 1950, The foraminiferal genus *Choffatella* Schlumberger, etc.: Eclog. géol. Helvet., v. 42, no. 2, p. 529–547.
——— 1959, Biocaractères et analyse morphometrique des éspèces Jurassiques du genre *Pseudocyclammina;* I. *P. lituus* (Yokoyama): Rev. de Micropal., v. 2, no. 3, p. 153–172.
——— 1960, Biocaractères et analyse morphometrique des éspèces Jurassiques du genre *Pseudocyclammina;* II. *P. jaccardi* (Schrodt): Rev. de Micropal., v. 3, no. 2, p. 103–118.
Morton, D. M., 1959, The geology of Oman: 5th World Petroleum Cong., Sec. 1, Paper 14.
Moullade, M., 1960, Les Orbitolinidae des microfacies Barremiens de la Drôme: Rev. de Micropal., v. 3, p. 188–198.
Owen, R. M. S., and Nasr, S. N., 1958, Stratigraphy of the Kuwait-Basra area, *in* Habitat of oil: Am. Assoc. Petroleum Geologists, p. 1252–1278.
Plumley, W. J., et al., 1962, Energy index for limestone interpretation and classification, *in* Classification of carbonate rocks—a symposium, Am. Assoc. Petroleum Geologists Mem. 1.
Reichel, M., 1956, Sur une trocholine du Valanginien d'Arzier: Schweiz. Pal. Gesell., Eclog. géol. Helvet., v. 48 (1955), p. 396–408.
Reiss, Z., 1961, Lower Cretaceous microfacies and microfossils from Galilee: Research Council Israel Bull., v. 10 G, no. 1–2, p. 223–242.
Rudik, V. A., 1960, O stratigraficheskoi granitse mezhdu otlozheniyami Yuri i Mela v Bukharo-Khivinskoi Oblasti (On the stratigraphic boundary between the Jurassic and Cretaceous in the Bukhara-Khivin region): Geologiya Nefti i Gaza, no. 5, p. 34–38.
Schnorf, A., 1961, Les Milleporidiidae des Marnes Valanginiennes d'Arzier: Schweiz. Pal. Gesell., Eclog. géol. Helvet., v. 53 (1960), p. 716–727.
Smout, A. H., and Sugden, W., 1962, New information on the foraminiferal genus *Pfenderina:* Palaeontology, v. 4, pt. 4, p. 581–591.
Steineke, M., and Bramkamp, R. A., 1952, Mesozoic rocks of eastern Saudi Arabia: Am. Assoc. Petroleum Geologists Bull., v. 36, no. 5, p. 909.
Steineke, M., Bramkamp, R. A., and Sander, N. J., 1958, Stratigraphic relations of Arabian Jurassic oil, *in* Habitat of oil: Am. Assoc. Petroleum Geologists, p. 1294–1329.
Sugden, W., ms., International stratigraphical lexicon: Qatar.
Vaughan, T. W., and Wells, J. W., 1943, Revision of the suborders, families and genera of the Scleractinia: Geol. Soc. America Special Paper 44.
Wells, J. W., and Hill, D., 1956, Anthozoa, *in* Part F, Coelenterata, Treatise on invertebrate palaeontology: Geol. Soc. America.
Zhukovski, L. G., Dudova, N. F., Kaesh, Yu. V., Petrov, I. V., 1957, Novie gazo-neftyanie mestorozhdeniya Bukharo-Khivinskoi Depressii (New oil and gas fields of the Bukhara-Khivin Depression): Geologiya Nefti, no. 11.